Nanotechnology in Aerospace and Structural Mechanics

Noureddine Ramdani
Research and Development Institute of Industry and Defense Technologies, Algeria & Harbin Engineering University, China

A volume in the Advances in Chemical and Materials Engineering (ACME) Book Series

IGI Global
DISSEMINATOR OF KNOWLEDGE

Published in the United States of America by
IGI Global
Engineering Science Reference (an imprint of IGI Global)
701 E. Chocolate Avenue
Hershey PA, USA 17033
Tel: 717-533-8845
Fax: 717-533-8661
E-mail: cust@igi-global.com
Web site: http://www.igi-global.com

Library of Congress Cataloging-in-Publication Data

Names: Ramdani, Noureddine, 1986- editor.
Title: Nanotechnology in aerospace and structural mechanics / Noureddine Ramdani, editor.
Description: Hershey, PA : Engineering Science Reference, [2019] | Includes bibliographical references and index.
Identifiers: LCCN 2018041945| ISBN 9781522579212 (hardcover) | ISBN 9781522579229 (ebook)
Subjects: LCSH: Aerospace engineering. | Nanocomposites (Materials)
Classification: LCC TL545 .N25 2019 | DDC 629.1/20284--dc23 LC record available at https://lccn.loc.gov/2018041945

This book is published in the IGI Global book series Advances in Chemical and Materials Engineering (ACME) (ISSN: 2327-5448; eISSN: 2327-5456)

British Cataloguing in Publication Data
A Cataloguing in Publication record for this book is available from the British Library.

Advances in Chemical and Materials Engineering (ACME) Book Series

ISSN:2327-5448
EISSN:2327-5456

Editor-in-Chief: J. Paulo Davim, University of Aveiro, Portugal

MISSION

The cross disciplinary approach of chemical and materials engineering is rapidly growing as it applies to the study of educational, scientific and industrial research activities by solving complex chemical problems using computational techniques and statistical methods.

The **Advances in Chemical and Materials Engineering (ACME) Book Series** provides research on the recent advances throughout computational and statistical methods of analysis and modeling. This series brings together collaboration between chemists, engineers, statisticians, and computer scientists and offers a wealth of knowledge and useful tools to academics, practitioners, and professionals through high quality publications.

COVERAGE

- Numerical Techniques
- Heat Treatments
- Multifuncional and Smart Materials
- Fracture Mechanics
- Thermo-Chemical Treatments
- Ductility and Crack-Resistance
- Composites
- Electrochemical and Corrosion
- Computational methods
- Coatings and surface treatments

IGI Global is currently accepting manuscripts for publication within this series. To submit a proposal for a volume in this series, please contact our Acquisition Editors at Acquisitions@igi-global.com or visit: http://www.igi-global.com/publish/.

The Advances in Chemical and Materials Engineering (ACME) Book Series (ISSN 2327-5448) is published by IGI Global, 701 E. Chocolate Avenue, Hershey, PA 17033-1240, USA, www.igi-global.com. This series is composed of titles available for purchase individually; each title is edited to be contextually exclusive from any other title within the series. For pricing and ordering information please visit http://www.igi-global.com/book-series/advances-chemical-materials-engineering/73687. Postmaster: Send all address changes to above address.

Titles in this Series

For a list of additional titles in this series, please visit:
https://www.igi-global.com/book-series/advances-chemical-materials-engineering/73687

Process Analysis, Design, and Intensification in Microfluidics and Chemical Engineering
Harrson Silva Santana (University of Campinas, Brazil) João Lameu da Silva Jr (Federal Institute of Education, Science, and Technology of South of Minas Gerais, Brazil) and Osvaldir Pereira Taranto (University of Campinas, razil)
Engineering Science Reference • ©2019 • 367pp • H/C (ISBN: 9781522571384) • US $215.00

Applications of Nanofluid Transportation and Heat Transfer Simulation
Mohsen Sheikholeslami (Babol Noshirvani University of Technology, Iran)
Engineering Science Reference • ©2019 • 692pp • H/C (ISBN: 9781522575955) • US $245.00

Strategic Applications of Measurement Technologies and Instrumentation
Soubantika Palchoudhury (University of Tennessee at Chattanooga, USA)
Engineering Science Reference • ©2019 • 223pp • H/C (ISBN: 9781522554066) • US $195.00

The Geometry of Higher-Dimensional Polytopes
Gennadiy Vladimirovich Zhizhin (Russian Academy of Sciences, Russia)
Engineering Science Reference • ©2019 • 286pp • H/C (ISBN: 9781522569688) • US $195.00

Composites and Advanced Materials for Industrial Applications
K. Kumar (Birla Institute of Technology, India) and J. Paulo Davim (University of Aveiro, Portugal)
Engineering Science Reference • ©2018 • 402pp • H/C (ISBN: 9781522552161) • US $225.00

Emerging Synthesis Techniques for Luminescent Materials
Ratnesh Tiwari (Bhilai Institute of Technology, India) Vikas Dubey (Bhilai Institute of Technology, India) and Sanjay J. Dhoble (Rashtrasant Tukadoji Maharaj Nagpur University, India)
Engineering Science Reference • ©2018 • 505pp • H/C (ISBN: 9781522551706) • US $215.00

For an entire list of titles in this series, please visit:
https://www.igi-global.com/book-series/advances-chemical-materials-engineering/73687

IGI Global
DISSEMINATOR OF KNOWLEDGE

701 East Chocolate Avenue, Hershey, PA 17033, USA
Tel: 717-533-8845 x100 • Fax: 717-533-8661
E-Mail: cust@igi-global.com • www.igi-global.com

Table of Contents

Detailed Table of Contents

Section 1
Nanotechnology for Aerospace Construction

Development of materials always plays a key role in the civilization of the societies. After the industrial revolution, material-based technologies received attention. Nanotechnology has a revolutionary part in the development of industries. Developing technologies cannot be considered without the usage of nanomaterials. Nano-sized materials have different properties than their bulk forms basically because of the increased surface area, surface reactivity, and energy according to decreased size. Carbon-based nanomaterials have a variety of possible application areas from biosensors to aviation. This compact review put a great deal of emphasis on the position of carbon nanomaterials: CNTs, graphene, and carbon nanofibers together with their main synthesis methods and their application areas. The reader can get a quick idea about the basics of CNTs, graphene, and carbon nanofibers and their fabrication techniques.

The formation processes of epoxy nanocomposites with carbon (nanotubes, graphene, and graphite), metal-containing, and aluminosilicate (montmorillonite and halloysite tubes) fillers are considered. A high reactivity of epoxy groups and a thermodynamic miscibility of epoxy oligomers with many substances make it possible to use diverse curing agents and to accomplish curing reactions under various technological conditions. Epoxy nanocomposites are designed to realize to the same extent the unique functional properties of nanoparticles: electric, magnetic, optical, chemical, and biological. The mutual effect of both a matrix and nanoparticles on the composite formation is discussed.

Epoxy resin has been employed as an important matrix for aerospace composite and nanocomposite. In this chapter, latent and essential features of carbon nanotube (CNT) nanofiller have been considered with reference to aeronautical application. Consequently, epoxy/carbon nanotube nanocomposite are conversed here for space competency. Inclusion of CNT in epoxy resin affected the prerequisite features of space nanocomposite. Dispersion of nanotube has been altered using suitable processing technique. Uniform nanotube network formation affects mechanical, electrical, and other physical properties of nanocomposite. Key application areas in this regard are flame and thermal stability, strength, lightning strike resistance, and radiation shielding of space vehicles. Further investigations to optimize structure and properties of multiscale epoxy/CNT nanocomposite are needed for future success in the field. Hence, towards the end, challenges and future prospects of epoxy/CNT nanocomposite have also been deliberated for the improvement of nanomaterial properties for aerospace relevance.

The cryogenic treatment of material has been known to motivate structural stability by rearranging its crystallographic structure in metals and by promoting intermolecular as well as intramolecular rearrangements in polymers. Additionally, in case of polymers reinforced with micro fillers, the structural changes brought about by cryogenic treatment are still largely governed by the polymer matrix itself. Thus, when investigated for their mechanical and tribological properties, the response of polymer/MWCNT nano-composites after cryogenic treatment was found to be depending on the cryo-structural modifications in the polymer matrix, followed by the MWCNT interaction to some extent. The enhancement in the mechanical properties of the polymer/MWCNT nano-composites is attributed to the increasing % crystallinity, changes in crystal structure, conversion of less stable phases into more stable phases, change in the nature of bonding and strengthening of interphase between polymer and MWCNT. Thus, for the cryogenic treatment temperature of -185 °C, the optimum soaking period for PA and PA/MWCNT nano-composite was 24 hrs, whereas for PBT and PBT/MWCNT nano-composite it was 12 hrs and 16 hrs, respectively. This agrees well with the popular claim that each polymer has a specific functional group and/or structural characteristic that readily responds to the cryogenic treatments conditions (irrespective of the filler type, content and/or interaction), thereby, modifying the structure and giving superior properties, which makes cryogenic treatment a material specific process.

Section 2
Modeling and Applications of Nanotechnology in Aerospace

Polymeric nanocomposites are significant engineering materials predominantly due to their enormous potential to meet a spectrum of applications, particularly in improving the strength-stiffness properties, thermal properties, optical and electrical properties. The exploitation of polymer nanocomposites in the aerospace industry is found to be attractive in recent times, since they can provide significant strength to the components with lightweight characteristics. In addition, a wide variety of polymers can be tuned with carbon and non-carbon-based nanomaterials and deployed as archetypes in the structural components of aerospace applications. Accordingly, this chapter consider the key properties of different nanomaterials in polymers as a function of nano-scale approach. Furthermore, this chapter is also dealing with the challenges that need to surmount the technological enduring of the polymer nanocomposites for advancements in the aerospace structural applications in the coming future.

Electromagnetic waves can have serious effects on human health by long-term exposure. Developing lightweight materials with good electromagnetic radiation shielding (EMS) that could prevent interference is a high desire for protection. Nanocomposites polymers have a wide range of potential applications and offers suggested solutions in environmental and aerospace applications. This chapter will cover the current challenge in the reduction of electromagnetic wave by developing lightweight absorber material with a wide absorption frequency. A wide range of different nanocomposites polymers contain conductive fillers such as metal or magnetic nanoparticles and carbon-based materials will be discussed. In addition, EMS mechanisms of reflection, absorption, and multiple reflections will be discussed. The unique of the chemical and physical properties of nanocomposites polymers are promising for shielding with low-cost environmentally friendly material.

This chapter deals with the modelling of nanomaterial and nanocomposite mechanical and thermal properties. Enrichment in the technology requires materials having higher thermal properties or higher structural properties. Nanomaterials and nanocomposites can serve this purpose accurately for aerospace or thermal applications and structural applications respectively. The thermal system requires materials having high thermal conductivity while structural system requires materials having high strength. Selection of the material for particular application is very critical and requires knowledge and experience. Al, Cu, TiO2, Al2O3, etc. are considered for thermal applications while epoxy-glass, FRP, etc. are considered for structural applications. Modelling of these nanomaterials and nanocomposites is done with the help of different mathematical models available in the literature. Results show that addition of the nanoparticle/composite in the base material can enhance the thermal and structural properties. Results also show that amount of weight percentage added also affects the properties.

The design methods based on aerospace model have been widely used in aircraft conceptual design for decades and proven very effective when restricted to simple problems with very approximate analyses. These monolithic, large, design and analysis codes are genuinely multidisciplinary, but as analyses become more complex, such codes have grown so large as to be incomprehensible and hence difficult to maintain. This chapter deals with the computational modeling of nanoparticles. Nanomaterials constitute a prominent sub-discipline in the materials and chemical sciences. Conventional materials like glass, ceramic, metals, polymers, or semiconductors can be acquired with nanoscale proportions. Nanomaterials have various microstructural distinctive attributes such as nanodiscs, nanotubes, nanocoatings, quantum dots, nanocomposites, and nanowires. The unique properties of nanoparticle-based materials and devices depend directly on size and structure dependent properties.

Preface

Nanotechnology has a revolutionary part in the development of various industries. However, developing nanotechnologies cannot be achieved without the utilization of nanomaterials. These nanomaterials have different tremendous properties compared to their bulky forms basically due to their higher surface area, surface reactivity, and energy according to reduced dimension. As an example, carbon nanomaterials have been widely used in industries from biosensors, actuators, to aviation. The development in the aerospace relates to many factors ranging from reduction of weight, less or no repairs, extended and safest life, availability of materials with multifunctional properties, bio-fuels, low fuel consumption, faster and highly responsive communication systems, decreased time frame of developing cycle from concept to implementation and others. In addition, the polymeric nanocomposites are important engineering materials predominantly due to their enormous potential to fit a wide range of applications, particularly in enhancing the mechanical strength-stiffness characteristics, thermal properties, optical and electrical properties. The use of polymer nanocomposites, such as polymer/Carbon nanomaterials, in the aerospace industry is increasing in the recent few years, since they guarantee a higher strength to the components with lightweight features. One famous useful polymer is epoxy resin, which has been utilized as an important matrix for producing aerospace composite and nanocomposite. Key application areas of these nanomaterials in this regard are thermal stability, mechanical strength, lightning strike resistance, and radiation shielding of space vehicles as the electromagnetic waves can engender serious effects on human health by long term exposure. Thus, producing of lightweight nanomaterials having enhanced electromagnetic radiation shielding could prevent interference is high desire for protection.

ORGANIZATION OF THE BOOK

The book is organized into eight chapters. A brief description of each of the chapters follows:

Chapter 1 put a great deal of emphasis on the position of carbon nanomaterials; CNTs, Graphene and Carbon Nanofibers together with their main synthesis methods and their application areas. The reader can get a quick idea about the basics of CNTs, Graphene and Carbon Nanofibers and their fabrication techniques.

Chapter 2 describes the formation processes of epoxy nanocomposites with carbon (nanotubes, graphene, and graphite), metal-containing, and aluminosilicate (montmorillonite and halloysite tubes) fillers are considered. It also discusses the design of epoxy nanocomposites to realize the unique functional properties of nanoparticles including electric, magnetic, optical, chemical, and biological.

Chapter 3 deals with latent and essential features of carbon nanotube (CNT) nanofiller have been considered with reference to aeronautical application. Consequently, epoxy/carbon nanotube nanocomposite are conversed here for space competency. Inclusion of CNT in epoxy resin affected the prerequisite features of space nanocomposite. Dispersion of nanotube has been altered using suitable processing technique. Uniform nanotube network formation affects mechanical, electrical, and other physical properties of nanocomposite. Key application areas in this regard are flame and thermal stability, strength, lightning strike resistance, and radiation shielding of space vehicles. Further investigations to optimize structure and properties of multiscale epoxy/CNT nanocomposite are needed for future success in the field. Hence, towards the end challenges and future prospects of epoxy/CNT nanocomposite have also been deliberated for the improvement of nanomaterial properties for aerospace relevance.

Chapter 4 describes the latent and essential features of carbon nanotube (CNT) nanofiller have been considered with reference to aeronautical application. Consequently, epoxy/carbon nanotube nanocomposite are conversed here for space competency. Inclusion of CNT in epoxy resin affected the prerequisite features of space nanocomposite. Dispersion of nanotube has been altered using suitable processing technique. Uniform nanotube network formation affects mechanical, electrical, and other physical properties of nanocomposite. Key application areas in this regard are flame and thermal stability, strength, lightning strike resistance, and radiation shielding of space vehicles. Further investigations to optimize structure and properties of multiscale epoxy/CNT nanocomposite are needed for future success in the field. Hence, towards the end challenges and future prospects of epoxy/CNT nanocomposite have also been deliberated for the improvement of nanomaterial properties for aerospace relevance.

Chapter 5 consider the key properties of different nanomaterials in polymers as a function of nano-scale approach. Furthermore, this chapter is also dealing with the challenges that need to surmount the technological enduring of the polymer nanocomposites for advancements in the aerospace structural applications in the coming future.

Chapter 6 covers the current challenge in the reduction of electromagnetic wave by developing lightweight absorber material with a wide absorption frequency. A wide range of different nanocomposites polymers contain conductive fillers such as metal or magnetic nanoparticles, carbon-based materials will be discussed. In addition, EMS mechanisms of reflection, absorption and multiple reflections will be discussed. The unique of the chemical and physical properties of nanocomposites polymers are promising for shielding with low-cost environmentally-friendly material.

Chapter 7 considers the key properties of different nanomaterials in polymers as a function of nano-scale approach. Furthermore, it is also dealing with the challenges that need to surmount the technological enduring of the polymer nanocomposites for advancements in the aerospace structural applications in the coming future.

Chapter 8 summarizes the computational modeling of nanoparticles. Nanomaterials constitute a prominent sub-discipline in the materials and chemical sciences. Conventional materials like glass, ceramic, metals, polymers or semiconductors can be acquired with a nanoscale proportion. Nanomaterials have various microstructural distinctive attributes such as nanodiscs, nanotubes, nanocoatings, quantum dots, nanocomposites, and nanowires. The unique properties of nanoparticle-based materials and devices depend directly on size and structure dependent properties.

Noureddine Ramdani
Research and Development Institute of Industry and Defense Technologies, Algeria & Harbin Engineering University, China

Acknowledgment

The editor would like to acknowledge the help of all the people involved in this project and, more specifically, to the authors and reviewers that took part in the review process. Without their support, this book would not have become a reality.

First, the editor would like to thank each one of the authors for their contributions. My sincere gratitude goes to the chapter's authors who contributed their time and expertise to this book.

Second, the editor wishes to acknowledge the valuable contributions of the reviewers regarding the improvement of quality, coherence, and content presentation of chapters. Most of the authors also served as referees; I highly appreciate their double task.

Noureddine Ramdani
Research and Development Institute of Industry and Defense Technologies, Algeria & Harbin Engineering University, China

Section 1

Nanotechnology for Aerospace Construction

Chapter 1
Carbon Nanomaterials:
Carbon Nanotubes, Graphene, and Carbon Nanofibers

Ezgi Ismar
https://orcid.org/0000-0002-4612-416X
Istanbul Technical University, Turkey

A.Sezai Sarac
Istanbul Technical University, Turkey

ABSTRACT

Development of materials always plays a key role in the civilization of the societies. After the industrial revolution, material-based technologies received attention. Nanotechnology has a revolutionary part in the development of industries. Developing technologies cannot be considered without the usage of nanomaterials. Nano-sized materials have different properties than their bulk forms basically because of the increased surface area, surface reactivity, and energy according to decreased size. Carbon-based nanomaterials have a variety of possible application areas from biosensors to aviation. This compact review put a great deal of emphasis on the position of carbon nanomaterials: CNTs, graphene, and carbon nanofibers together with their main synthesis methods and their application areas. The reader can get a quick idea about the basics of CNTs, graphene, and carbon nanofibers and their fabrication techniques.

DOI: 10.4018/978-1-5225-7921-2.ch001

INTRODUCTION

Materials have always had an important place in the improvement of societies since the beginning. The prehistoric ages were named according to the commonly used material such as; stone, bronze and iron ages which reflects the importance of the materials in the development of civilizations. When we look at the today's technological developments, composite materials gain a huge impact due to their unique and combined properties in the same product. As represented in Figure 1 our century is called the Nano age due to the developments on the engineered nano materials. Not only their neat forms but also their cooperation with the previously improved polymeric matrixes take attention and have a pivotal role in the material science.

When Richard Feynman emphasized the importance of nanoparticles by saying "There's plenty of room at the bottom." nanomaterials start to draw attention. However, characterization tools had not been powerful enough to examine the submicron systems, thus; technological developments needed to wait until they let the examine nano sized materials. Nano materials have distinctively different properties than their bulk forms in terms of surface area and dominance of quantum effects,

Figure 1. The materials time-line with a nonlinear scale

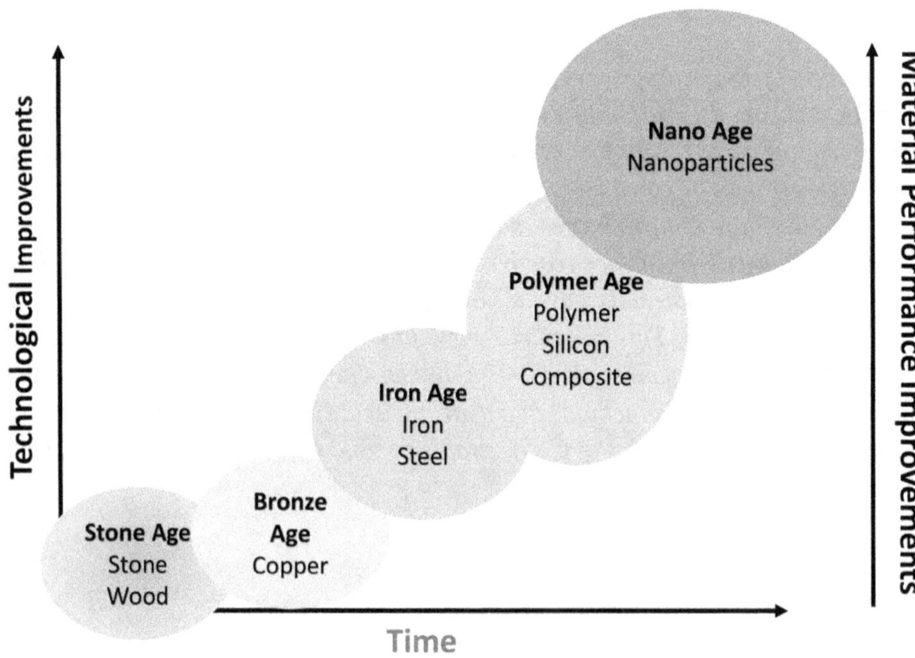

while increased surface area play a role on the improved chemical reactivity of the material, quantum effects are resultant with the change a material's optical, magnetic or electrical properties (Britain, 2004). Thus, polymer composites with nanoparticle addition display different role than the bulk size of polymeric composites. Total investment in the nanotechnology is getting rise day by day due to the proliferation product range (Navarro et al., 2008).

Carbon is not the most abundant element in the world however, its role is fundamental and Earth's 0.2% of mass consists of carbon (Delhaes, 2013).

Different carbon atom bonding causes to obtain allotropes of carbon which are comprised of unique properties. Most well-known carbon allotropes are; diamond and graphite (Figure 2). Besides, graphene, fullerenes, carbon nanotubes are the new generation of carbon allotropes (Krüger, 2010; Delhaes, 2013). The ideal crystal structure of graphite (see Figure) consists of layers in which the carbon atoms are arranged in an open honeycomb network containing two atoms per unit cell in each layer, a carbon nanotube is a graphene sheet appropriately rolled into a cylinder of nanometer size diameter. Therefore, it is expected to the planar sp^2 bonding is characteristic of graphite to play a significant role in carbon nanotubes (Ajayan, 1999). Each carbon atom has six electrons which occupy $1s^2$, $2s^2$, and $2p^2$ atomic orbitals. Three possible hybridizations occur in carbon: sp, sp^2, and sp^3 while other group IV elements such as Si and Ge exhibit primarily sp^3 hybridization (Dresselhaus

Figure 2. (A) Graphene, (B) CNT, (C)Fullerene, (D)Carbon Nanofiber

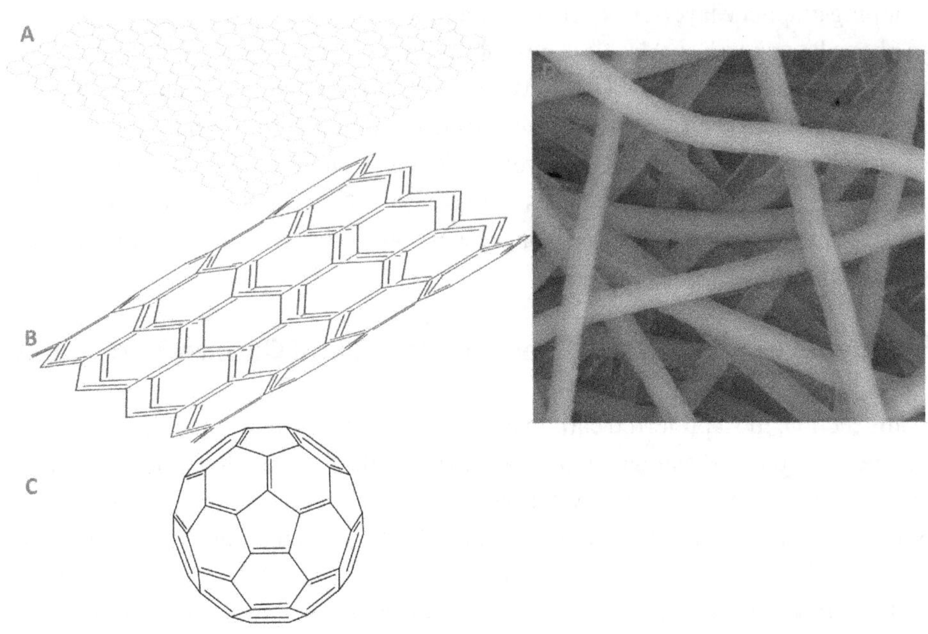

& Avouris, 2001). The discovery more than two decades ago of the allotrope of carbon referred to Buckminsterfullerene which was a pivotal event at the beginning of nanotechnology. This key discovery showed that; how intense energy sources may be used to vaporize ordinary carbon and convert it into the nanomaterials such as fullerenes and carbon nanotubes (Weisman & Subramoney, 2006). Multi-walled carbon nanotube variants were synthesized initially in 1991 (Iijima, 1991), while graphene is the unwrapped form of the carbon nanotubes.

Polymer nanocomposites consist of nanoparticles such as, carbon nanotubes, carbon nanofibers, and graphene, which embedded inside the polymeric matrix and different application areas are possible according to homogeneous dispersion of nanoparticles inside the host matrix without losing their performance and preventing the agglomeration (Njuguna & Pielichowski, 2004). Nanoparticles have a huge variety of application areas from sports to the aviation industries (Cao & Wang, 2011). In this chapter, carbon based nanoparticles; carbon nanotubes, graphene, and carbon nanofibers were considered.

CARBON NANOTUBES

Carbon nanotubes (CNTs) are novel carbonaceous materials and their composites have been widely studied since their discovery in 1991. CNTs have a significant potential as a building block, due to their high aspect-ratio together with their nano-sized tubular structure which leads to strong confinement effect. Principally, CNTs are created by wrapping graphene layers. CNTs can be singular or multi-walled according to folded graphene layer number. Also, it is possible to increase the variety of CNTs by using doping techniques. Doped CNTs have numerous advantages and can be useful for nanoscale devices such as; miniaturized electronics and electromechanical, chemical and scanning probe devices. Moreover, CNTs can be used as filler for macroscopic composites. In this review, synthesis methods for CNTs and doping techniques are briefly introduced together with their present application areas to give an approach to basic knowledge about CNTs.

Synthesis of Multi-Walled and Single-Walled Carbon Nanotubes

In this section, most preferred carbon nanotube fabrication techniques are presented together with their advantages and drawbacks. Synthesis of single-walled and multi-walled carbon nanotubes mainly followed by arc-discharge (Iijima, 1991; Journet et al., 1997; Moravsky, Wexler, & Loutfy, 2004), laser ablation (Moravsky et al., 2004; Rinzler et al., 1998) and chemical vapor deposition (CVD) (Z. P. Huang et al., 1998) methods and schematic presentations were given in the Figure 3. Past decades,

Figure 3. Schematic description of (a) electric arc-discharge, (b) laser ablation, (c) chemical vapor deposition methods

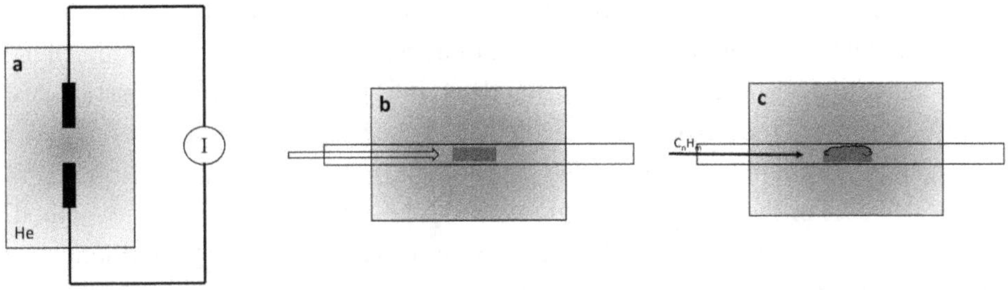

arc-discharge and laser ablation methods had more interest to build nanotubes, and common principle of these methods is to condensation of carbon atoms which are obtained from evaporation of solid carbon sources and temperature during these methods are nearby to the melting temperature of graphite which is around 3000-4000 °C (Dai, 2001). During the growth of SWNTs via laser ablation or arc-discharge methods, by products of the process can be; graphitic polyhedrons with enclosed metal particles, fullerenes and amorphous carbon as particles or over coating on the sidewalls of nanotubes (Dai, 2001).For scale-up productions laser ablation and arc discharge method have limitations. Methods require further purification steps due to the creation of impurities in the form of catalyst particles, amorphous carbon during the process. Moreover, these methods make the production of CNTs expensive (Thostenson, Ren, & Chou, 2001). On the other hand, gas phase techniques, CVD can be a potential for the scale-up production of CNTs with fewer impurities (W. Z. Li et al., 1996; Thostenson et al., 2001).

Electric Arc Discharge

A basic of the arc-discharge method is explained as; plasma of helium gas ignited by high currents passed through carbon anode and cathode aids to evaporate carbon atoms. The arc-discharge method has been developed to fabricate high quality multi-walled nanotubes (MWNTs) and single-walled nanotubes (SWNTs). MWNTs can be obtained by adjusting the growth conditions such as the arcing current and the pressure of an inert gas in the discharge chamber. In 1992, a revolution in MWNT growth by arc-discharge was first made by Ebbesen and Ajayan who have been achieved the growth and purification of high quality MWNTs at the gram level (Ebbesen & Ajayan, 1992).

MWNTs are synthesized by arc discharge, have lengths around ten microns and diameters in between the range of 5-30 nm. The nanotubes are naturally connected together by strong van der Waals interactions and form tight bundles. MWNTs, which are created by the arc-discharge method, are very straight, due to their high crystallinity. For as grown materials, there are a few defects such as pentagons or heptagons existing on the sidewalls of the nanotubes. Moreover, the by-product of the arc-discharge growth process is multi-layered graphitic particles in polyhedron shapes. It is also possible to the purification of MWNTs by heating the grown material under an oxygen atmosphere to eliminate the graphitic particles by oxidation (Ebbesen & Ajayan, 1992). For the growth of single-walled tubes via the arc-discharge method, a metal catalyst is required. The first success in producing substantial amounts of SWNTs by arc-discharge was achieved by Bethune and coworkers in 1993 (Bethune et al., 1993).

When the arc-discharge is carried on keeping the gap between the carbon electrodes about 1mm, cylindrical deposits forms on the surface of the cathode. The diameter of this cathode deposit is the same as that of the anode stick (Yumura, 1999).

To obtain a high yield of tubes via carbon arc technique is challenging and requires careful control of experimental conditions as opposed to its production simplicity. Generally, in laboratory scale production operation is carried with direct current (DC) and arc operates 1 to 4 mm wide gap between two graphite electrodes which have a diameter in 6 to 12 mm and can be placed vertically or horizontally in a water cooled chamber which is complete with Helium gas at sub atmospheric pressure (Moravsky et al., 2004). During the fabrication of MWNT, quality or quantity is not affected from the position of the electrode axis however, Helium gas and DC current are more remarkably affect the fabrication yield and the high yield arc process requires a constant feed rate and arc current for the duration of synthesis for the stable synthesis (Kiselev, Moravsky, Ormont, & Zakharov, 1999).Graphite cathode and anode holder rods are generally equipped with the water cooling system to prevent the damage which can be happened when the arc is closed to one of the holders. The cooling has effects on the properties of the cathode deposit merely a first few millimeters from its origin; the deposit up to 10 to 15 cm total length (rest of the deposit) has a stable cross-sectional structure and composition (Moravsky et al., 2004). The cathode deposit growth in the first few millimeters, the arc process stabilizes and a steady electrode surface temperature is set. Then, the cooling of the cathode is not affecting the deposit working surface temperature. Thus, it is beneficial to use a cathode of adequately small diameter (for instance; equal or smaller than the anode rod diameter) that is long enough (a few centimeters) to more quickly reach a steady temperature at the cathode working surface and in the arc. Furthermore, the cathode deposit is more strongly connected to a smaller diameter cathode compared to larger ones. A strong bond has a key role to prevent the cathode deposit from breaking off the cathode surface in the sequence of the production process. Moreover, excessive cooling of the cathode holder is undesirable property and, same is also true for the anode holder. Because when the arc zone approaches to cold anode holder, the anode working surface becomes cooler, which affects the plasma properties, and deteriorating the uniformity of the deposit structure and composition. Advisable holders are generally low-heat conductive anode holders (Moravsky et al., 2004). Another production parameter is, Helium gas and it can be a static or dynamic atmosphere during the process and there is not any product quality difference is recorded. For MWCNTs, requirements for Helium gas purity are quite less rigid than for fullerene or SWNT synthesis; to be exact; 99.5% and 99.999% pure gases, respectively produce the same amount of MWNTs. The MWNT synthesis is much more flexible for the presence of oxygen based impurities in the buffer gas due to the formation of MWNTs is occurred in the inter electrode gap, where the gas does not penetrate, while fullerenes and SWNTs are formed outside the gap (Moravsky et al., 2004).

Laser Ablation

Laser ablation method was used when C_{60} was first discovered in 1985 (Kroto et al., 1985). This method has also been applied for the synthesis of CNT, but the length of MWCNT is much shorter than that by the arc-discharge method. So, this method does not seem adequate to the synthesis of MWCNT. However, in the synthesis of SWCNT marvelously high yield has been obtained by this method, hence, the laser ablation method has become another important technology in this respect (Yumura, 1999).

During fullerene production experiments using a laser vaporization apparatus with an ablated graphite sample positioned in an oven, it was found that closed-ended MWNTs were produced in the gas phase through homogeneous carbon-vapor condensation in a hot argon atmosphere (Jiang et al., 2004).

MWNTs which are produced with the laser ablation method are similar to arc produces ones, due to their relatively short size (ca. 300 nm), but the number of layers which are ranging between 4 and 24 nm, and the inner diameter, varying between 1.5 and 3.5 nm. MWNTs with an even number of layers was generally obtained, however there is not any SWNTs were observed. The yield and quality of MWNTs are dropped at oven temperatures below 1200°C, and at 200°C there is not any nanotubes were detected. Which is explained by the need for the sufficiently high temperature to rapidly anneal the imperfect structures into a closed form through the incorporation and rearrangement of pentagons (Smalley, 1992).

How an open-end nanotube would be grown rather than close off into a short spheroidal structure at 1200°C is unclear due to the applied temperature range and in the absence of extrinsic stabilizing factors (such as the high electric fields) proposed earlier in the literature (Smalley, 1993). It was suggested that adequate temperature and rapid annealing play a key role, stabilization of the open-end interaction between adjacent tube layers is provided by carbon "adatom" bridges that saturate some of the dangling bonds (Guo et al., 1995). Such cohesive lip-lip interactions may explain the tendency of MWNT layers to close in pairs, as well as occasionally observed (Kwon et al., 1997) "turnaround" growth in arc-produced MWNTs. To keep the growing MWNT end open, it is essential to have lip-lip interaction between at least two inner layers. The next layers may grow with some delay, for example, by island nucleation (Kwon et al., 1997) and anneal on the underlying nanotube template. Nanotube thickening by additional layer growth terminates when carbon-vapor density falls below a threshold value corresponding to the saturated vapor pressure of small carbon clusters adsorbed on the tube surface. This threshold value decreases with the increasing tube radius thus, may be imposed a limit on the MWNT diameter (Guo et al., 1995).

Chemical Vapor Deposition

Chemical vapor deposition (CVD) methods have been successfully used for creating carbon materials such as; fibers, filaments, and nanotubes for more than decades (Morgan, 2005). With arc-discharge or laser ablation techniques, only tangled nanotubes mixed randomly with various impurities are obtainable.

Recent research activities in CVD nanotube growth have also been sparked by the idea that aligned and ordered nanotube structures can be grown on surfaces with control. Growth process explained as; catalyst material is heated to high temperature inside the tubular furnace and time to time hydrocarbon gas flows through the furnace and materials grown over the catalyst are taken with the cooling of the system at room temperature (Dai, 2002).The key parameters for CVD growth of nanotubes are the hydrocarbons, catalysts and growth temperature. The active catalytic species are typically transition-metal nanoparticles formed on a support material such as alumina (Dai, 2001).

CVD has been extensively used to grow a variety of CNT structures, a single-walled nanotube is rolled-up tubular shell of the graphene sheet, which is made up of benzene type hexagonal rings of carbon atoms and, a multi-walled nanotube is a rolled-up stack of graphene sheets in concentric cylinders. In contrast, a stacked cone arrangement in the central core region is also seen in some CVD samples (Nolan, Lynch, & Cutler, 1998).

Doping Techniques for Carbon Nanotubes

Substitutional doping of carbon nanotubes was first attempted in 1994 (Stephan et al., 1994). They synthesized, boron-nitrogen doped MWNTs using the arc-discharge method. Nitrogen doped nanotubes are less stable than their pure carbon counterparts, breaking easily under the electron beam in the transmission electron microscope (TEM) and oxidizing at lower temperatures than undoped tubes. This is understandable when one considers the nitrogen atoms as localized defects, which will be energetically less stable than a pure carbon lattice. This also makes the tube surfaces chemically active, facilitating chemical functionalization (Jiang et al., 2004).

The possibility of attaching functional groups to the nanotube surface also allows the combination of the properties of the nanotubes and the additional unit.

Besides the chemical modification of defect groups, a significant improvement for single walled carbon nanotubes have sidewall functionalization via addition reactions to use reactive organic species such as; nitrenes, carbenes, lithium alkyls, fluorine and diazonium salts (Holzinger et al., 2001).

Nanotube doping can be achieved in two different ways, either directly during the synthesis or by post treatment of pre-synthesized carbon nanotubes. For both approaches, various methods have successfully been employed for making nitrogen doped carbon nanotubes. Arc-discharge method; the tubes are grown through evaporation of a graphite rod via an electric-arc, in a Helium gas atmosphere. Another high temperature growth technique is laser ablation, whereby a catalyst containing a graphite target is ablated using a pulsed laser, with the vaporization products rapidly removed by flowing gas such as N_2. Laser ablation is also commonly used to produce nitrogen doped fullerenes. Arc-discharge, laser ablation and the solar oven all belong to the so-called high temperature synthesis methods. The high temperature is used in all these cases to evaporate the carbon source, normally graphite (Journet et al., 1997). Nitrogen doped multi-walled nanotubes have been synthesized both by classical chemical vapor deposition techniques (CVD) and by aerosol assisted CVD methods (Ewels, Glerup, Krstic, Basiu, & Basiuk, 2007). Post treatment techniques; apart from nitrogen incorporation during growth, it is also possible to nitrogen doped nanotubes through post treatments of pristine carbon nanotubes. Chemical substitution has been used to nitrogen doped SWNTs, through partial substitution reactions of undoped tubes, using B_2O_3 vapor and N_2 gas at 1500-1700 °K. This resulted in irregular damaged nanotube walls with less than 1% N substitution, when it is comparable with arc-discharge techniques. D.C. magnetron sputtering has successfully been applied by Suenaga et al, achieving high nitrogen concentrations of 15-30% nevertheless, mainly in the carbon nanoturbolites (Suenaga et al., 1999).

Based on theoretical studies, ion implantation has been proposed as a way to selectively dope carbon nanotubes with nitrogen. This approach was used experimentally to implant nitrogen in MWNTs (Morant et al., 2006). They quantified the nitrogen concentration to be 0.47% by using XPS. In particular, spatially resolved doping should allow both homogenous dopant distribution and spatially localized doping as required. Notably, concentration variances in nanotubes doped during synthesis can arise from inhomogeneous synthesis conditions, e.g. impurities in the catalyst, catalyst size, an uneven gas flow, or temperature gradients in the reaction zone (Ewels et al., 2007).

Properties and Applications

CNT is a folded layer of graphene which is rolled around and according to rolling angle it is named such as arm chair, chiral and zigzag (Figure 4). Moreover, If CNT is only one layer it is named as single-walled carbon nanotubes (SWCNTs) and if it has multi layers than it is named as multi-walled carbon nanotubes (MWCNTs) which is basically centric SWCNT with outer rolled graphene layers (Thostenson et al., 2001).

CNTs have a low density with one-dimensional properties in nano range, extraordinary mechanical and electrical properties with high aspect ratios which makes them a good reinforcement material for the polymeric composites (Spitalsky, Tasis, Papagelis, & Galiotis, 2010). Their highlighted properties are mechanical, electrical, magnetic, optical and thermal properties (Xie, Mai, & Zhou, 2005).

CNTs can be used in a variety range of products in the field of aerospace applications. Application of CNTs can take a place in commercial aircrafts,

Figure 4. Schematic representation of CNT type according to rolling angle

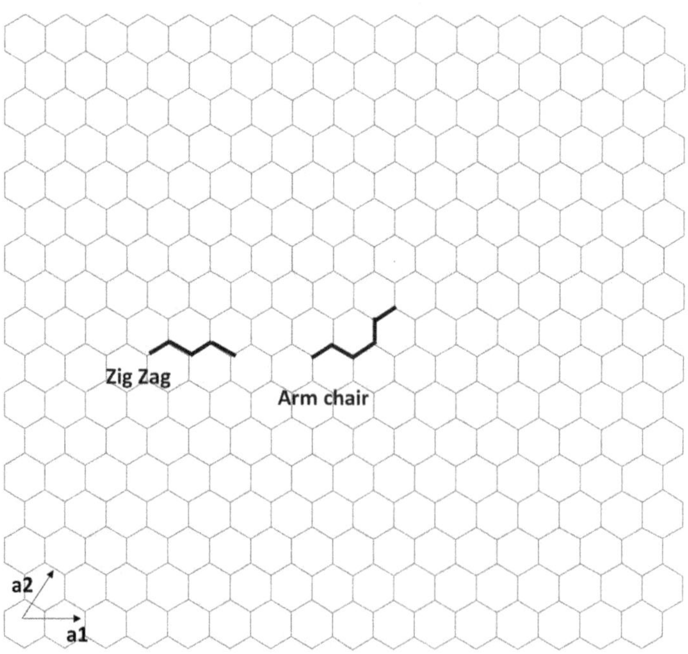

satellites, space launch vehicles moreover, their stipulated applications such as hydrogen storage encapsulation, lightning protection for aircraft, reduced weight of satellites are breathtaking (Gohardani, Elola, & Elizetxea, 2014). For the high-tech application areas such as aerospace, aviation, automobile, etc. polymeric composite structures should have anti-wearability, load carrying capacity together with the low density, in that perspective nanoparticle addition to composite structures are studied (Su, Zhang, & Liu, 2006). CNTs are considered as the new area for new the enhancement of the polymeric composite structures for tougher and/or fatigue resistant materials which is highly depended to their good dispersion on the matrix (Bal & Samal, 2007; Cooper, Ravich, Lips, Mayer, & Wagner, 2002). The aim of creating the polymer/CNT composites is to combine ease of processability of the polymer with the excellent properties of the CNTs (Xie et al., 2005). Extent and efficiency of stress transfer through the interface between nanotubes and polymers makes their dispersion difficult for the nanocomposite (Njuguna & Pielichowski, 2004). Different than the conventional filled polymer composites, CNT composites are required low filling ratios to achieve desired property enhancements, thus they are a strategic candidate for the aerospace applications (Njuguna, Pielichowski, & Fan, 2012). Polymer and CNT cooperation creates composite structures which can be used in a variety of applications. Composite fabrication of CNTs via conventional composite processing have a wide range of challenges thereby the uniform dispersion or difficulty of precise control. Layer by layer assembly of CNTs offers advantages such as; uniform dispersion of CNTs into a composite which is enabled by direct adsorption of CNTs from a solution to a solid state without phase segregation, tailor-made multifunctionality, and ease of fabrication (Podsiadlo, Shim, & Kotov, 2009).

Depend on the dispersion and functionalization of carbon nanotubes they have fruitful mechanical and multifunctional properties (Ma, Siddiqui, Marom, & Kim, 2010). Their mechanical properties are highly dependent on the cooperating polymer type, filling ratio of CNTs and together with the single or multi walled structure of the carbon nanotubes. Moreover, CNTs exhibit an improvement on the electrical properties of the composites whereas it is highly fluctuating in accordance with the function of the polymer matrix, processing method and CNT type (Spitalsky et al., 2010).

NASA also has a research on the composite structures with CNTs for the Mars mission applications (Dresselhaus, Dresselhaus, & Avouris, 2001). Likewise, CNTs are appropriate candidates to work as a pressure sensor, and durable interfaces with electronics as a form of semiconductor membranes with the combination of textiles (Gugliuzza & Drioli, 2013). Polymer composites of nanotubes can be used for transparent conductive coatings, electromagnetic interference shielding applications (Spitalsky et al., 2010).

Another most studied polymer/CNT combination is based on fibers. Addition of CNT to fiber improves the properties of the fiber even though they have superiorities than the high performance fibers in terms of tensile strength values. Direct dispersion of CNTs into the chopped fiber creates a polymeric composite whereas alignment of the CNT and fibers have a significant role (Iwahori, Ishiwata, Sumizawa, & Ishikawa, 2005). CNT and fibers are expected to be used in a wide range of application areas such as; reinforcements for composites, supercapacitors, transmission lines and electrochemical sensors (Yadav, Dasgupta, Patwardhan, & Joshi, 2017). For all cases, the huge amount of CNT is needed to create combined structures like polymer composites or fiber/CNT combinations. This is come together with the scale-up limitation issues of CNTs and the need for further purifications. Thus, fabricating the highest amount of CNTs, the CVD method is the most promoted one (Thostenson et al., 2001).

GRAPHENE

Graphene is a two dimensional layer of sp^2 bonded carbon atoms and has only a one atom thickness (Figure 5), which displays improved mechanical and thermal properties (H. Kim, Abdala, & Macosko, 2010) (Stankovich et al., 2006) (Ramanathan et al., 2008). Graphene exhibits a superior mechanical and electrical features which makes it ideal for potential applications and it is possible to obtain polymeric composite structures with a lower amount of loadings (Du & Cheng, 2012).

Synthesis of Graphene

Single atom thickness is the most significant reason which makes graphene the rising star of nanoscience. Primarily, the graphene layer is obtained via mechanical exfoliation of graphite with the assistance of scotch tape (Geim & Novoselov, 2009). Different synthesis techniques for graphene layers are available with bottom up and top down approaches. Most common methods are; mechanical exfoliation, chemical exfoliation and chemical vapor deposition (Allen, Tung, & Kaner, 2010; Bhuyan, Uddin, Islam, Bipasha, & Hossain, 2016; Novoselov et al., 2004; Reina et al., 2009). Principally mechanical exfoliation of graphite is based on a mechanical force and adhesive material to remove a graphene layer from graphite (Bhuyan et al., 2016). Chemical exfoliation is explained as a solution with alkali ions dispersed to graphite and large alkali ions help to open up the graphite layers (Stankovich et al., 2007). As described in the previous part, the catalytic thermal CVD method is highly suitable to produce scale up graphene layers. Different than those methods, chemical synthesis methods are also available in the literature (Pei & Cheng, 2012).

Figure 5. Schematic representation of graphene layers in a form of graphite

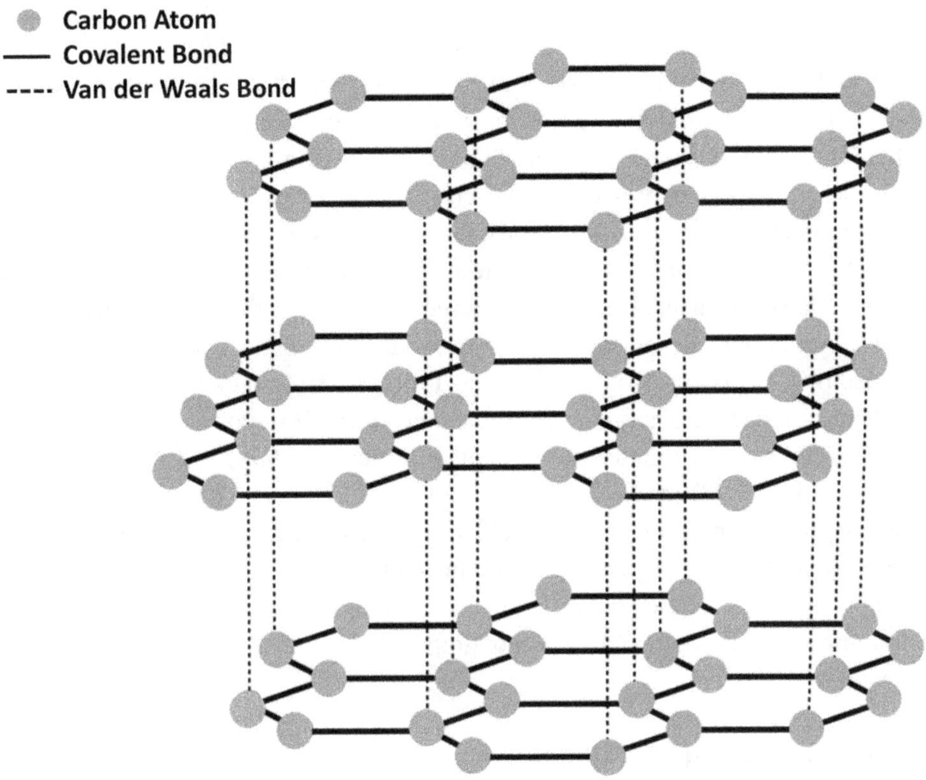

In terms of fabricating the graphene, exfoliation means a separation/peel off graphite layers or its derivatives such as graphene oxide (GO) to obtain a graphene layer. For mechanical exfoliation single graphene layer is extracted on the preferred substrate with the applied external force to remove a single graphene layer on the graphite (Bhuyan et al., 2016). External force should be around 300 nN/μm^2 to separate mono layer of graphene which is consist of the top down approach, opposite of the stacking mono layers (Zhang, Small, Pontius, & Kim, 2005). This mechanical process can be conducted with the assistance of several products; such as scotch tape (Geim & Novoselov, 2009), the electric field (X. Liang et al., 2009), and ultra-sonication (Ci et al., 2009). Graphene flakes can be characterized via optical microscopy, Raman spectroscopy and AFM to determine the thickness and the amount of layers (Z. H. Ni et al., 2007; Z. Ni, Wang, Yu, & Shen, 2008). By this method high quality graphene layers without defects are achieved however scale up issues are existing due to the usage of the wafer as a plate to collect graphene layers and which is labor exhaustive process (Bhuyan et al., 2016).

Chemical exfoliation method is based on a separation of graphite to graphene layers via the assistance of the chemical substrates. Chemical exfoliation of graphite has two main stages. Firstly, colloidal suspension helps to reduce Van der Waals forces to increase the interlayer spaces between graphene layers than, rapid heating or sonication helps to exfoliate graphene with single to few layers (Wu, Yu, & Shen, 2010). This method is quite appropriate to obtain easily graphene layers.

As described in the previous parts, chemical vapor deposition method helps to vaporize heated material to gaseous state, which is named as a precursor and thanks to the high temperature substrate is diffused on the precursor. Deposition of high quality graphene layers are achieved via CVD on the different transition metal substrates and Ni was the first substrate where the large area of graphene was obtained by CVD (Obraztsov, Obraztsova, Tyurnina, & Zolotukhin, 2007).

Chemical reduction of graphite oxide is another method to produce graphene flakes. Mainly methods are Hummer's (Hummers & Offeman, 1958) and, Staudenmaier (Staudenmaier, 1898) methods. They are followed to chemically reduce GO. For reduction of GO, sonication and electrochemical reduction, and, pyrolysis of graphene are also available. It is also possible to obtain graphene layer via, unzipping of CNTs. Unzipping process is based on a chemical and plasma etched method of CNT. As a bottom-up approach, epitaxial growth of graphene is another way to build graphene (Bhuyan et al., 2016).

Properties and Applications

The twenty-first century has an impact on the development of nanomaterials and graphene is a 2-D new generation material with lots of unique properties. These thin and strong materials provide cutting-edge technologies for the development of light and strong materials. Nanoparticle applications in engineered structural composites have an important role. Among the different structural composite systems aerospace structural composites should provide tremendous physical, chemical, electrical, thermal and mechanical properties. Thus, nanoparticles especially carbon nanoparticle combinations serve a good opportunity for the structural aerospace applications. Whereas they need high property demands for bringing out higher performance materials. While creating the high performance structural composites for the aviation industry, performance criteria such as weight, processability, life cycle and risks assessment of the new material should be considered and balance with the performance expectation (Baur & Silverman, 2007). Nanocomposites are getting attention due to the enhanced properties and application areas of engineering polymers (Njuguna & Pielichowski, 2004).

Carbon nanomaterials are used as effective fillers to produce high performance structural composites. Enhanced electrical, chemical and mechanical properties make them considered as nanocomposite materials for a variety of applications such as aviation, electronics and biotechnology applications (Iwahori et al., 2005; Lim, Guzman, Schaefer, & Minaie, 2013; Zhou, Pervin, Rangari, & Jeelani, 2006). Flame retardancy and thermal stability is one of the required material properties for aviation applications and, desired properties are achieved with the usage of the nanocomposites (Njuguna & Pielichowski, 2004). Carbon nanoparticles are also considered as an additive for aerospace applications (Rettenbacher, Elliott, Hudson, Amirkhanian, & Echegoyen, 2006).

Graphene has a variety of application areas from optoelectronics to polymeric composites. Mainly electronic and photonic applications are emerging technologies (W. Choi, Lahiri, Seelaboyina, & Kang, 2010). There is also gas and biosensor and battery applications are available for the products which contain graphene. Graphene can be used as a gas and bioelectronic sensors and working principle is based on a change in the electrical conductivity, sigma bonds (σ) for the reason of adsorption of molecules on the graphene sheet. The absorbed molecule can act as a donor or acceptor and the change in the conductivity aids to change the carrier concentration of graphene (C. Lee, Wei, Kysar, & Hone, 2008). Also, graphene is an attractive material for the membrane applications thanks to its atomic thickness, mechanical strength, and chemical stability and there are also some challenges about adjusting the pore size for the graphene added membranes (Editorial, 2015). There are also some market products are available, which includes graphene. Wheel producer Vittoria built a wheel which contains graphene and expressed that, they are the best wheels in their products and the fastest in the world. Another study released the Unmanned aerial vehicles (UAV) which were constructed with the graphene and it is the point of that graphene can be integrated within the aerospace industry ("Media Innovation Studio," 2018).

CARBON NANOFIBERS

Different than the carbon nanotubes and graphene, carbon nanofibers have several advantages on their web form. Which makes their fabrication easier and moreover, let their processability in the polymer matrix easily while CNTs and graphene have to overcome the agglomeration issues. Besides, CNTs and Graphene have scale up difficulties (Y.-H. Lee et al., 2013).

Through the decades carbon fibers are intensively used in the composite applications. Different than the carbon fiber's micro size, carbon nanofibers offer a huge surface area due to their reduced fiber diameter together with the porous surface structure. Moreover, carbon nanofibers can be available in activated carbon nanofiber form, which makes them feasible to use as a membrane.

Furthermore, CNFs are lighter than the micro sized carbon fibers while serving an improved surface area. Different precursors are available to fabricate carbon nanofibers. Most studies precursor types are PAN, pitch and cellulose based ones (Edie, 1998; Morgan, 2005). Acrylonitrile unit has a 67.9% carbon content and carbon yield of PAN precursors is around 50–55%, for cellulosic precursors $(C_6H_{10}O_5)_n$ carbon contents is 44.4% however, they have a complex carbonization process which limits the carbon yield in between 25–30% when it comes to pitch, it has a higher yield of 85% however its graphitic nature resultant with poorer compression and transverse properties compared to PAN based carbon fibers (Morgan, 2005). Thus, among the others PAN gets more attention due to its carbonization process advantages. Figure 6 represents the chemical changes of PAN through the carbonization which is formed as a nanofiber. Not only homopolymer PAN is used as a precursor but

Figure 6. Carbonization mechanism of polyacrylonitrile

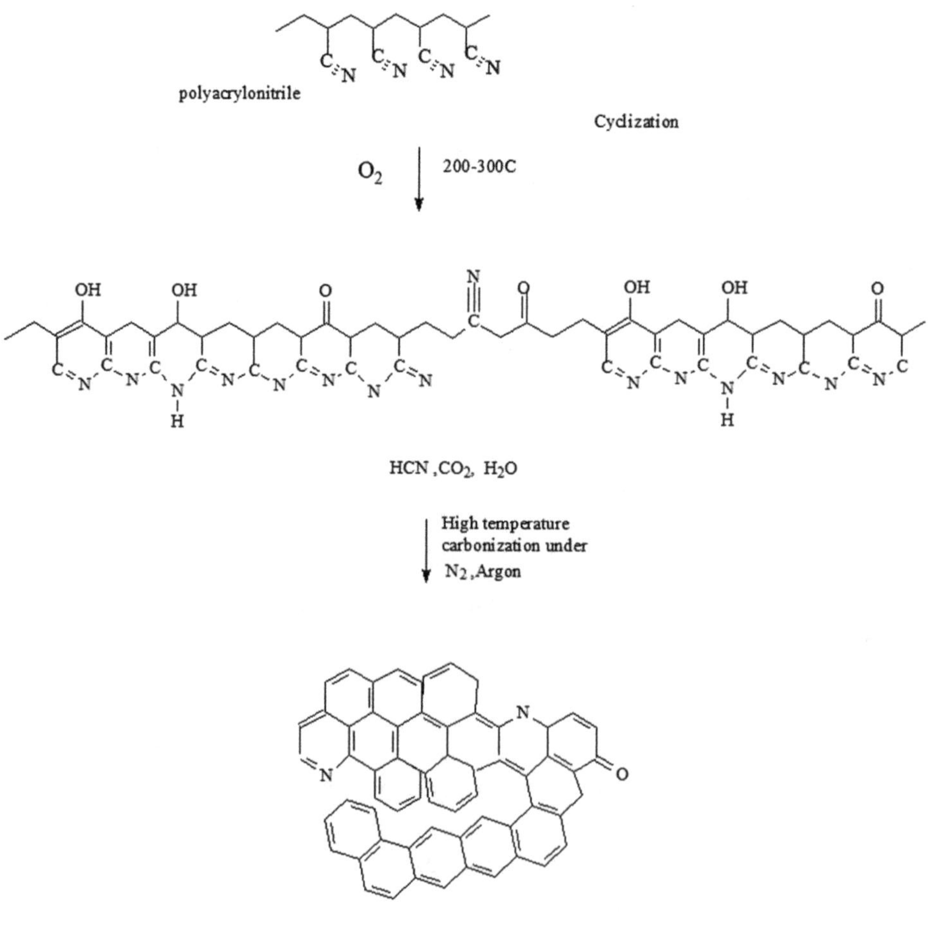

also its copolymers are suitable for the fabrication of the carbon fibers whereas they have advantages to control the carbonization and oxidation process easily (Bashir, 1991; Ismar & Sarac, 2016; Faraji, Yardim, Can, & Sarac, 2017).

Fabrication of CNFs

For fabricating the carbon nanofibers, several methods are presented in the literature such as electrospinning of polymeric precursors (Ali & El-Hamid, 2006; Ismar & Sarac, 2016), and vapor growth method (VGCNF) (Tibbetts, Lake, Strong, & Rice, 2007; Al-Saleh & Sundararaj, 2009). Generally, fabricating PAN based carbon nanofibers, electrospinning method is selected and electrospun PAN nanofibers are converted to the carbon nanofiber through the oxidative stabilization and carbonization processes. Vapor grown carbon nanofibers are fabricated via catalytic chemical vapor deposition are different than the PAN based CNFs in terms of production method, physical properties and structure of the CNF. During this fabrication process, raw material can be hydrocarbons (natural gas, propane, acetylene, benzene, and ethylene, etc.) or carbon monoxide onto a metals (Au, Fe, Co, Ni,) or metal alloy such as; Fe-Ni, Ni-Cu with a catalysts at a high temperature range (500–1500 °C) (Sattar, Kausar, & Siddiq, 2015; Zhou, Pervin, & Jeelani, 2007). Microstructural properties of VGCNFs have some impurities thus, elimination of the defects and side groups on the VGCNFs, graphitization method is the one of the effective way (Endo et al., 2003). VGCNFs structures can be a potential for polymer composite structures as a filler due to their light weight and high strength and good chemical resistance to creating an inexpensive semiconducting composite, however agglomeration of the VGCNFs one of the issues to overcome (Cortés, Lozano, Barrera, & Bonilla-Rios, 2003; Zeng, Saltysiak, Johnson, Schiraldi, & Kumar, 2004; Higgins & Brittain, 2005; Y. K. Choi, Sugimoto, Song, & Endo, 2005).

Electrospinning of Carbon Nanofibers

Nanofibers cannot be produced via conventional spinning techniques; such as dry, wet, melt and gel spinning methods, with those methods polymeric fibers can be in the micrometer range. For reducing the fiber diameter from micro to nano scale electrospinning method is preferable for polymeric precursors (Gu, Ren, & Vancso, 2005). The spinning of polymers with the help of electrostatic force is a well-known process for more than a hundred years. First patents using the electrostatic force to produce filaments go to the 1930s by Formhals (Z.-M. Huang, Zhang, Kotaki, & Ramakrishna, 2003). Moreover, in nano scale; aspect ratio of the fibers is increased and surface functionalities and better mechanical performance is achieved (Afshari, Kotek, Tonelli, & Jung, 2007). Furthermore, desired properties of electrospun nanofibers can be achieved through the adjustments on the electrospinning process (D. Li & Xia, 2004). Major electrospinning fabrication parameters are divided into three main group; solution parameters, ambient parameters, process parameters and for all have a different effect on the fiber diameter and homogeneity of the web (Bognitzki et al., 2001; Deitzel, Kleinmeyer, Harris, & Beck Tan, 2001; Doshi & Reneker, 1995; Sarac, 2015).

Indeed, it is not possible to achieve carbon nanofibers only by the finalization of the electrospinning. Electrospun fibers should be thermally treated to obtain carbon nanofibers. PAN based carbon nanofibers can be simply obtained via electrospinning technique, if those electrospun nanofibers are heat treated, PAN based carbon nanofibers are achieved (Ismar & Sarac, 2016, 2018). The oxidation mechanism should be conducted preciously to prevent the fiber damages and achieve the desired properties. During the stabilization, cyclization and dehydrogenation reactions occurs which makes PAN fibers denser and stable and prevents its fibrous structure for further high temperature steps; carbonization (Edie, 1998; Rahaman, Ismail, & Mustafa, 2007). Carbonization step can be dived into two according to applied temperature; low temperature and high temperature carbonization. Carbonization temperature has a critical role as much as the oxidation step and applied temperature determine the mechanical properties of the material. As a concept, increased temperature assists to increase the carbon content of the nanofiber thus, high temperature (1000-1500 °C) carbonization resultant with the increment on the tensile strength. According to researches the maximum strength is achieved at 1500 and 1700 °C where the elastic modulus increases monotonically with temperature due to the increased volume fraction and crystallite size, especially at temperatures between 2000 and 3000 °C which is also another phase called graphitization (Arshad, Naraghi, & Chasiotis, 2011; Edie, 1998; Liu, Wang, Xue, Fan, & Zhu, 2008).

Vapor Grown Carbon Nanofibers

Vapor Grown Carbon Nanofibers (VGCNFs) can be considered as an inexpensive alternative to CNTs, in terms of their ease fabrication when it is compared with the CNTs (Al-Saleh & Sundararaj, 2009). CNTs have improved mechanical properties than the VGCNFs, while they have a smaller diameter and lower density (Al-Saleh & Sundararaj, 2011). VGCNFs are fabricated via catalytic chemical vapor deposition which is basically similar with the previous CVD methods. During this fabrication process, precursor can be hydrocarbons (natural gas, propane, acetylene, benzene, and ethylene, etc.) or carbon monoxide onto a metal metals (Au, Fe, Co, Ni,) or metal alloy such as; Fe-Ni, Ni-Cu as a catalysts at a high temperature range (500–1500 °C) (Sattar et al., 2015; Zhou et al., 2007). In other words, the catalyst can be collected on a substrate or directly fed with the gas phase (Mordkovich, 2003).

According to the operating conditions, feedstock and catalyst type it is possible to obtain different VGCNFs in shapes (helical, twisted) and characteristics (B. O. Lee, Woo, & Kim, 2001). Not only their shape but also surface area and electrical conductivity features of VGCNFs are dependent to catalyst and feedstock type (Al-Saleh & Sundararaj, 2009).

Properties and Applications

CNFs are in a nanometer scale in diameter, and their morphology and dimensions are highly dependent on the fabrication technique together with the process conditions (Cooper et al., 2002). Applied oxidation and carbonization process temperatures, durations, temperature ramps, a cooling process, inert gas content are all affect the mechanical and surface structure of the nanofibers (Ismar & Sarac, 2018). Carbon fiber structure can be crystalline, amorphous, or partly crystalline. The crystalline form consists of a crystal structure of graphite with sp^2 hybridized carbon atoms which are arranged two-dimensionally in a honeycomb structure in the x-y plane (Chung, 1994). It is possible to fabricate self-standing, flexible supercapacitors from carbon nanofiber webs (McDonough, Choi, & Yang, 2009; Ismar, Karazehir, Ates, & Sarac, 2018). Porous carbon nanofiber webs are a good candidate for the electrochemical energy applications as an electrode (Ra, Raymundo-Piñero, Lee, & Béguin, 2009). Alignment of the nanofibers in the bundle also has an effect on the capacitive behavior of the carbon nanofiber electrode, one way alignment enhances the electrochemical performance of the web (M. Kim et al., 2016). Nanoscale carbon reinforcements, for example, CNFs are taking attention due to their improved matrix dominated properties and functionalities of the structure via their unique properties (Lee et al., 2011) which is essential for creating advanced composite structures. Due to the remarkable properties; high tensile strength and modulus of CNFs have some potential applications in polymer reinforcement composite structures (Al-Saleh & Sundararaj, 2011; Zhou et al., 2006).

Nanofibers have possible application areas from composite materials, reinforcements to tissue engineering applications. Moreover, combined structures with carbon nanofibers and the carbon nanotubes and graphene are also studied as an alternative to macroscopic composite structures. Graphene oxide and carbon nanofiber composite structure is studied in detail to improve the capacitive behavior of the webs (Gergin, Ismar, & Sarac, 2017). There is also CNF/polymer composite combinations are available for a variety of applications. Carbon nanofibers are also a good candidate for the electromagnetic shielding applications (Tibbetts et al., 2007). Carbon nanofiber and epoxy composites have stunning properties such as high strength, high toughness and low weight, which makes them appropriate candidate for the aerospace, automobile, coating, electronics sport and etc. applications

(Chisholm, Mahfuz, Rangari, Ashfaq, & Jeelani, 2005; Rana, Alagirusamy, & Joshi, 2011; Shalin, 1995). Carbon nanofibers, especially vapor grown ones used as nanofillers in different type of polymers (nylon, epoxy, polypropylene, etc.) to reinforce the polymeric matrix and improve the mechanical and thermal properties of the structure (Pogue, Ye, Klosterman, Glass, & Chartoff, 1998; Patton, Pittman, Wang, & Hill, 1999; Kumar, Doshi, Srinivasarao, Park, & Schiraldi, 2002; Pervin, Zhou, Rangari, & Jeelani, 2005). Polymer based CNF filled composite structures exhibit enhanced properties with low filler amounts (such as lower than 3wt.%) thanks to the increased surface to volume ratio and the high aspect ratio of the nanofibers (Zhou, Pervin, Jeelani, & Mallick, 2008).

CNT and CNF enhanced composite structures are a new class of materials and improved performance of these composite structures are highly depended to their well dispersion inside the composite structure and feeding weight percentage of the nanomaterials (Rana et al., 2011). In the literature; CNF/epoxy composites are studied moreover combination of CNFs and CNTs as nanofillers with epoxy resin or other polymeric matrices as a form of hybrid composites are suggested to get an absolute potential of each nanoparticle to obtain improved mechanical, chemical and thermal properties for the structural composites (Y. L. Liang & Pearson, 2010; Shokrieh, Esmkhani, Haghighatkhah, & Zhao, 2014). Besides, it is possible to create tailor-made polymeric composite structures with enhanced properties by using CNFs and CNFs are more economical than the CNTs which makes composite structures

Figure 7. Schematic description of carbon nanoparticle composites

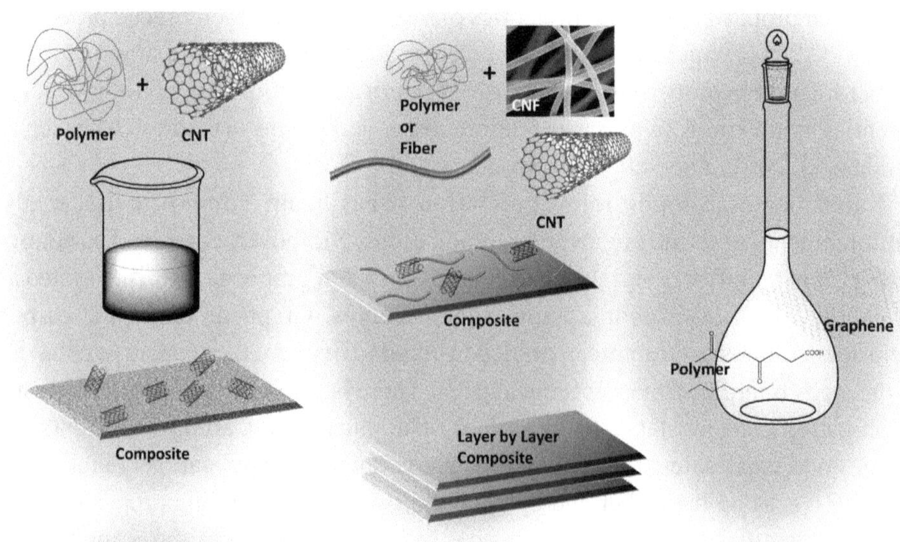

more cost effective, and multifunctional (Miyagawa, Rich, & Drzal, 2006; Wang, Sihn, Roy, Baek, & Tan, 2010). Layer by layer composite structures with CNF filling were prepared as a film for significant application areas such as radiation shielding, thermal interfaces, space and aviation (Shim, Starkovich, & Kotov, 2006).

CONCLUSION

In the aerospace applications, weight is the primary factor to be considered together with the strength, and thermal and chemical stability of the material. The nanotechnology has a key role on the engineering composites due to the extraordinary reduced weight of the nanoparticles. When material size is reduced to nano level, energy becomes more remarkable and small amount of loadings of those carbonaceous materials give a chance to create unique polymeric composite structures. The future is promising with all these carbon based nanoparticles. State of the art technologies nanoparticles work as a back bone of technological developments. Polymeric nanocomposite structures offer superior features. Thus,in this chapter selected carbon nanoparticles; carbon nanotubes, graphene, and carbon nanofibers were shortly enlightened from their synthesis to applications by taking into account their outstanding polymeric composites structures.

REFERENCES

Afshari, M., Kotek, R., Tonelli, A. E., & Jung, D.-W. (2007). Producing polyamide nanofibers by electrospinning. In P. J. Brown & K. Stevens (Eds.), *Nanofibers and Nanotechnology in Textiles* (pp. 71–89). Woodhead Publishing. doi:10.1533/9781845693732.1.71

Ajayan, P. M. (1999). Nanotubes from Carbon. *Chemical Reviews*, *99*(7), 1787–1800. doi:10.1021/cr970102g PMID:11849010

Al-Saleh, M. H., & Sundararaj, U. (2009). A review of vapor grown carbon nanofiber/polymer conductive composites. *Carbon*, *47*(1), 2–22. doi:10.1016/j.carbon.2008.09.039

Al-Saleh, M. H., & Sundararaj, U. (2011). Review of the mechanical properties of carbon nanofiber/polymer composites. *Composites. Part A, Applied Science and Manufacturing*, *42*(12), 2126–2142. doi:10.1016/j.compositesa.2011.08.005

Ali, A. A., & El-Hamid, M. A. (2006). Electro-spinning optimization for precursor carbon nanofibers. *Composites. Part A, Applied Science and Manufacturing*, *37*(10), 1681–1687. doi:10.1016/j.compositesa.2005.10.008

Allen, M. J., Tung, V. C., & Kaner, R. B. (2010). Honeycomb Carbon: A Review of Graphene. *Chemical Reviews*, *110*(1), 132–145. doi:10.1021/cr900070d PMID:19610631

Arshad, S. N., Naraghi, M., & Chasiotis, I. (2011). Strong carbon nanofibers from electrospun polyacrylonitrile. *Carbon*, *49*(5), 1710–1719. doi:10.1016/j.carbon.2010.12.056

Bal, S., & Samal, S. S. (2007). Carbon nanotube reinforced polymer composites—A state of the art. *Bulletin of Materials Science*, *30*(4), 379–386. doi:10.100712034-007-0061-2

Bashir, Z. (1991). A critical review of the stabilisation of polyacrylonitrile. *Carbon*, *29*(8), 1081–1090. doi:10.1016/0008-6223(91)90024-D

Baur, J., & Silverman, E. (2007). Challenges and Opportunities in Multifunctional Nanocomposite Structures for Aerospace Applications. *MRS Bulletin*, *32*(4), 328–334. doi:10.1557/mrs2007.231

Bethune, D. S., Klang, C. H., de Vries, M. S., Gorman, G., Savoy, R., Vazquez, J., & Beyers, R. (1993). Cobalt-catalysed growth of carbon nanotubes with single-atomic-layer walls. *Nature*, *363*(6430), 605–607. doi:10.1038/363605a0

Bhuyan, M. S. A., Uddin, M. N., Islam, M. M., Bipasha, F. A., & Hossain, S. S. (2016). Synthesis of graphene. *International Nano Letters*, *6*(2), 65–83. doi:10.100740089-015-0176-1

Bognitzki, M., Czado, W., Frese, T., Schaper, A., Hellwig, M., Steinhart, M., ... Wendorff, J. H. (2001). Nanostructured Fibers via Electrospinning. *Advanced Materials*, *13*(1), 70–72. doi:10.1002/1521-4095(200101)13:1<70::AID-ADMA70>3.0.CO;2-H

Cao, G., & Wang, Y. (2011). *Nanostructures and Nanomaterials: Synthesis, Properties, and Applications*. World Scientific. doi:10.1142/7885

Chisholm, N., Mahfuz, H., Rangari, V. K., Ashfaq, A., & Jeelani, S. (2005). Fabrication and mechanical characterization of carbon/SiC-epoxy nanocomposites. *Composite Structures*, *67*(1), 115–124. doi:10.1016/j.compstruct.2004.01.010

Choi, W., Lahiri, I., Seelaboyina, R., & Kang, Y. S. (2010). Synthesis of Graphene and Its Applications: A Review. *Critical Reviews in Solid State and Material Sciences*, *35*(1), 52–71. doi:10.1080/10408430903505036

Choi, Y. K., Sugimoto, K. I., Song, S. M., & Endo, M. (2005). Mechanical and thermal properties of vapor-grown carbon nanofiber and polycarbonate composite sheets. *Materials Letters*, *59*(27), 3514–3520. doi:10.1016/j.matlet.2005.05.082

Chung, D. D. L. (1994). *Carbon Fiber Composites*. Butterworth-Heinemann. doi:10.1016/B978-0-08-050073-7.50009-9

Ci, L., Song, L., Jariwala, D., Elías, A. L., Gao, W., Terrones, M., & Ajayan, P. M. (2009). Graphene Shape Control by Multistage Cutting and Transfer. *Advanced Materials*, *21*(44), 4487–4491. doi:10.1002/adma.200900942

Cooper, C. A., Ravich, D., Lips, D., Mayer, J., & Wagner, H. D. (2002). Distribution and alignment of carbon nanotubes and nanofibrils in a polymer matrix. *Composites Science and Technology*, *62*(7), 1105–1112. doi:10.1016/S0266-3538(02)00056-8

Cortés, P., Lozano, K., Barrera, E. V., & Bonilla-Rios, J. (2003). Effects of nanofiber treatments on the properties of vapor-grown carbon fiber reinforced polymer composites. *Journal of Applied Polymer Science*, *89*(9), 2527–2534. doi:10.1002/app.12309

Dai, H. (2001). Nanotube Growth and Characterization. In M. S. Dresselhaus, G. Dresselhaus, & P. Avouris (Eds.), *Carbon Nanotubes* (pp. 29–53). Springer Berlin Heidelberg; doi:10.1007/3-540-39947-X_3

Dai, H. (2002). Carbon nanotubes: Opportunities and challenges. *Surface Science*, *500*(1-3), 218–241. doi:10.1016/S0039-6028(01)01558-8

Deitzel, J. M., Kleinmeyer, J., Harris, D., & Beck Tan, N. C. (2001). The effect of processing variables on the morphology of electrospun nanofibers and textiles. *Polymer*, *42*(1), 261–272. doi:10.1016/S0032-3861(00)00250-0

Delhaes, P. (2013). *Carbon Based Solids and Materials*. John Wiley & Sons. doi:10.1002/9781118557617

Doshi, J., & Reneker, D. H. (1995). Electrospinning process and applications of electrospun fibers. *Journal of Electrostatics*, *35*(2), 151–160. doi:10.1016/0304-3886(95)00041-8

Dresselhaus, M. S., & Avouris, P. (2001). Introduction to Carbon Materials Research. In M. S. Dresselhaus, G. Dresselhaus, & P. Avouris (Eds.), *Carbon Nanotubes* (pp. 1–9). Springer Berlin Heidelberg; doi:10.1007/3-540-39947-X_1

Dresselhaus, M. S., Dresselhaus, G., & Avouris, P. (Eds.). (2001). *Carbon Nanotubes: Synthesis, Structure, Properties, and Applications*. Berlin, Heidelberg: Springer-Verlag. doi:10.1007/3-540-39947-X

Du, J., & Cheng, H.-M. (2012). The Fabrication, Properties, and Uses of Graphene/Polymer Composites. *Macromolecular Chemistry and Physics, 213*(10–11), 1060–1077. doi:10.1002/macp.201200029

Ebbesen, T. W., & Ajayan, P. M. (1992). Large-scale synthesis of carbon nanotubes. *Nature, 358*(6383), 220–222. doi:10.1038/358220a0

Edie, D. D. (1998). The effect of processing on the structure and properties of carbon fibers. *Carbon, 36*(4), 345–362. doi:10.1016/S0008-6223(97)00185-1

Editorial. (2015). Graphene opens up to new applications. *Nature Nanotechnology, 10*(5), 381. doi:10.1038/nnano.2015.110

Endo, M., Kim, Y. A., Hayashi, T., Yanagisawa, T., Muramatsu, H., Ezaka, M., ... Dresselhaus, M. S. (2003). Microstructural changes induced in "stacked cup" carbon nanofibers by heat treatment. *Carbon, 41*(10), 1941–1947. doi:10.1016/S0008-6223(03)00171-4

Ewels, C., Glerup, M., Krstic, V., Basiu, V. A., & Basiuk, E. V. (2007). Nitrogen and boron doping in carbon nanotubes. *Doped Nanomaterials and Nanodevices, 3*. Retrieved from http://ewels.info/science/publications/papers/2007.NitrogenChapter.pdf

Faraji, S., Yardim, M. F., Can, D. S., & Sarac, A. S. (2017). Characterization of polyacrylonitrile, poly(acrylonitrile-co-vinyl acetate), and poly(acrylonitrile-co-itaconic acid) based activated carbon nanofibers. *Journal of Applied Polymer Science, 134*(2). doi:10.1002/app.44381

Geim, A. K., & Novoselov, K. S. (2009). The rise of graphene. In *Nanoscience and Technology* (Vol. 1–0, pp. 11–19). Co-Published with Macmillan Publishers Ltd. doi:10.1142/9789814287005_0002

Gergin, İ., Ismar, E., & Sarac, A. S. (2017). Oxidative stabilization of polyacrylonitrile nanofibers and carbon nanofibers containing graphene oxide (GO): a spectroscopic and electrochemical study., Oxidative stabilization of polyacrylonitrile nanofibers and carbon nanofibers containing graphene oxide (GO): a spectroscopic and electrochemical study. *Beilstein Journal of Nanotechnology, 8*(16), 1616–1628. doi:10.3762/bjnano.8.161, 10.3762/bjnano.8.161

Gohardani, O., Elola, M. C., & Elizetxea, C. (2014). Potential and prospective implementation of carbon nanotubes on next generation aircraft and space vehicles: A review of current and expected applications in aerospace sciences. *Progress in Aerospace Sciences*, *70*, 42–68. doi:10.1016/j.paerosci.2014.05.002

Gu, S. Y., Ren, J., & Vancso, G. J. (2005). Process optimization and empirical modeling for electrospun polyacrylonitrile (PAN) nanofiber precursor of carbon nanofibers. *European Polymer Journal*, *41*(11), 2559–2568. doi:10.1016/j.eurpolymj.2005.05.008

Gugliuzza, A., & Drioli, E. (2013). A review on membrane engineering for innovation in wearable fabrics and protective textiles. *Journal of Membrane Science*, *446*, 350–375. doi:10.1016/j.memsci.2013.07.014

Guo, T., Nikolaev, P., Rinzler, A. G., Tomanek, D., Colbert, D. T., & Smalley, R. E. (1995). Self-assembly of tubular fullerenes. *Journal of Physical Chemistry*, *99*(27), 10694–10697. doi:10.1021/j100027a002

Higgins, B. A., & Brittain, W. J. (2005). Polycarbonate carbon nanofiber composites. *European Polymer Journal*, *41*(5), 889–893. doi:10.1016/j.eurpolymj.2004.11.040

Holzinger, M., Vostrowsky, O., Hirsch, A., Hennrich, F., Kappes, M., Weiss, R., & Jellen, F. (2001). Sidewall Functionalization of Carbon Nanotubes This work was supported by the European Union under the 5th Framework Research Training Network 1999, HPRNT 1999-00011 FUNCARS. *Angewandte Chemie International Edition*, *40*(21), 4002–4005. doi:10.1002/1521-3773(20011105)40:21<4002::AID-ANIE4002>3.0.CO;2-8 PMID:12404474

Huang, Z.-M., Zhang, Y.-Z., Kotaki, M., & Ramakrishna, S. (2003). A review on polymer nanofibers by electrospinning and their applications in nanocomposites. *Composites Science and Technology*, *63*(15), 2223–2253. doi:10.1016/S0266-3538(03)00178-7

Huang, Z. P., Xu, J. W., Ren, Z. F., Wang, J. H., Siegal, M. P., & Provencio, P. N. (1998). Growth of highly oriented carbon nanotubes by plasma-enhanced hot filament chemical vapor deposition. *Applied Physics Letters*, *73*(26), 3845–3847. doi:10.1063/1.122912

Hummers, W. S. Jr, & Offeman, R. E. (1958). Preparation of Graphitic Oxide. *Journal of the American Chemical Society*, *80*(6), 1339–1339. doi:10.1021/ja01539a017

Iijima, S. (1991). Helical Microtubules of Graphitic Carbon. *Nature*, *354*(6348), 56–58. doi:10.1038/354056a0

Ismar, E., Karazehir, T., Ates, M., & Sarac, A. S. (2018). Electrospun carbon nanofiber web electrode: Supercapacitor behavior in various electrolytes. *Journal of Applied Polymer Science, 135*(4), 45723. doi:10.1002/app.45723

Ismar, E., & Sarac, A. S. (2016). Synthesis and characterization of poly (acrylonitrile-co-acrylic acid) as precursor of carbon nanofibers. *Polymers for Advanced Technologies, 27*(10), 1383–1388. doi:10.1002/pat.3807

Ismar, E., & Sarac, A. S. (2018). Oxidation of polyacrylonitrile nanofiber webs as a precursor for carbon nanofiber: Aligned and non-aligned nanofibers. *Polymer Bulletin, 75*(2), 485–499. doi:10.100700289-017-2043-x

Iwahori, Y., Ishiwata, S., Sumizawa, T., & Ishikawa, T. (2005). Mechanical properties improvements in two-phase and three-phase composites using carbon nano-fiber dispersed resin. *Composites. Part A, Applied Science and Manufacturing, 36*(10), 1430–1439. doi:10.1016/j.compositesa.2004.11.017

Jiang, K., Schadler, L. S., Siegel, R. W., Zhang, X., Zhang, H., & Terrones, M. (2004). *Protein immobilization on carbon nanotubes via a two-step process of diimide-activated amidation*. Academic Press. doi:10.1039/B310359E

Journet, C., Maser, W. K., Bernier, P., Loiseau, A., de la Chapelle, M. L., Lefrant, S., ... Fischer, J. E. (1997). Large-scale production of single-walled carbon nanotubes by the electric-arc technique. *Nature, 388*(6644), 756–758. doi:10.1038/41972

Kim, H., Abdala, A. A., & Macosko, C. W. (2010). Graphene/Polymer Nanocomposites. *Macromolecules, 43*(16), 6515–6530. doi:10.1021/ma100572e

Kim, M., Kim, Y., Lee, K. M., Jeong, S. Y., Lee, E., Baeck, S. H., & Shim, S. E. (2016). Electrochemical improvement due to alignment of carbon nanofibers fabricated by electrospinning as an electrode for supercapacitor. *Carbon, 99*, 607–618. doi:10.1016/j.carbon.2015.12.068

Kiselev, N. A., Moravsky, A. P., Ormont, A. B., & Zakharov, D. N. (1999). SEM and HREM study of the internal structure of nanotube rich carbon arc cathodic deposits. *Carbon, 37*(7), 1093–1103. doi:10.1016/S0008-6223(98)00309-1

Kroto, H. W., Heath, J. R., O'Brien, S. C., Curl, R. F., & Smalley, R. E. (1985). C 60: Buckminsterfullerene. *Nature, 318*(6042), 162–163. doi:10.1038/318162a0

Krüger, A. (2010). *Carbon Materials and Nanotechnology*. John Wiley & Sons. doi:10.1002/9783527629602

Kumar, S., Doshi, H., Srinivasarao, M., Park, J. O., & Schiraldi, D. A. (2002). Fibers from polypropylene/nano carbon fiber composites. *Polymer*, *43*(5), 1701–1703. doi:10.1016/S0032-3861(01)00744-3

Kwon, Y.-K., Lee, Y. H., Kim, S.-G., Jund, P., Tománek, D., & Smalley, R. E. (1997). Morphology and Stability of Growing Multiwall Carbon Nanotubes. *Physical Review Letters*, *79*(11), 2065–2068. doi:10.1103/PhysRevLett.79.2065

Lee, B. O., Woo, W. J., & Kim, M.-S. (2001). EMI Shielding Effectiveness of Carbon Nanofiber Filled Poly(vinyl alcohol) Coating Materials. *Macromolecular Materials and Engineering*, *286*(2), 114–118. doi:10.1002/1439-2054(20010201)286:2<114::AID-MAME114>3.0.CO;2-8

Lee, C., Wei, X., Kysar, J. W., & Hone, J. (2008). Measurement of the Elastic Properties and Intrinsic Strength of Monolayer Graphene. *Science*, *321*(5887), 385–388. doi:10.1126cience.1157996 PMID:18635798

Lee, Y.-H., Kim, J.-S., Noh, J., Lee, I., Kim, H. J., Choi, S., ... Choi, J. W. (2013). Wearable Textile Battery Rechargeable by Solar Energy. *Nano Letters*, *13*(11), 5753–5761. doi:10.1021/nl403860k PMID:24164580

Li, D., & Xia, Y. (2004). Electrospinning of Nanofibers: Reinventing the Wheel? *Advanced Materials*, *16*(14), 1151–1170. doi:10.1002/adma.200400719

Li, W. Z., Xie, S. S., Qian, L. X., Chang, B. H., Zou, B. S., Zhou, W. Y., ... Wang, G. (1701–1703). … Wang, G. (1996). Large-Scale Synthesis of Aligned Carbon Nanotubes. *Science*, *274*(5293), 1701–1703. doi:10.1126cience.274.5293.1701

Liang, X., Chang, A. S. P., Zhang, Y., Harteneck, B. D., Choo, H., Olynick, D. L., & Cabrini, S. (2009). Electrostatic Force Assisted Exfoliation of Prepatterned Few-Layer Graphenes into Device Sites. *Nano Letters*, *9*(1), 467–472. doi:10.1021/nl803512z PMID:19072062

Liang, Y. L., & Pearson, R. A. (2010). The toughening mechanism in hybrid epoxy-silica-rubber nanocomposites (HESRNs). *Polymer*, *51*(21), 4880–4890. doi:10.1016/j.polymer.2010.08.052

Lim, C.-S., Guzman, M., Schaefer, J., & Minaie, B. (2013). Fabrication and properties of dense thin films containing functionalized carbon nanofibers. *Thin Solid Films*, *534*, 111–115. doi:10.1016/j.tsf.2013.02.010

Liu, F., Wang, H., Xue, L., Fan, L., & Zhu, Z. (2008). Effect of microstructure on the mechanical properties of PAN-based carbon fibers during high-temperature graphitization. *Journal of Materials Science*, *43*(12), 4316–4322. doi:10.100710853-008-2633-y

Ma, P.-C., Siddiqui, N. A., Marom, G., & Kim, J.-K. (2010). Dispersion and functionalization of carbon nanotubes for polymer-based nanocomposites: A review. *Composites. Part A, Applied Science and Manufacturing*, *41*(10), 1345–1367. doi:10.1016/j.compositesa.2010.07.003

McDonough, J. R., Choi, J. W., Yang, Y., La Mantia, F., Zhang, Y., & Cui, Y. (2009). Carbon nanofiber supercapacitors with large areal capacitances. *Applied Physics Letters*, *95*(24), 243109. doi:10.1063/1.3273864

Media Innovation Studio. (2018, August 7). Retrieved August 7, 2018, from http://mediainnovationstudio.org/graphene-uav-first-for-uclan-and-university-of-manchester/

Miyagawa, H., Rich, M. J., & Drzal, L. T. (2006). Thermo-physical properties of epoxy nanocomposites reinforced by carbon nanotubes and vapor grown carbon fibers. *Thermochimica Acta*, *442*(1), 67–73. doi:10.1016/j.tca.2006.01.016

Morant, C., Andrey, J., Prieto, P., Mendiola, D., Sanz, J. M., & Elizalde, E. (2006). XPS characterization of nitrogen-doped carbon nanotubes. *Physica Status Solidi. A, Applications and Materials Science*, *203*(6), 1069–1075. doi:10.1002/pssa.200566110

Moravsky, A., Wexler, E., & Loutfy, R. (2004). Growth of Carbon Nanotubes by Arc Discharge and Laser Ablation. In *Carbon Nanotubes* (Vol. 1–0, pp. 65–97). CRC Press. doi:10.1201/9780203494936.ch3

Mordkovich, V. Z. (2003). Carbon Nanofibers: A New Ultrahigh-Strength Material for Chemical Technology. *Theoretical Foundations of Chemical Engineering*, *37*(5), 429–438. doi:10.1023/A:1026082323244

Morgan, P. (2005). *Carbon Fibers and Their Composites*. Taylor & Francis. doi:10.1201/9781420028744

Navarro, E., Baun, A., Behra, R., Hartmann, N. B., Filser, J., Miao, A.-J., ... Sigg, L. (2008). Environmental behavior and ecotoxicity of engineered nanoparticles to algae, plants, and fungi. *Ecotoxicology (London, England)*, *17*(5), 372–386. doi:10.100710646-008-0214-0 PMID:18461442

Ni, Z., Wang, Y., Yu, T., & Shen, Z. (2008). Raman spectroscopy and imaging of graphene. *Nano Research*, *1*(4), 273–291. doi:10.100712274-008-8036-1

Ni, Z. H., Wang, H. M., Kasim, J., Fan, H. M., Yu, T., Wu, Y. H., ... Shen, Z. X. (2007). Graphene Thickness Determination Using Reflection and Contrast Spectroscopy. *Nano Letters, 7*(9), 2758–2763. doi:10.1021/nl071254m PMID:17655269

Njuguna, J., & Pielichowski, K. (2004). Polymer Nanocomposites for Aerospace Applications: Fabrication. *Advanced Engineering Materials, 6*(4), 193–203. doi:10.1002/adem.200305111

Njuguna, J., Pielichowski, K., & Fan, J. (2012). Polymer nanocomposites for aerospace applications. In F. Gao (Ed.), *Advances in Polymer Nanocomposites* (pp. 472–539). Woodhead Publishing. doi:10.1533/9780857096241.3.472

Nolan, P. E., Lynch, D. C., & Cutler, A. H. (1998). Carbon Deposition and Hydrocarbon Formation on Group VIII Metal Catalysts. *The Journal of Physical Chemistry B, 102*(21), 4165–4175. doi:10.1021/jp980996o

Novoselov, K. S., Geim, A. K., Morozov, S. V., Jiang, D., Zhang, Y., Dubonos, S. V., ... Firsov, A. A. (2004). Electric Field Effect in Atomically Thin Carbon Films. *Science, 306*(5696), 666–669. doi:10.1126cience.1102896 PMID:15499015

Obraztsov, A. N., Obraztsova, E. A., Tyurnina, A. V., & Zolotukhin, A. A. (2007). Chemical vapor deposition of thin graphite films of nanometer thickness. *Carbon, 45*(10), 2017–2021. doi:10.1016/j.carbon.2007.05.028

Patton, R. D., Pittman, J. C. U. Jr, Wang, L., & Hill, J. R. (1999). Vapor grown carbon fiber composites with epoxy and poly(phenylene sulfide) matrices. *Composites. Part A, Applied Science and Manufacturing, 30*(9), 1081–1091. doi:10.1016/S1359-835X(99)00018-4

Pei, S., & Cheng, H.-M. (2012). The reduction of graphene oxide. *Carbon, 50*(9), 3210–3228. doi:10.1016/j.carbon.2011.11.010

Pervin, F., Zhou, Y., Rangari, V. K., & Jeelani, S. (2005). Testing and evaluation on the thermal and mechanical properties of carbon nano fiber reinforced SC-15 epoxy. *Materials Science and Engineering A, 405*(1), 246–253. doi:10.1016/j.msea.2005.06.012

Podsiadlo, P., Shim, B. S., & Kotov, N. A. (2009). Polymer/clay and polymer/carbon nanotube hybrid organic–inorganic multilayered composites made by sequential layering of nanometer scale films. *Coordination Chemistry Reviews, 253*(23), 2835–2851. doi:10.1016/j.ccr.2009.09.004

Pogue, R. T., Ye, J., Klosterman, D. A., Glass, A. S., & Chartoff, R. P. (1998). Evaluating fiber–matrix interaction in polymer–matrix composites by inverse gas chromatography. *Composites. Part A, Applied Science and Manufacturing*, *29*(9), 1273–1281. doi:10.1016/S1359-835X(98)00074-8

Ra, E. J., Raymundo-Piñero, E., Lee, Y. H., & Béguin, F. (2009). High power supercapacitors using polyacrylonitrile-based carbon nanofiber paper. *Carbon*, *47*(13), 2984–2992. doi:10.1016/j.carbon.2009.06.051

Rahaman, M. S. A., Ismail, A. F., & Mustafa, A. (2007). A review of heat treatment on polyacrylonitrile fiber. *Polymer Degradation & Stability*, *92*(8), 1421–1432. doi:10.1016/j.polymdegradstab.2007.03.023

Ramanathan, T., Abdala, A. A., Stankovich, S., Dikin, D. A., Herrera-Alonso, M., Piner, R. D., ... Brinson, L. C. (2008). Functionalized graphene sheets for polymer nanocomposites. *Nature Nanotechnology*, *3*(6), 327–331. doi:10.1038/nnano.2008.96 PMID:18654541

Rana, S., Alagirusamy, R., & Joshi, M. (2011). Development of carbon nanofibre incorporated three phase carbon/epoxy composites with enhanced mechanical, electrical and thermal properties. *Composites. Part A, Applied Science and Manufacturing*, *42*(5), 439–445. doi:10.1016/j.compositesa.2010.12.018

Reina, A., Jia, X., Ho, J., Nezich, D., Son, H., Bulovic, V., ... Kong, J. (2009). Large Area, Few-Layer Graphene Films on Arbitrary Substrates by Chemical Vapor Deposition. *Nano Letters*, *9*(1), 30–35. doi:10.1021/nl801827v PMID:19046078

Rettenbacher, A. S., Elliott, B., Hudson, J. S., Amirkhanian, A., & Echegoyen, L. (2006). Preparation and Functionalization of Multilayer Fullerenes (Carbon Nano-Onions). *Chemistry (Weinheim an der Bergstrasse, Germany)*, *12*(2), 376–387. doi:10.1002/chem.200500517 PMID:16189840

Rinzler, A. G., Liu, J., Dai, H., Nikolaev, P., Huffman, C. B., Rodríguez-Macías, F. J., ... Smalley, R. E. (1998). Large-scale purification of single-wall carbon nanotubes: Process, product, and characterization. *Applied Physics. A, Materials Science & Processing*, *67*(1), 29–37. doi:10.1007003390050734

Sarac, A. S. (2015). *Nanofibers of Conjugated Polymers*. Pan Stanford.

Sattar, R., Kausar, A., & Siddiq, M. (2015). Advances in thermoplastic polyurethane composites reinforced with carbon nanotubes and carbon nanofibers: A review. *Journal of Plastic Film & Sheeting*, *31*(2), 186–224. doi:10.1177/8756087914535126

Shalin, R. E. (Ed.). (1995). *Polymer Matrix Composites*. Springer Netherlands. doi:10.1007/978-94-011-0515-6

Shim, B. S., Starkovich, J., & Kotov, N. (2006). Multilayer composites from vapor-grown carbon nano-fibers. *Composites Science and Technology*, *66*(9), 1174–1181. doi:10.1016/j.compscitech.2005.11.004

Shokrieh, M. M., Esmkhani, M., Haghighatkhah, A. R., & Zhao, Z. (2014). Flexural fatigue behavior of synthesized graphene/carbon-nanofiber/epoxy hybrid nanocomposites. *Materials & Design (1980-2015)*, *62*, 401–408. doi:10.1016/j.matdes.2014.05.040

Smalley, R. E. (1992). Self-assembly of the fullerenes. *Accounts of Chemical Research*, *25*(3), 98–105. doi:10.1021/ar00015a001

Smalley, R. E. (1993). From dopyballs to nanowires. *Materials Science and Engineering B*, *19*(1), 1–7. doi:10.1016/0921-5107(93)90156-H

Spitalsky, Z., Tasis, D., Papagelis, K., & Galiotis, C. (2010). Carbon nanotube–polymer composites: Chemistry, processing, mechanical and electrical properties. *Progress in Polymer Science*, *35*(3), 357–401. doi:10.1016/j.progpolymsci.2009.09.003

Stankovich, S., Dikin, D. A., Dommett, G. H. B., Kohlhaas, K. M., Zimney, E. J., Stach, E. A., ... Ruoff, R. S. (2006). Graphene-based composite materials. *Nature*, *442*(7100), 282–286. doi:10.1038/nature04969 PMID:16855586

Stankovich, S., Dikin, D. A., Piner, R. D., Kohlhaas, K. A., Kleinhammes, A., Jia, Y., ... Ruoff, R. S. (2007). Synthesis of graphene-based nanosheets via chemical reduction of exfoliated graphite oxide. *Carbon*, *45*(7), 1558–1565. doi:10.1016/j.carbon.2007.02.034

Staudenmaier, L. (1898). Verfahren zur Darstellung der Graphitsäure. *Berichte der Deutschen Chemischen Gesellschaft*, *31*(2), 1481–1487. doi:10.1002/cber.18980310237

Stephan, O., Ajayan, P. M., Colliex, C., Redlich, P., Lambert, J. M., Bernier, P., & Lefin, P. (1994). Doping graphitic and carbon nanotube structures with boron and nitrogen. *Science*, *266*(5191), 1683–1685. doi:10.1126cience.266.5191.1683 PMID:17775629

Su, F., Zhang, Z., & Liu, W. (2006). Mechanical and tribological properties of carbon fabric composites filled with several nano-particulates. *Wear*, *260*(7), 861–868. doi:10.1016/j.wear.2005.04.015

Suenaga, K., Johansson, M. P., Hellgren, N., Broitman, E., Wallenberg, L. R., Colliex, C., ... Hultman, L. (1999). Carbon nitride nanotubulite – densely-packed and well-aligned tubular nanostructures. *Chemical Physics Letters*, *300*(5–6), 695–700. doi:10.1016/S0009-2614(98)01425-0

Thostenson, E. T., Ren, Z., & Chou, T.-W. (2001). Advances in the science and technology of carbon nanotubes and their composites: A review. *Composites Science and Technology*, *61*(13), 1899–1912. doi:10.1016/S0266-3538(01)00094-X

Tibbetts, G. G., Lake, M. L., Strong, K. L., & Rice, B. P. (2007). A review of the fabrication and properties of vapor-grown carbon nanofiber/polymer composites. *Composites Science and Technology*, *67*(7–8), 1709–1718. doi:10.1016/j.compscitech.2006.06.015

Wang, D. H., Sihn, S., Roy, A. K., Baek, J.-B., & Tan, L.-S. (2010). Nanocomposites based on vapor-grown carbon nanofibers and an epoxy: Functionalization, preparation and characterization. *European Polymer Journal*, *46*(7), 1404–1416. doi:10.1016/j.eurpolymj.2010.04.025

Weisman, R. B., & Subramoney, S. (2006). Carbon nanotubes. *The Electrochemical Society Interface*, 43.

Wu, Y. H., Yu, T., & Shen, Z. X. (2010). Two-dimensional carbon nanostructures: Fundamental properties, synthesis, characterization, and potential applications. *Journal of Applied Physics*, *108*(7), 071301. doi:10.1063/1.3460809

Xie, X.-L., Mai, Y.-W., & Zhou, X.-P. (2005). Dispersion and alignment of carbon nanotubes in polymer matrix: A review. *Materials Science and Engineering R Reports*, *49*(4), 89–112. doi:10.1016/j.mser.2005.04.002

Yadav, M. D., Dasgupta, K., Patwardhan, A. W., & Joshi, J. B. (2017). High Performance Fibers from Carbon Nanotubes: Synthesis, Characterization, and Applications in Composites—A Review. *Industrial & Engineering Chemistry Research*, *56*(44), 12407–12437. doi:10.1021/acs.iecr.7b02269

Yumura, M. (1999). Synthesis and Purification of Multi-Walled and Single-Walled Carbon Nanotubes. In K. Tanaka, T. Yamabe, & K. Fukui (Eds.), *The Science and Technology of Carbon Nanotubes* (pp. 2–13). Oxford, UK: Elsevier Science Ltd.; doi:10.1016/B978-008042696-9/50002-1

Zeng, J., Saltysiak, B., Johnson, W. S., Schiraldi, D. A., & Kumar, S. (2004). Processing and properties of poly(methyl methacrylate)/carbon nano fiber composites. *Composites. Part B, Engineering*, *35*(2), 173–178. doi:10.1016/S1359-8368(03)00051-9

Zhang, Y., Small, J. P., Pontius, W. V., & Kim, P. (2005). Fabrication and electric-field-dependent transport measurements of mesoscopic graphite devices. *Applied Physics Letters*, *86*(7), 073104. doi:10.1063/1.1862334

Zhou, Y., Pervin, F., & Jeelani, S. (2007). Effect vapor grown carbon nanofiber on thermal and mechanical properties of epoxy. *Journal of Materials Science*, *42*(17), 7544–7553. doi:10.100710853-007-1618-6

Zhou, Y., Pervin, F., Jeelani, S., & Mallick, P. K. (2008). Improvement in mechanical properties of carbon fabric–epoxy composite using carbon nanofibers. *Journal of Materials Processing Technology*, *198*(1), 445–453. doi:10.1016/j.jmatprotec.2007.07.028

Zhou, Y., Pervin, F., Rangari, V. K., & Jeelani, S. (2006). Fabrication and evaluation of carbon nano fiber filled carbon/epoxy composite. *Materials Science and Engineering A*, *426*(1), 221–228. doi:10.1016/j.msea.2006.04.031

KEY TERMS AND DEFINITIONS

Carbon Nanotubes: Rolled graphene layers in 2-D.

Electrospinning: The method for producing submicron fibers via electrical force.

Graphene: Single layer of graphite.

Nano: Unit which equals to 10^{-9}.

Polyacrylonitrile: Synthetic semi-crystalline organic polymer, generally used as a precursor for carbon fiber production.

Chapter 2
Synthesis of Epoxy Nanocomposites

Tamara F. Irzhak
Institute of Problems of Chemical Physics of RAS, Russia

Vadim I. Irzhak
Institute of Problems of Chemical Physics of RAS, Russia

ABSTRACT

The formation processes of epoxy nanocomposites with carbon (nanotubes, graphene, and graphite), metal-containing, and aluminosilicate (montmorillonite and halloysite tubes) fillers are considered. A high reactivity of epoxy groups and a thermodynamic miscibility of epoxy oligomers with many substances make it possible to use diverse curing agents and to accomplish curing reactions under various technological conditions. Epoxy nanocomposites are designed to realize to the same extent the unique functional properties of nanoparticles: electric, magnetic, optical, chemical, and biological. The mutual effect of both a matrix and nanoparticles on the composite formation is discussed.

INTRODUCTION

Since around the mid-1990s, polymer nanocomposites have become the subject of considerable attention, as evidenced by monographs, and a great number of reviews: since 2010, more than two dozen papers have been published (Irzhak, 2017). The application of nanocomposites is associated with their unique properties related to a huge specific surface and high surface energy of nanoparticles. Nanosized particles, as opposed to microinclusions and coarser inclusions, are not stress concentrators,

DOI: 10.4018/978-1-5225-7921-2.ch002

and this circumstance facilitates a marked improvement in the mechanical properties of nanocomposites. Compared with the respective polymers, the transparency of nanocomposites does not decrease, because nanoparticles do not scatter light because of their small sizes. Depending on the type of nanoparticles introduced in polymeric materials even at a low concentration, nanocomposites acquire remarkable chemical (primarily catalytic), electrophysical, and biomedical properties, thereby opening wide potential for their use.

Among polymer nanocomposites (Figure 1, curve *1*), it would appear that composites based on an epoxy matrix occupy an insignificant place—nearly 10% as to the number of publications—but an ever increasing number of papers appear

Figure 1. The number of publications on (1) polymeric and (2) epoxy nanocomposites and (3) the number of their citations according to the Web of Science Reprinted with permission from IAPC "Nauka" (2017, Irzhak)

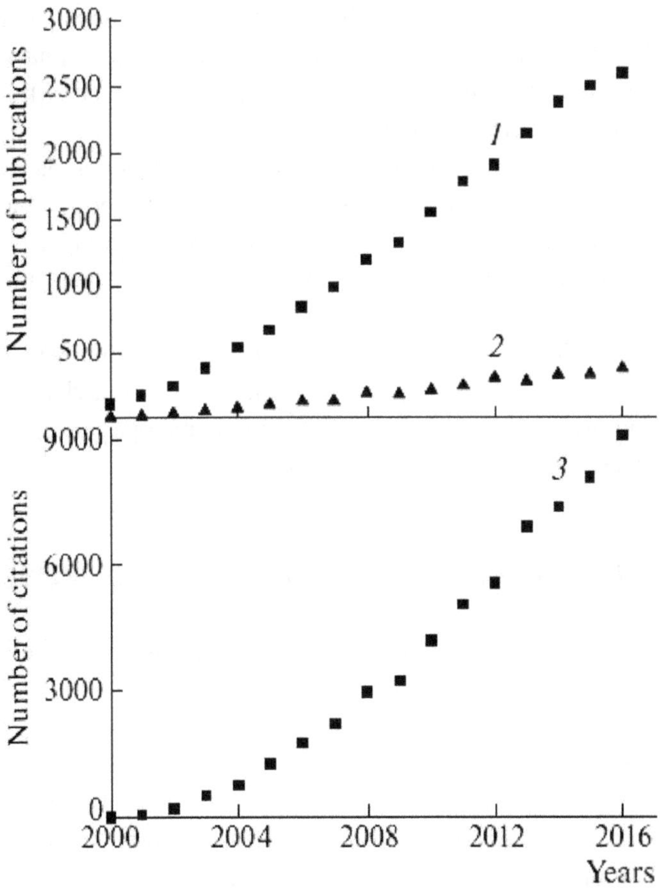

annually (curve *2*). Moreover, the interest in them grows almost exponentially, as evidenced by the number of citations (curve *3*).

Epoxy polymers in terms of a set of properties stand out among other polymeric materials and play an important role in aerospace, automotive, shipbuilding, and other industries. Their wide application in engineering is associated, firstly, with a high processability of epoxy resins and, secondly, with the unique combination of performance characteristics of their curing products (Paquin, 1958, Lee, 1967, Chernin, 1982, Khozin, 2004).

A high reactivity of epoxy groups and a thermodynamic miscibility of epoxy oligomers with many substances make it possible to use diverse curing agents and to accomplish curing reactions under various technological conditions (Irzhak, 1979, Rozenberg, 1986, Okabe, 2013). Of no small importance are the features of synthesis processes, such as the absence of volatile products and low level of shrinkage.

Epoxy polymers have high values of static and shock strength, hardness, and wear resistance. They possess marked thermal stability and heat resistance. Many solid surfaces form strong adhesive bonds with epoxy polymers (Paquin, 1958, Lee, 1967). This circumstance determines their use as compounds, glues, paint and lacquer materials, and coatings, including in aerospace engineering. Epoxy nanocomposites are designed to realize the unique functional properties of nanoparticles: electric, magnetic, optical, chemical, and biological.

Information about epoxy nanocomposites is contained most completely in (Irzhak, 2017). Reviews (Irzhak, 2011, Kablov, 2013) are devoted to epoxy nanocomposites with carbon nanotubes (CNTs). Some aspects of epoxy nanocomposites containing graphene were covered in (Song, 2013). Other generalizing papers are unavailable in this field.

Taking into consideration the urgency of the issue and the presence of recent studies not covered in the mentioned reviews, authors took a chance to consider epoxy nanocomposites containing metal and mineral nanoparticles, graphene, and CNTs and to discuss the processes of their formation.

FEATURE OF SYNTHESIS OF EPOXY NANOCOMPOSITES

The generally accepted scheme of the amino-epoxy cure involves two main reactions of the glycidyl ether (Box 1).

In principle, a reaction of esterification is possible – the resulting hydroxyls with epoxy groups.

The ratio of the rate constant of the secondary amine addition to the corresponding constant of the primary one depends on electron-donating properties of an amine and may vary with temperature. Normally this ratio varies within 0.1–0.6. It is because

Box 1. Epoxy-amine cure reactions scheme

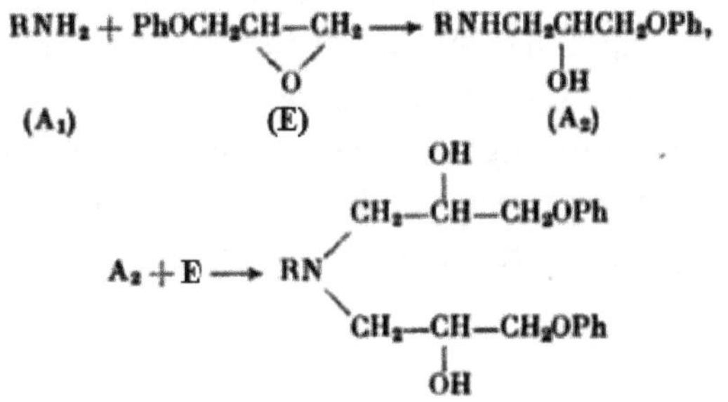

of accumulation of hydroxyl groups during the reaction that the process of curing epoxy oligomers under the action of amines has autocatalytic character. Kinetics is often analyzed using empirical Eq. (1)

$$\frac{d\alpha}{dt} = \left(k_1 + k_2\alpha^m\right)\left(1-\alpha\right)^n \quad (1)$$

$$\frac{d\alpha}{dt} = \left(k_1 + k_2\alpha\right)\left(1-\alpha\right)^n \quad (1a)$$

where α is the degree of conversion of epoxy groups; the sum of exponents $m + n$ defines the overall reaction order, which is usually two. Sometimes it is assumed that $m = 1$ (Eq. (1a)). Constants k_1 and k_2 reflect the autocatalytic character of the process.

From the standpoint of formation of epoxy nanocomposites, there are two types of fillers: fillers belonging to the first type are chemically unchanged and should be reduced to the desired size in one way or another. An example is provided by CNTs, diverse minerals, and graphene. Compounds whose chemical nature changes during composite formation belong to the second type. These are, in particular, metal salts, in which cations should be reduced to the zero-valence state.

In contrast to linear polymers, epoxy nanocomposites are obtained only via the curing of epoxy oligomers in the presence of a filler or the corresponding precursor. It is evident that these additives may affect the kinetics of the process and the properties of the resulting matrix. The mechanism of curing of epoxy resins is fairly complicated, because many reactions occur simultaneously and depend on such phenomena as gelation and glass transition (Irzhak, 1979, Rozenberg, 1986). After incorporation of nanoparticles into the epoxy resin, the process of its curing becomes even more complicated. A considerable number of papers concern the kinetics of curing of epoxy systems with various types of nanoparticles.

Most of kinetic studies are using DSC. DSC analysis has shown that etherification occurs at elevated temperatures once all the primary amines are exhausted. In this case, the data are analyzed as a rule in terms of the generalized formula

$$\frac{d\alpha}{dt} = k(T)f(\alpha) \tag{2}$$

For nonisothermal conditions the temperature varies with time with a constant heating rate, $\beta = dT/dt$.

Isoconversional kinetic analysis has shown that the reaction rate at constant conversion depends only on the temperature. In other words,

$$E_a = -Rd\ln\left(\frac{d\alpha}{dt}\right) / \, dT^{-1} \tag{3}$$

where E_α is the effective activation energy at a given conversion. The energy of activation was determined by varying the rate of scanning. Multistage processes show the dependence of E_α on α, the analysis of which helps not only to reveal the complexity of the process, but also to identify its kinetic scheme. So, for example, starting from equation (1), we get:

$$E_a = -R\left(k_1E_1 + \alpha^m k_2E_2\right) / \left(k_1 + \alpha^m k_2\right) \tag{4}$$

In Figure 2 shows the data (Vyazovkin, 1996) experimental, processed according to (3), and calculated by the equation (4). The calculation was carried out under the condition $E_1 = 120$ and $E_2 = 50$ kJ/mol and a variation of the exponent m. The first value was obtained from the experiment as the limiting value of the effective activation energy, since $E_\alpha = E_1$ as $\alpha \rightarrow 0$.

As can be seen, the calculation can be reconciled with the experiment by specifying the values of E_2 and m. Thus, we can assume that Eq. (1) adequately describes the experimental data. Namely this approach is used to study the effect of various additives on the kinetics of epoxy nanocomposites formation.

Figure 2. Dependencies of the activation energy upon conversion computed by eq. 4 for different m: (1) 0.7, (2) 0.9, (3) 1.1, (4) 1.3, and (5) 1.5, as compared to the actual value (•). Curve 6 represents the model eq. 4 dependence with E2 = 53 kJ/mol and m = 1.3. Reprinted with permission from Am. Chem. Soc. (1996, Vyasovkin).

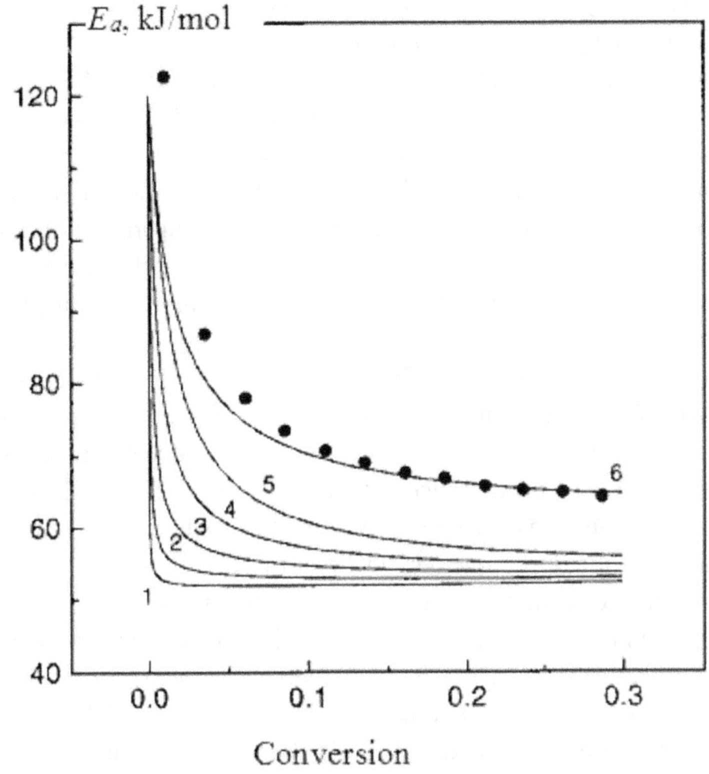

INFLUENCE OF THE FILLERS ON THE NANOCOMPOSITE SYNTHESIS PROCESS

Carbon Fillers: CNTs, Graphene, Graphite, and Carbon Fiber

Taking into consideration the molecular structure of carbon nanoparticles, in particular, graphene and CNTs (D'yachkov, 2006, Grayfer, 2011, Ivanovskii, 2012, Rakov, 2006, 2013) it may be stated that their influence on the kinetics of the process of curing of epoxy oligomers will be similar. Indeed, graphene, CNTs, and other compounds with the *sp*^{2}-hybridized carbon can catalyze diverse organic reactions (Schaetz, 2012, Navalon, 2014, Chua, 2015). Their surface energy is fairly high; therefore, the adsorption of various molecules is typical of them (Liang, 2004,

Britz, 2006, Shen, 2013, Xue, 2015). The components of epoxy binders are not exceptional in this respect. Adsorbed molecules may be involved in the process of matrix formation in a certain manner.

In work (Xie et al., 2004) it was investigated the high-temperature isothermal curing of tetraglycidyl-4,4′-diamino-diphenylmethane with 4,4′-diaminodiphenyl sulfone (DDS) in the presence of multi-wall CNTs.

The typical technique used to prepare the reaction mixture for kinetic studies of the process will be described below, and the techniques described in other papers may differ from the typical one only in some details.

A mixture of an epoxy resin with preliminarily purified multi-wall CNTs was sonificated for 2 h and placed in an oil bath at temperature of 120°C, and the stoichiometric amount of a curing agent was slowly added under continuous mechanical stirring until formation of a homogeneous mixture. This process took nearly 10 min.

The kinetics was analyzed using Eq. (1). It was found that with the increase in concentration of CNTs, constant k_1, which defines the initial rate of reaction, grows, while the corresponding activation energy drops. The autocatalytic constant k_2 is practically unaffected by the presence of tubes. (Xie et al., 2004) believe that these effects may be attributed to the catalytic effect of surface hydroxyl groups that arise as a result of oxidation during the purification of CNTs.

The initial acceleration of the reaction under the action of single-wall CNTs was also observed (Tao, 2006), but the magnitude of this effect was insignificant. At the same time, the glass-transition temperature T_g decreased, thus suggesting a reduction in the degree of crosslinking of the matrix (Irzhak, 1979).

The effect of concentration and type of CNTs (single-, double-, and multi-wall) on the kinetics of reaction between low viscosity mixture of epoxy oligomers and amine curing agents was studied (Esmizadeh, 2015). Analysis was conducted in terms of the equation

$$\frac{d\alpha}{dt} = k(T)\alpha^m \left(1-\alpha\right)^n \tag{5}$$

It was shown that the type of CNTs exerts almost no effect on kinetic parameters possibly because of their low concentration (0.01%). At a concentration of 0.1, 0.2, and 0.5%, the rate constant changes nonmonotonically, but on the whole it is lower than that in the absence of CNTs. The energy of activation increases from ~6 to ~9 kJ/mol. The heat of reaction decreases, indicating the incompletion of the process. This is also evidenced by a decrease in the high-elasticity modulus, i.e., matrix network density. In accordance with (Esmizadeh, 2015), the observed effects may be explained by increase in the viscosity and thermal conductivity of the system in the presence of CNTs, although these properties were not estimated.

Multi-wall CNTs, when taken at low concentrations (0.5%), slightly accelerate the reaction of amine curing of bisphenol F diglycidyl ether, while at higher concentrations (1.5%) they decelerate this reaction (Visco et al., 2009). They believe that the rate of the process is controlled by viscosity of the system.

The curing of an epoxy resin is regarded as a heterogeneous phase formation process and the role of multi-wall CNTs is discussed from this viewpoint (Susin et al., 2014). They believe that the tubes restrict the local free volume and assist in development of the heterogeneous morphology in resin, especially at high content of multi-wall CNTs. At the same time, with the increase in their concentration (to 1%), the ultimate heat of reaction increases, while the energy of activation decreases.

Most of kinetic studies are aimed at gaining insight into the effect of functionalized CNTs using DSC. Rahaman and Mohanty (2014) presented the data of the influence of multi-wall CNTs carrying COOH groups on the process of curing of epoxy resin EPOLAM with anhydride of 1,2,3,6-tetrahydroxymethyl-1,3,6-methanephthalic acid. According to these authors, this dependence provides evidence that in the presence of multi-wall CNTs the degree of crosslinking increases; as a result, the mobility of unreacted groups declines. The closeness of the curves at the beginning of the curing process indicates that the addition of CNTs does not affect the value of E_1. From the form of the divergence of the curves in the course of the reaction, it follows that the entrainment of the filler leads to an increase in E_2.

Note that methods using the dependence of a change in the exothermal peak on the rate of heating cannot provide reasonable isothermal predictions for the kinetics of curing of epoxy compounds because of the autocatalytic character of the process (Irzhak, 1979, Rozenberg, 1986). The presence of at least two kinetic constants, as in Eq. (1), should be taken into account.

In this context, noteworthy are other methods that may be successfully used depending on the type of investigation or the nature of epoxy resins (Cividanes, 2014). Among them, for example, the method of luminescence spectroscopy makes it possible to determine the degree of conversion with high accuracy at the final stage of reaction. This is hardly achievable by other methods. Analysis with the aid of a rheometer provides information about the time of gelation which cannot be obtained by another method.

The DSC studies (Zhou, 2009) showed that multi-wall CNTs act as catalysts, with COOH functionalization stimulating the initial stage of the curing of bisphenol A diglycidyl ether (BADGE). This accelerating effect is noticeable even at a 1% content of multi-wall CNTs. Nonfunctionalized multi-wall CNTs decrease the degree of crosslinking, as evidenced by a lower overall heat of reaction and lower glass-transition temperatures of the nanocomposites compared with the neat epoxy resin. At the same time, the functionalization of multi-wall CNTs increases the degree of crosslinking.

Compared with the neat epoxy resin, 1% of carboxylated multi-wall CNTs decreases the heat of reaction and increases the energy of activation (Abdalla, 2008). Fluorinated tubes insignificantly influence the value of E_a but lower the ultimate degree of conversion Q_{lim}.

The grafting of butylamine onto plasma- and CF_4-treated single-wall CNTs markedly improves the ultimate conversion, whereas clean tubes, when the reaction is accomplished in the nonisothermal regime, have no effect on this parameter, while in the case of the isothermal regime (30°C), they decrease it (Valentini, 2008).

There was another conclusion (Qiu, 2010): clean and aminated single-wall CNTs reduce the ultimate heat of reaction. Tubes with the grafted epoxy groups give almost the same value of heat as that obtained in the curing of resin without any filler: 355 versus 362 J/g. It is possible that different results may be attributed to different concentrations of single-wall CNTs.

The kinetic analysis of the curing process in terms of Eq. (1) does not reveal marked differences in the values of constants and exponents for the systems of interest. However, it was shown (Choi, 2009) that the introduction of 3% of clean multi-wall CNTs does not influence the kinetics of reaction, but tubes with grafted amino groups decrease the constant k_1 by factor of almost 2.5, increase k_2 by factor of 3, and decrease exponent m from 0.53 to 0.27.

Multi-wall CNTs with amino groups decelerate the curing of BADGE with 2-ethyl-4-methylimidazole at concentrations of 0.5 and 1 wt % (Yang, 2008), but the deceleration effect vanishes at a concentration of nanoparticles of 3%. However, in this case the value of ultimate heat decreases as well. Note that clean multi-wall CNTs accelerate curing of the same reaction system (Zhou, 2009).

The impact of multi-wall CNTs carrying acidic and amino groups on the process of curing was studied by Raman and luminescence spectroscopy (Cividanes, 2013). Throughout the reaction conducted in the presence of nanoparticles the rates were higher compared with the neat resin. The difference in the rates of curing was explained by homogeneity of the sample and the presence of chemical groups.

The kinetics of the amine curing of BADGE in the presence of multi-wall CNTs functionalized by oxygen-containing groups using viscometry and transmission electron microscopy along with calorimetry was analyzed (Grachev, 2014). The samples of multi-wall CNTs had different values of specific surface S. This factor was found to be decisive in the kinetic study of reaction: if at the onset the rates were equal, then, by the time of attaining the maximum rate W_{max}, the higher the value of S, the more pronounced the deceleration of the process, so that the time of attaining W_{max} increased. Afterwards, the inverse effect was observed, namely, acceleration, so that the higher the value of S, the more pronounced the final heat release. The rheokinetic study demonstrated that the time of a sharp gain in viscosity of the system (the gel point) also shifts to longer times with an increase in S. However, variation in the concentration of multi-wall CNTs (to 5%) insignificantly influences the kinetics of reaction.

The kinetic features of the process may become understandable if the micrograph of the epoxy composition which is measured at the initial stage of curing performed in the presence of multi-wall CNTs is examined (Figure 3). It is seen that compact polymer structures grow along the tube. Evidently, hydroxyl groups grafted on the surface catalyze the reaction of epoxy groups with amine (Rozenberg, 1986) to give rise to new hydroxyl groups accelerating this reaction. In doing so, the process of polymer formation is localized and the reaction assumes the frontal character. The natural consequence of the localization process is the formation of ineffective crosslinks. Therefore, the value of critical conversion increases and correlation is observed between the time of a sharp gain in viscosity and the specific surface of multi-wall CNTs.

There was shown that graphene oxide (GO) accelerates the curing of tetraglycidyl-4,4′-diaminodiphenylmethane with 4,4′-diaminodiphenyl sulfone (Qiu, 2011). According to this work, this effect is related to the presence of hydroxyl and carboxyl groups on the surface of GO.

The kinetics of the nonisothermal curing of BADGE with liquid poly(amido amine) in the presence of amine-functionalized GO using Eq. (5) was investigated (Ryu, 2014). It is seen that for the systems without any filler and in the presence of GO containing NH_2 groups, the parameters of this equation are similar. At the same time, GO slightly decelerates the process.

The mixture of BADGE and 1,1,2,2-tetra(*p*-hydroxyphenyl)ethane tetraglycidyl ether with diethyltoluenediamine in the presence of graphene was cured (Park, 2014a). There was showed that the latter somewhat accelerates the reaction and increases T_g by 15– 25°C. Previously, the same authors showed using Eq. (1) that this effect is associated with an increase in constant k_1 and the functionalization of graphene by amine enhances the effect of the filler (Park, 2014b).

The IR spectral analysis of the curing of BADGE with 4,4'-diaminodiphenylmethane in the presence of GO permited to obtain kinetic curves for both epoxy groups and

Figure 3. Initial stage of curing of the epoxy matrix in the presence of COOH-functionalized CNTs. Reprinted with permission from IAPC "Nauka" (2014, Grachev).

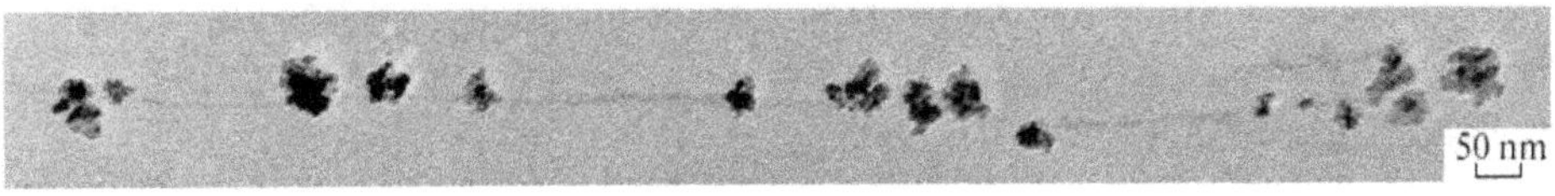

primary, secondary, and tertiary amino groups (Galpaya, 2015). These studies made a considerable contribution to gaining insight into the mechanism of reaction. It was shown that the original GO has no effect on the overall kinetics of the process and even decelerates the consumption of primary amino groups. But after autoclave purification, GO increases the rate of reaction of epoxy groups by more than factor of 2 and the rate of reaction of primary amino groups by factor of 1.8. As was shown by X-ray photoelectron spectroscopy, purification leads to a marked reduction in the amount of oxygen-containing groups on the surface of GO. The glass-transition temperature of a nanocomposite based on the crude GO is much lower than that of the epoxy matrix, whereas the purification of GO causes an increase in this parameter, although T_g does not attain the T_g value of the matrix.

The influence of various graphite fillers (graphite with a high surface area, graphite oxide, and exfoliated graphite oxide) on the reaction of epoxy ring opening of BADGE by amines—primary (benzylamine and cyclohexylamine) and secondary (dibenzylamine)—and hydroxyl (benzyl alcohol) was studied (Acocella, 2016). These data indicate the strong catalytic effect of fillers on the reaction with amines. This effect is the most pronounced for the exfoliated graphite oxide. In the case of benzyl alcohol, interaction with epoxy groups was observed only for graphite oxide. A similar effect was coused by fillers on the process of matrix formation: in the presence of graphite fillers, the rate and heat effect of the reaction grow and the gel point shifts to smaller times.

In essence, analogous data were reported (Mauro at al., 2014). These authors believe that a marked rise in the T_g of nanocomposites compared with the neat matrix is evidence of the catalytic effect of graphite with a high surface area (308 m^2/g) and graphite oxide. The catalytic effect of graphite was confirmed using the reaction of epoxystyrene with benzylamine.

On the other hand, it was found that the samples of graphite oxide with carboxyl or amino groups have almost no effect on the kinetics of nonisothermal curing of BADGE with 4,4'-diaminodiphenylmethane (Li, 2015). Possibly, this is related to a low concentration of the fillers (0.5%).

The effect of carbon nanofibers (diameter of 100–200 nm and length of 30–100 μm) oxidized in a solution of nitric acid and then treated with 3-glycidoxypropy ltrimethoxysilane on the kinetics of curing of epoxy resin Cycom 977 described by Eq. (1) was studied (Seyhan, 2009). It was shown that all kinds of fibers exert a catalytic effect which manifests itself as an increase in ultimate conversion and growth of kinetic constants k_1 and k_2. Note that E_{a1} decreases, while E_{a2} increases. In terms of catalytic efficiency, the fibers may be arranged in the following sequence: untreated, oxidized (COOH groups on the surface), and treated with silane (epoxy groups on the surface).

The catalytic effect of a carbon nanofiber was observed: the ultimate conversion and the kinetic constants in Eq. (1) increase, while the respective activation energies decrease (Cai, 2010). The higher activity is exhibited by fiber whose surface is modified through the oxidative polymerization of aniline (in accordance with the authors, the "nanograssy" coating).

The above results are contradictory. Most likely, this circumstance may be explained by ambiguity with respect to the filler concentration and uncertainty in the degree of dispersion and in the value and structure of its surface. The above results, summarized in Table 1, are rather contradictory, which is probably caused by ambiguity in the concentration of the filler, uncertainty in the degree of dispersion, the magnitude and structure of its surface.

Table 1. Kinetic features of curing epoxy oligomers in the presence of carbon nanoparticles

Filler	Concentration, wt %	Method of Experiments	Reaction Results	T_g	References
Multi-wall CNT-COOH	1	Isothermal calorimetry, eq. (5)	Decrease of Q_{lim}, increase of E_a	No effect	Abdalla, 2008
Multi-wall CNT-F	1	Isothermal calorimetry, eq. (5)	Decrease of Q_{lim}	No effect	Abdalla, 2008
Multi-wall CNT	3	Isothermal calorimetry, eq. (1)	No effect	Increase	Choi, 2009
Multi-wall CNT-NH_2	3	Isothermal calorimetry, eq. (1)	Decrease of k_1 and m, increase of k_2	Increase	Choi, 2009
Multi-wall CNT-COOH Multi-wall CNT-NH_2	0.2	Raman- and luminescence spectroscopy	Increase of initial reaction rate	–	Cividanes, 2013
Single-, Double- and Multi-wall CNT	0.1 – 0.5	Isothermal calorimetry, eq. (5)	Decrease of $k(T)$, increase of E_a	–	Esmizadeh, 2015
Multi-wall CNT-COOH, Different values of S	0.1 – 5	Isothermal calorimetry, viscometry	Increase of Q_{lim} with decrease of S, no concentrate effect	Increase	Grachev, 2014
Single-wall CNT-NH_2	–	Nonisothermal calorimetry	Decrease of Q_{lim}	–	Qiu, 2010
Multi-wall CNT-COOH	0.1 – 1	Nonisothermal calorimetry	Increase of Q_{lim}	Increase	Rahaman, 2014
Multi-wall CNT	0.25 – 1	Nonisothermal calorimetry	Increase of Q_{lim}, decrease of E_a	–	Susin, 2014

continued on following page

Table 1. Continued

Filler	Concentration, wt %	Method of Experiments	Reaction Results	T_g	References
Single-wall CNT	1	Isothermal calorimetry	Weak initial increase of reaction rate	Decrease	Tao, 2006
Single-wall CNT-NH_2	0.5 – 1.5	Nonisothermal calorimetry	Reaction rate depends on concentration	–	Visco, 2009
Multi-wall CNT-NH_2	0.5, 1, 3	Nonisothermal calorimetry, eq. (3)	Decrease of reaction rate and $Q_{\lim}$	–	Yang, 2008
Multi-wall CNT	1, 5	Isothermal calorimetry, eq. (1)	Increase of k_1	–	Xue, 2015
Multi-wall CNT	1, 3, 5	Nonisothermal calorimetry	Initial increase of reaction rate, decrease of $Q_{\lim}$	Decrease	Zhou, 2009
Multi-wall CNT-COOH	1, 3, 5	Nonisothermal calorimetry	Increase of reaction rate, increase of $Q_{\lim}$	Decrease	Zhou, 2009
GO	–	Fourier IR spectroscopy	Increase of reaction rate	Decrease	Navalon, 2014
GO	1 – 5	Nonisothermal calorimetry	Increase of reaction rate, weak increase of $Q_{\lim}$	–	Qiu, 2011
GO	1	Nonisothermal calorimetry, eq. (5)	Decrease of reaction rate	–	Ryu, 2014
GO-NH_2	1	Nonisothermal calorimetry, eq. (5)	No effect	–	Ryu, 2014
Graphene	0.5 – 2	Nonisothermal calorimetry	Weak increase of reaction rate	Increase	Galpaya, 2015
Graphene-NH_2	1	Isothermal calorimetry, eq. (1)	Increase of k_1	Increase	Park, 2014a
Graphite	3	Nonisothermal calorimetry, viscometry	Increase of reaction rate, increase of $Q_{\lim}$	–	Acocella, 2016
Graphite	0.5	Nonisothermal calorimetry, eq. (5)	Decrease of reaction rate, increase of E_a	–	Li, 2015
Graphite	3	Model reaction	Catalytic effect	Increase	Mauro, 2014
Carbon nanofibers, polyaniline	1	Isothermal calorimetry, eq. (1)	Increase of k_1 and k_2, decrease of E_{a1} and E_{a2}, increase of $Q_{\lim}$	–	Cai, 2010
Carbon nanofibers -COOH, epoxy-	1, 3, 5	Isothermal calorimetry, eq. (1)	Increase of $Q_{\lim}$, decrease of E_{a1}, increase of E_{a2}	–	Seyhan, 2009

Noncarbon Fillers: Oxides of Metals and Silicon

The effect of Al_2O_3 nanoparticles on the kinetics of polycondensation of BADGE under the action of diethylenetriamine using modulated DSC was explored (Sanctuary, 2009). As a result, not only the rate of the process was registered, but also variation in the heat capacity of the system during the process was monitored. It was shown that the filler increases the rate of reaction but decreases the ultimate heat. Viscosity measurements confirmed that the formation of the polymer network accelerates in the presence of nanoparticles and the gel point shifts not only with time but also with conversion. This implies that nanoparticles are directly involved in the formation of intermolecular bonds. At the same time, experiments with water additives revealed that nanoparticles affect the kinetics of the curing reaction in qualitatively the same manner as Al_2O_3 nanoparticle (Baller, 2011). The authors inferred that water adsorbed by nanoparticles is responsible for the catalytic effect.

DSC was used to examine the effect of additives of Al_2O_3 and ZnO nanoparticles on the curing of BADGE with *o*-tolylbiguanidine (Karasinski, 2013). Both oxides decelerate the reaction but increase the ultimate limiting degree of conversion. Exponents m and n in Eq. (5) remain almost unchanged, whereas the activation energy decreases. Note that in the case of ZnO this decrease is considerable.

At a fairly low concentration (1 and 5%), ZnO nanoparticles accelerate the reaction of BADGE with 2,2'-diamino-1,1'-binaphthyl; at concentration of 10%, their catalytic efficiency declines, while at content of 15% retardation is observed (Zabihi, 2011). Compared with the neat matrix, nanocomposites feature higher values of ultimate heat and glass-transition temperature with the maximum values corresponding to 5% content of nanoparticles. Probably, reduction in the catalytic activity of nanoparticles with increasing concentration may be attributed to their aggregation; as a result, the effective surface decreases.

In work (Ghaffari, 2015) there was studied how the size of ZnO nanoparticles affects the kinetics of BADGE curing with poly(aminoamide). Nanoparticles were sheets with a thickness of nearly 20–40 nm, while microparticles were rods with a length of ~1 μm. Analysis was performed using Eq. (5). It was found that the autocatalysis of reaction is absent; that is, $m = 0$ and n is somewhat above unity. For both composites, compared with the neat matrix, the energy of activation decreases, but the rate constant slightly grows in the case of the microcomposite and declines in the case of the nanocomposite.

BADGE with the propylenimine dendrimer carrying eight end groups -NH_2 in the presence of Fe_2O_3 nanoparticles was cured (Zabihi, 2012). The latter manifested the catalytic effect. The higher the concentration of nanoparticles, the more pronounced the increase in ultimate conversion and glass-transition temperature. It is shown that the kinetics of formation of the nanocomposite containing 10% Fe_2O_3 is adequately described by Eq. (1a). No data are available for other systems, including the neat matrix. A similar result was reported in (Omrani, 2011). It was showen that the kinetics of curing of glycerol diglycidyl ether with 3,3'-dimethylglutaric anhydride in the presence of Al_2O_3 obeys Eq. (1a).

Nanoparticles of metal oxides are able to adsorb components of the reaction system to one extent or another (Hong, 2001). Possibly, their kinetic role is associated with this property.

Direct measurements of the complex specific heat capacity demonstrated that the interaction of SiO_2 nanoparticles and BADGE molecules is very weak (Baller, 2009). At all stages of polymer network formation, the interaction of nanoparticles and matrix is of a physical origin. The effect of the filler on the kinetics of curing was insignificant.

In contrast, acceleration of the process was observed (Ghaemy, 2007, Rosso, 2007). The kinetic studies revealed that the catalytic effect of SiO_2 nanoparticles is related to the presence of hydroxyl groups on their surface (Alzina, 2011). When the latter groups are changed for epoxy groups, the effect of nanoparticles on the kinetics of BADGE reaction with *m*-phenylenediamine is eliminated.

The kinetic results of this section are summarized in Table. 2.

Minerals

For polymer nanocomposites, the most popular fillers from the class of minerals are layered silicates (Pavlidou, 2008), which are sometimes called nanoclays, in particular, montmorillonite (MMT). The structure of its crystal lattice is such that it can adsorb various ions (mostly cations) and swell in polar liquids owing to their penetration into the interlayer space (Becker, 2005, Paiva 2008).

At the nanometer scale, MMT is composed of three-layer stacks ~0.7 nm in thickness and several hundred nanometers in length and width. At micron level, these stacks are united into primary particles with interlayer distance of about 1.35 nm. At higher level, they form aggregates. During formation of nanocomposites, stacks should be exfoliated in order to reach a high area of contact with the matrix. In order to facilitate exfoliation, the surface of stacks should be treated for the purpose of changing their hydrophilic nature to hydrophobic, because the hydrophilic character of the silicate surface hampers the dispersion of MMT. Neutral organic compounds may form complexes with interlayer cations; for example, alkylamines are transformed into alkylammonium cations.

These properties of MMT govern the kinetic features of formation of epoxy nanocomposites.

In the absence of the curing agent (1,3-phenylenediamine), the modified MMT and even the unmodified MMT promotes the homopolymerization of BADGE

Table 2. Kinetic features of curing epoxy oligomers in the presence of noncarbon nanoparticles

Filler	Concentration, wt %	Method of Experiments	Reaction Results	T_g	References
Al_2O_3	7, 12, 16	Isothermal calorimetry	Increase of reaction rate	–	Baller, 2011
ZnO	1	Isothermal calorimetry, eq. (5)	Decrease of reaction rate, decrease of E_a, $m = 0$, increase of n	–	Ghaffari, 2015
Al_2O_3, ZnO	2.5	Nonisothermal calorimetry, eq. (5)	Decrease of reaction rate, decrease of E_a, increase of Q_{lim}	Increase	Karasinski, 2013
Al_2O_3	2, 3.5, 4.5	Isothermal calorimetry, eq. (1), viscometry	Increase of reaction rate, decrease of Q_{lim}	–	Sanctuary, 2009
ZnO	1, 5, 10, 15	Nonisothermal calorimetry, eq. (5)	Increase of reaction rate depending on concentration, increase of Q_{lim}	Increase	Zabihi, 2011
Fe_2O_3	1, 5, 10	Isothermal calorimetry, eq. (1a)	Increase of Q_{lim}, $n = 2.75$	Increase	Zabihi, 2012
SiO_2	10, 20	Nonisothermal Calorimetry	Increase of reaction rate	Decrease	Alzina, 2010
SiO_2	5-20	Nonisothermal calorimetry, eq. (3)	Increase of reaction rate, increase of Q_{lim} depending on concentration	–	Ghaemy, 2007
SiO_2	1, 3, 5	Nonisothermal calorimetry, eq. (1)	Increase of reaction rate, decrease of Q_{lim}	Increase	Rosso, 2007

at a high temperature (Alzina 2012). Depending on the nature of the intercalated modifier (octadecyl-, trimethylstearyl-, methyldihydroxyethylammonium), MMT may either catalyze the reaction of the epoxy oligomer or react with a prepolymer or a curing agent.

At the same time, it was found that MMTs modified with alkylamines weakly accelerate (Ivankovic, 2006, Román, 2010, Alzina, 2010), retard (Li, 2014), or do not affect at all (Ferdosian, 2013) the kinetics of curing of epoxy oligomers with amines. A weak acceleration was also observed for the unmodified MMT and MMT with intercalated 3-aminopropylethoxysilane (Shanmugharaj, 2012).

Thus, it should be stated that the rate of curing of epoxy oligomers is almost insensitive to the presence of MMT. However, the kinetics of formation of polymers is not reduced only to a change in the concentration of reactants: structure formation should be taken into account. Even to a higher extent, this applies to a nanocomposite, whose properties are determined not only by structural levels of polymer matrix but also by the structure of nanoparticles and the character of their distribution in the material bulk.

As was noted above, MMT requires exfoliation. Namely this process occurs during the chemical reaction, and its efficiency depends on the reaction conditions. For example, it was shown (Shiravand, 2014a) that the cationic polymerization of triglycidyl-*p*-aminophenol occurs within the interlayer space, which entails the exfoliation of MMT, whereas with DDS the epoxy oligomer reacts outside the interlayer cavity. An increase in temperature is favorable for the former reaction: ultimate conversions inside and outside the cavity are 0.19 and 0.74, respectively, at 120°C or 0.76 and 0.77 at 180°C.

The optimum structure of epoxy composites was achieved when BADGE was cured with poly(ester diamine) in the nonisothermal regime at low rise in temperature (2.5 and 5 K/min) (Montserrat, 2008). Small-angle X-ray scattering studies revealed the exfoliation of MMT in the matrix. The authors of (Cortés, 2014) used hyperbranched polyethylenimine with end amino groups as a curing agent and attained effective exfoliation. A comparison of three systems, such as triglycidyl-*p*-aminophenol + DDS, BADGE + poly(ester diamine), and BADGE + hyperbranched polyethylenimine, showed (Shiravand, 2014b) that the exfoliation ability of MMT decreases in the mentioned sequence.

As was reported in (Shiravand, 2015), the nonisothermal curing of epoxy oligomers with amines in the presence of MMT includes four different reactions: formation of the matrix via the interaction of epoxy groups with the diamine curing agent, intracavity homopolymerization, and two homopolymerization reactions outside MMT which are catalyzed by onium ions of organically modified clay and tertiary amines.

X-ray diffraction was used to probe the exfoliation of MMT intercalated by octadecylammonium during the isothermal curing of BADGE with DDS (Kong, 2003). It was showed that this process may be divided into three different stages (Figure 4).

The first stage is related to the penetration of BADGE into the interlayer space of MMT; at the second stage the cationic polymerization of the epoxy resin catalyzed by ammonium takes place; and at the third stage BADGE sorbed by MMT is cured with amine.

In order to identify the intracavity polymerization, mixture of MMT and triglycidyl-*p*-aminophenol was held at various temperatures for tens of days in the absence of the amine curing agent. Afterwards, DDS was added and the curing process was

Figure 4. Change in distance d between MMT sheets during the isothermal curing of BADGE at (1) 140, (2) 130, and (3) 120°C. Roman numerals denote stages of the exfoliation process. Arrows indicate the expected tendency of exfoliation. Reprinted with permission from Am. Chem. Soc. (2003, Kong).

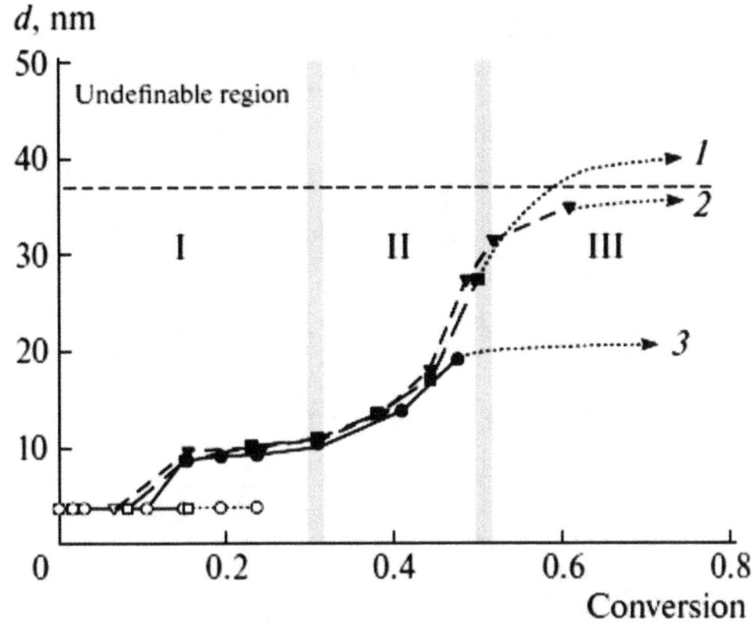

conducted in the nonisothermal regime (Hutchinson, 2013). A similar procedure was employed for the system BADGE–MMT–poly(ester diamine) (Pustkova, 2009). At the first stage the epoxy equivalent and the glass-transition temperature increased. This technology makes it possible to improve both the degree of dispersion of MMT in the epoxy resin and the subsequent exfoliation of the clay during formation of epoxy nanocomposites.

The period of the intracavity polymerization was reduced to tens of minutes by the use of complex $BF_3 \cdot C_2H_5NH_2$ as a catalyst (Hutchinson, 2014).

An attempt to accomplish the intracavity polycondensation was made (Jagtap 2015). In contrast to the commonly used MMT modifiers, i.e., polyalkylamines (see Table 3), they used half neutralized salt of poly(ester diamine), which was intercalated via ionic exchange in an aqueous–organic solution, assuming that these macromolecules, when fixed by the ionic end on the walls of the cavity, will react with epoxy groups of the binder via its free amine end. However, no direct evidence for this reaction is available (Jagtap 2015).

Table 3. Kinetic features of curing epoxy oligomers in the presence of MMT nanoparticles

Modifier	Concentration, wt %	Method of Experiments	Reaction Results	T_g	References
Different amines	5	Nonisothermal calorimetry	Reaction rate depends on modifier	Decrease	Alzina, 2011
Octadecyamine	5	Nonisothermal calorimetry, eq. (1)	Weak increase of reaction rate	Decrease	Alzina, 2012
Polyoxypropilenediamine	5, 10	Isothermal calorimetry, eq. (1a)	Weak increase of reaction rate	–	Ivankovic, 2006
Polyoxypropilenediamine	5	Nonisothermal calorimetry, eq. (5)	Decrease of reaction rate, increase of Q_{lim}, decrease of E_a	–	Li, 2014
Octadecyamine	2, 5	Nonisothermal calorimetry, eq. (5)	Weak increase of reaction rate	Depends on concentration	Rosso, 2007
3-aminopropilethoxysilane	2	Nonisothermal calorimetry, eq. (5)	Weak increase of reaction rate, decrease of Q_{lim}, weak increase of E_a	–	Shanmugharaj, 2012

Metal-Containing Nanoparticles Synthesized in Situ

The synthesis of metal-containing nanoparticles for producing nanocomposites may be accomplished by various physical processes on the preformed matrix containing molecules of appropriate precursors (Pomogailo, 2005, Wilcoxon, 2006, Gusev, 2007, Suzdalev, 2006, Faraji, 2010, Lu, 2012). Physical methods of obtaining metal-containing nanoparticles (photolysis, radiolysis, and thermolysis) are as a rule accompanied by chemical reactions leading to their formation. An important factor is the diffusion of preformed substances (metal atoms): the glassy state of the matrix provides a considerable obstacle to diffusion. For example, *N*-cetylpyridinium tetrachloroaurate was dissolved in methyl methacrylate and, after polymerization of the latter, was subjected to UV radiation. However, the formation of gold nanoparticles was registered only at temperatures above T_g of the polymer (Agareva, 2009).

A substantially different method involves the combination of processes of formation of a matrix and metal-containing nanoparticles, i.e., formation of nanocomposites in situ.

Main chemical method used at moderate temperatures includes the reduction of chemically bound metal atoms in nonpolar media. These methods of chemical reduction are the subject of most publications (Wilcoxon, 2006, Lu, 2012, Rozenberg, 2008, Wuithschick, 2013, Wang, 2014, Yan, 2014, Völkle, 2015). Mechanisms controlling formation of metal-containing nanoparticles in situ are highlighted in the review (Irzhak, 2016).

The transformation of the resulting single valence atoms, or monomers, into nanoparticles includes nucleation stages with formation of primary clusters or stable particles, their growth by addition of monomers, possible subsequent coagulation and/or Ostwald ripening. The kinetics of all these stages determines the size distribution function which strongly affects the ways and possibilities of the nanoparticles and corresponding nanocomposites application. An important role in this is played by a polymer medium in which chemical reactions take place, including the ability of its components or fragments of macromolecules to be adsorbed on the particles, as well as the possibility of formation of micelle-type structures.

With rapid decay of the precursor, there will be a supersaturation of the system with a monomer. In this case, the nucleation will, as a first approximation, describe by thermodynamic theory of Gibbs-Volmer-Frenkel (Volmer, 1939, Frenkel', 1975), according to which the magnitude of the radius of the critical nucleus is determined the formula:

$$r_c = \frac{2\sigma V_m}{RT \ln S}$$

where r_c, σ and V_m are radius, surface tension and molar volume of the critical nucleus, R is the gas constant, T is temperature, and S is the supersaturation. In this case, the nucleus will be unstable if $r < r_c$, and it stabilizes when $r \geq r_c$.

A qualitative picture of nucleation kinetics on the basis of the classic thermodynamic notions was proposed (LaMer, 1950). The pattern illustrating their idea is shown in Figure 5.

Rapid increase in the concentration of monomer will lead to the achievement the critical value of C_{min} (region I, the stage of prenucleation), and then to overcome C_{min}. Exceeding this value gives rise to proper nucleation. Because of the balance between the rate of formation of the monomer and its expenditure consumption on nucleation and growth of the generated nuclei, the concentration will reach a peak, C_{max}, and then begins to decrease due to an increase in consumption for growth and again reach the critical level of C_{min}, marking the end of the nucleation stage (region II). After that, the concentration of the monomer will continue to decrease to the equilibrium value of C_0, expending on the growth of nucleus without renucleation, due to the fact that the supersaturation is below the critical level (region III).

Although short spatial and temporal scales of the nucleation stage hamper the direct observation of the classic process, to trace the time-resolved formation of silver CNs using small-angle X-ray scattering *in situ* was managed (Polte, 2012). Silver perchlorate was reduced with sodium boron hydride in an aqueous solution. As shown in Figure 6, the nanoparticles nucleation kinetics corresponds to the LaMer pattern: the number of particles initially grows without any appreciable change in

Figure 5. LaMer pattern (see text for explanations)

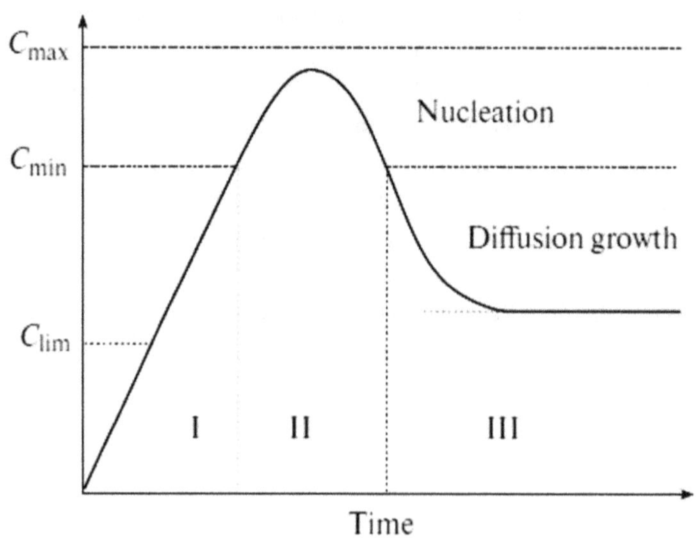

their sizes, attaining a maximum in fractions of a second. Further, the number of nanoparticles descends, and their radius increases. This corresponds to stage III in the pattern, and nanoparticles interact with each other alongside with their simple growth via the reaction with monomers, i.e., the aggregation mechanism is engaged.

However, under conditions of formation of composites, it is hardly possible to implement such a mechanism. An alternative variant was obtained in (Turkevich et al., 1951). When studying the synthesis of gold by reducing $HAuCl_4$ with sodium citrate, it was shown that rapid nucleation is not observed; on the contrary, the kinetic curve of accumulation of critical nuclei is S-shaped with a more or less extended induction period. The authors suggested that such feartures of the process are due to redox reactions leading to the conversion of gold cations to a zero-valent atom and citrate ion to acetone dicarboxylic acid. In this case, the supersaturation of the system by the monomer, Au(0), is absent. Obviously, in this case the concept of a critical nucleus becomes meaningless.

Later it was established that S-shaped kinetics is inherent in many metals with variable valence (see, for example, the review (Finney, 2008)). To describe such processes, a rather simple two-step scheme was proposed (Watzky, 1997):

$$A \xrightarrow{k_1} B$$
$$A + B \xrightarrow{k_2} B_2$$

Figure 6. Kinetics of change in the average radius and number of particles for first 2 s. Reprinted with permission from Am. Chem. Soc. (2012, Polte)

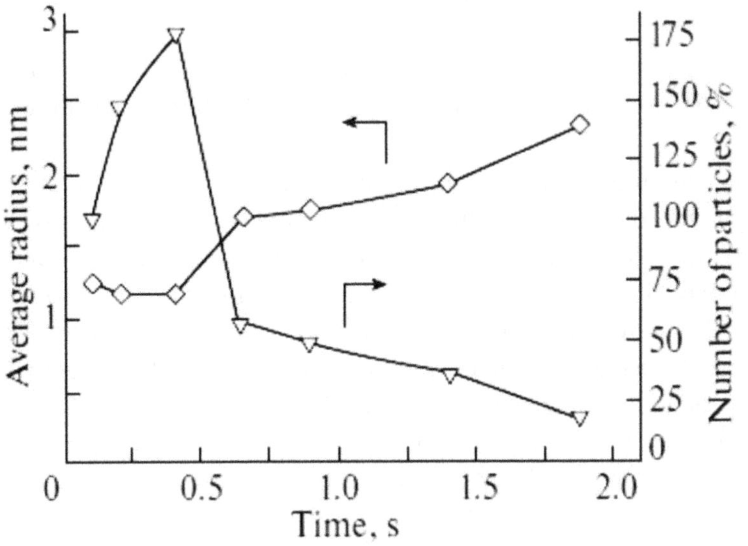

The first stage is the slow nucleation of "kinetically efficient" clusters *B* from precursor *A*, and the second stage is the particle's fast growth reaction. In the original studies, in particular in (Turkevich, 1951), *A* represented complex $[(n\text{-}C_4H_9)_4N]5Na_3[(1,5\text{-cyclooctadiene})Ir \cdot P_2W_{15}Nb_3O_{62}]$, and *B* was the catalytic surface of a $Ir_n(0)$ nanocluster.

Studies of a wide range of systems show (see (Irzhak, 2016)) that the value depends on the number of catalytic active nuclei, namely, their ratio with increasing ratios k_2A_0/k_1 for practically unchanged diameters of the order of 2 nm. The medium, active additives, temperature – these factors make it possible to achieve the formation of an almost monodisperse distribution (width not exceeding 15%) of the nanoparticles with a size determined, as a rule, by a "magic number" (number of atoms in a filled shell: 13, 55, 147, 309 and etc.).

The processes of formation of zero-valence atoms or other monomers inevitably lead to their clustering. To stabilize clusters in a nonpolar solution, it is necessary to have amphiphilic molecules capable of forming adsorption layers and thereby form inverse micelles from the nanoparticles.

At the same time, when the epoxy nanocomposite films are stored under light for a while, the optical density D_{max} in the region of the surface plasmon resonance of silver nanoparticles decreases. Similar changes are observed in the spectra and at storage of films in dark (Bogdanova, 2019) (Figure 7). The kinetic curves are described by the first-order equation (Figure 7*a*, straightening in Figure 7*b*).

$$D_{\max} = D_{\lim} + A\exp\left\{-kt\right\}$$

D_{lim} = 0.534 (in the dark) и 0.42 (in the light).

Accordingly, the value of the limiting conversion of nanoparticles under light is higher than in dark: the conversion under light is 0.29, while in dark, the conversion is 0.11.

The decrease in D_{max} in time means a drop in the total concentration of metallic silver, apparently due to its "dissolution". The relatively recent discovery of digestive ripening (DS), which represents the transfer of atoms from large metal nanoparticles to smaller ones (Irzhak, 2016), indicates the possibility of such a process.

Although the mechanism of the DS process is not fully understood, there are grounds to believe that the equation proposed in (Lifshits, 1958, Wagner, 1961) for describing the kinetics of Ostwald ripening (OS) is applicable to DS. The latter is one of the options for the growth of the nanoparticles, when large particles grow at the expense of small particles. According to the theory, the change in the radius *r* of a spherical particle in time will obey equation:

Figure 7. Kinetics of the decrease in the concentration of silver NP in the dark (1) and in light (2) in natural (a) and semi logarithmic (b) coordinates. Reprinted with permission from IAPC "Nauka" (2019, Bogdanova).

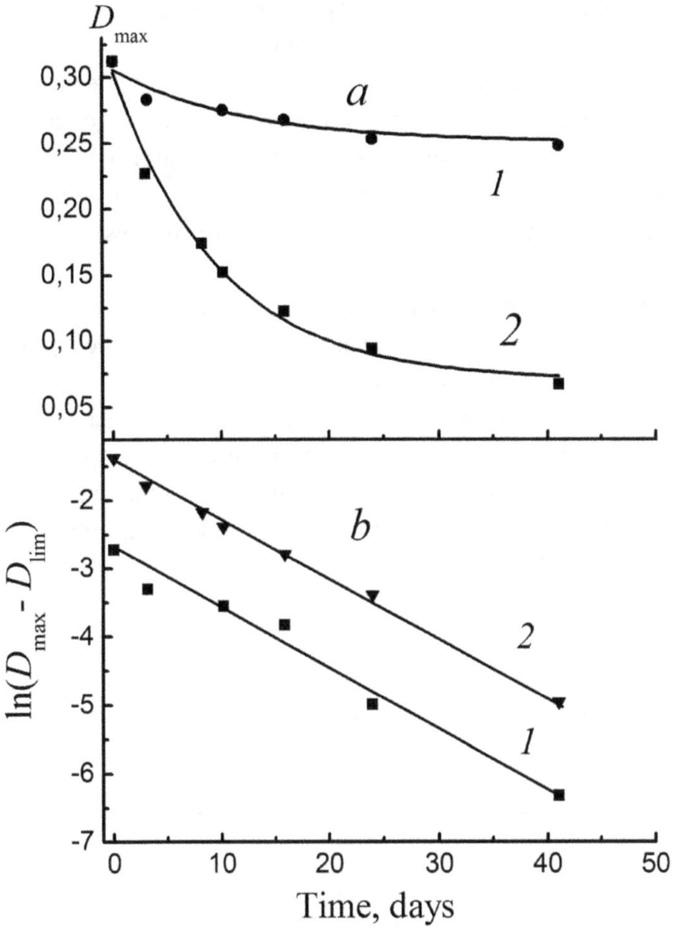

$$\frac{dr}{dt} = \frac{K_D}{r^2}\left(\frac{r}{r_{cr}} - 1\right)$$

where D is diffusion coefficient, V is molar volume, and C_0 is solubility of the monomer; C is monomer concentration; $S = C/C_0$; σ is interphase free energy; k is the Boltzmann constant; and T is temperature.

As can be seen, the particles grow if $r \geq r_{cr}$, their dimensions decrease otherwise. This is the physical meaning of the phenomenon of OS.

A vivid example of reversible OR and DR was demonstrated (Xin, 2012). They observed fluctuating growth of bismuth nanoparticles in the absence of precursor at 180°C. The reaction system consisted of a limited number of large Bi nanoparticles (80–150 nm in diameter) in a solution of oleylamine (surfactant) and dichlorobenzene. The Bi nanoparticles served as a source of monomeric Bi(0).

The process includes formation and growth of small particles due to large ones (DR) with simultaneous OR. The total number of particles increased for a while and then changed to the fluctuation mode. The total volume of content fluctuated near a certain level evidently given by the total volume of initial Bi nanoparticles. The size of each of the observed particles and their assemblies experienced similar fluctuations.

Figure 8 shows the trajectories of variation of the volume of close-lying particles *P*1 and *P*2. Their size changes in antiphase, demonstrating the simultaneous DR (growth of the first particle from the zero volume to the volume equal to that of the second particle) and OR (further growth of the first particle, leading to a decrease in the size of the second). The total volume of these particles is maximum when only one of them remains and minimum at equal sizes. In other words, the monomer concentration in the diffusion layer around the particles is minimum in the former case and maximum in the latter.

Figure 8. Volume trajectory of the P1 and P2 particles. Reprinted with permission from Am. Chem. Soc. Am. Chem. Soc. (2012, Xin).

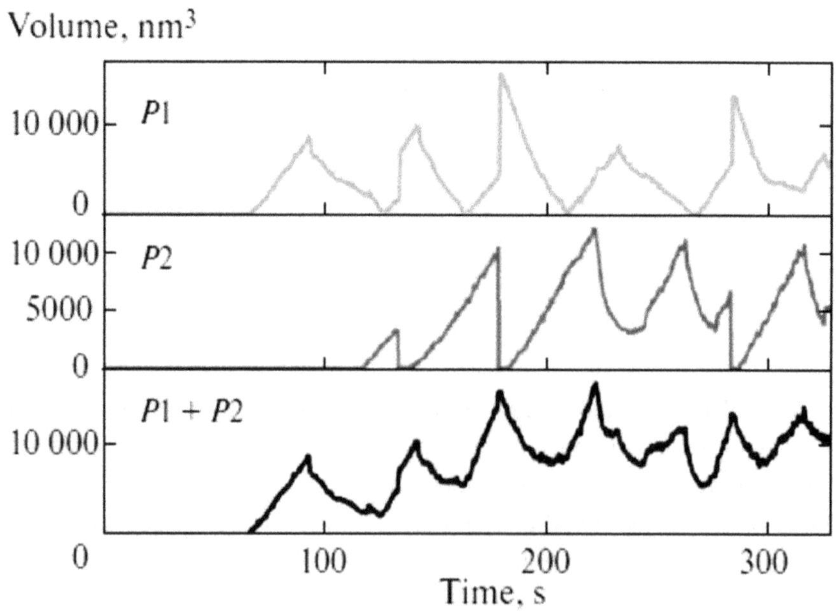

Obviously, the first stage of the DS process is the disassembly of small nanoparticles on zero-valence atoms and metal clusters. Under normal conditions, they subsequently form nanoparticles, which requires their diffusion displacement. But under the conditions of a glassy matrix, diffusion is difficult, if not forbidden at all. Therefore, the entire process is reduced to the first stage, i.e. "dissolution" of large nanoparticles with the formation of zero-valence atoms and silver clusters. The presence of a limit, apparently, is due to the saturation of the boundary zone surrounding the particle by the zero-valence atoms and clusters of silver formed.

As precursors soluble in organics, complexes of univalent gold $[O(AuPR_3)_3](CF_3SO_3)$ (R = Ph or CH_3) (Bardaji, 2002), $[RN(CH_3)_3][Au(SC_{12}H_{25})_2]$ (R = C_8H_{17}, $C_{12}H_{25}$, and $C_{14}H_{29}$), and $[(C_{18}H_{37})_2N(CH_3)_2][Au(SC_{12}H_{25})_2]$ were proposed (Nakamoto, 2005); salts of organic acids with a fairly bulky (even high-molecular-mass (Tkachuk, 2013, Kuzub, 2015)) radical, such as silver myristate $C_{13}H_{27}COOAg$, copper oleate $(C_{18}H_{33}COO)_2Cu$, silver oleate, silver octanoate $C_7H_{15}COOAg$, silver stearate $C_{17}H_{35}COOAg$, silver 2-hexyldecanoate *p*-$C_8H_{17}CH$(*n*-C_6H_{13})$COOAg$, *cis*9-octadecanoate *p*-$C_8H_{17}CH{=}CH(CH_2)_7COOAg$, and silver neodecanoate $CH_3(CH_2)_3C(CH_3)_2COOAg$, have enjoyed popularity (Bardaji, 2002, Yamamoto, 2003, 2006, Kashiwagi, 2006, Khanna, 2008, Clary, 2011, 2012).

However, the nonideal state of solutions of these compounds should be taken into account. This implies that, when a certain concentration is exceeded, precursors in solution are united into associates, i.e., are clustered. Evidently, this circumstance cannot be disregarded when considering feasible mechanisms governing formation of metal-containing nanoparticles.

The silver nanoparticles by the reduction of alkyl carboxylates in trimethylamine at 78°C were synthesized (Kuzub, 2009, 2010). It was found that the induction period grows and the maximum rate decreases in the following sequence: decanoate, myristate, and stearate. But in this sequence, the length of the hydrocarbon radical of carboxylates (C_9, C_{13}, and C_{17}) increases. It is reasonable to assume that solubility grows in the same sequence and, hence, the possibility of formation of clusters decreases. Thus, there is a direct relationship between the rate of formation of nanoparticles and the concentration of precursor clusters.

This idea underlies the theory of formation of metal-containing nanoparticles from precursors of the silver carboxylates type via their reduction which was put forth (Solov'ev, 2015).

The model adopted for formation of metal-containing nanoparticles may be presented as follows. Carboxylates reversibly form triangular and tetrahedral clusters. The development of larger clusters is not allowed because of steric reasons. The reduction of cation in them occurs. As a result, the adsorption of new salt molecules becomes possible. Indeed, if for carboxylates the tetrahedral structure is limiting, then the metal atom in the limit may be surrounded by 12 molecules (the icosahedron structure). It is assumed that the concentration of the reducing agent is high, so that the corresponding reaction is pseudo first order. Thus, the kinetic scheme may be written as follows:

$$\begin{array}{l} 3R_{01} \xrightarrow{k_1} R_{03} \xrightarrow{k_2} 3R_{01} \\ R_{01} + R_{03} \xrightarrow{k_3} R_{04} \xrightarrow{k_4} R_{01} + R_{03} \\ R_{01} + R_{ji} \xrightarrow{k_3} R_{j,i+1} \\ R_{ji} \xrightarrow{k_4} R_{01} + R_{j,i-1} \\ R_{ji} \xrightarrow{k_5} R_{j+1,i-1} \end{array}, \quad (6)$$

$i = 0, 1, 2, \ldots; j = 3, 4, \ldots$).

Here R_{ji} denote clusters composed of i carboxylate molecules and j atoms of the zero-valence metal. Accordingly, R_{01} is the initial carboxylate, R_{03} is the cluster of the triangular carboxylate, and R_{04} is the cluster of the tetrahedral carboxylate. Reactions with constants k_1 and k_2 are responsible for the formation and dissociation of associates consisting of three carboxylate molecules, and reactions with constants k_3 and k_4 correspond to the addition of one molecule to cluster R_{ji} and its detachment. The reaction with constant k_5 involves reduction of the bound metal in a cluster.

A system of equations corresponding to scheme (6) was analyzed with a wide variation in kinetic constants. It was shown that the values of k_2, k_4, and k_5 have a slight effect on kinetics of the process. The decisive role is played by constants k_1 and k_3, that is, those constants that determine reactions giving rise to clusters, including mixed clusters.

The kinetics of the process is characterized by existence of the induction period in the reaction of carboxylate consumption and an almost linear growth of average sizes of metal-containing nanoparticles with conversion. With increase in constant k_1, the maximum rate increases, the induction period shortens, and the sizes of resulting particles decrease After a certain period of growth, the number of particles reaches a limiting value which is lower the greater the value of k_1 (Figure 9). At the same time the mass of NPs varies in proportion to conversion, irrespective of the constant (insert in Figure 9). At the same time, the narrow size distribution is typical of these particles.

Papers addressing methods of in situ obtainment of epoxy nanocomposites with metal nanoparticles are scarce. Below we will concern studies published during the last decade.

Figure 9. Variation in N-value and mass M (insert) of nanoparticles; k1, L2·mol−2·s−1 = (1) 10, (2) 30, (3) 50, (4) 100, and (5) 200. Reprinted with permission from IAPC "Nauka" (2016, Solov'ev).

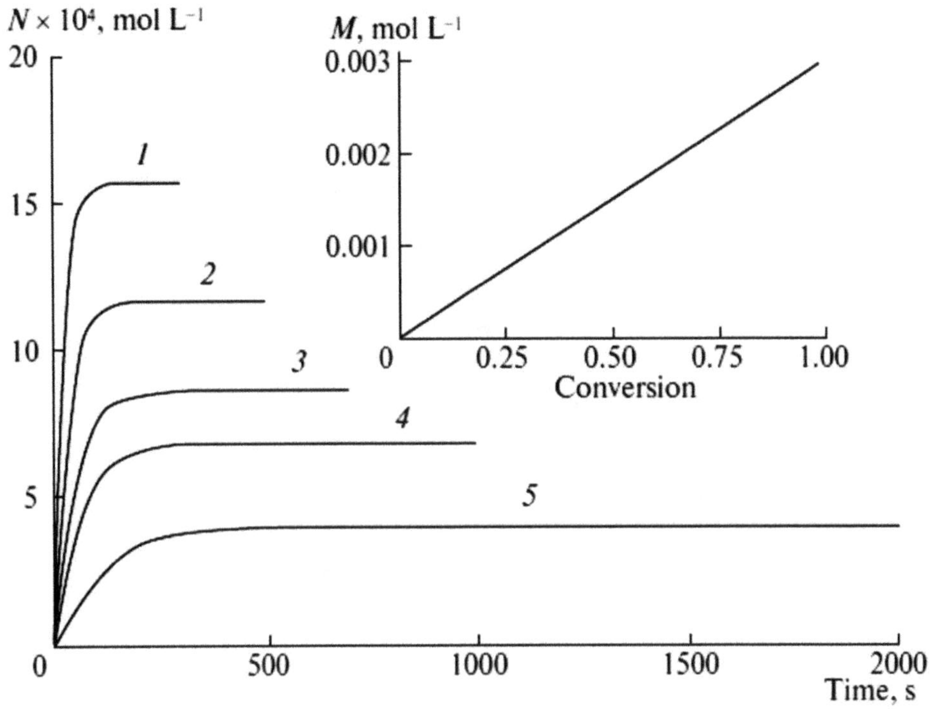

Under UV radiation 2,2'-dimethoxy-2-phenylacetophenone decomposes into radicals. A dimethoxyphenyl-carbonium radical reacts with $AgSbF_6$ and reduces a silver cation to Ag(0) via the transfer of electron, and then the radical transforms into the carbonium cation able to initiate the polymerization of diepoxide. Hence, silver nanoparticles and network matrix are formed simultaneously (Sangermano, 2007). With increasing concentration of silver salt, the rate of polymerization and the ultimate conversion decrease, but the glass-transition temperature increases.

The concentration of $AgSbF_6$ in a wide range varied and the yield of silver nanoparticles determined (Vescovo, 2010) (Figure 10). As it is seen, an almost full conversion was attained (the amount of Ag is 31% of the precursor mass). In part, these results were proved in (Kortaberria, 2012).

The silver nanoparticles with the same precursor, $AgSbF_6$ were obtained, but in the presence of degradable under irradiation with visible light 3,5-bis(4-methoxyphenyl)-dithieno[3,2-b;2,3d]-thiophene (Yagci, 2011). A similar technique was employed to synthesize epoxy nanocomposites with nanoparticles of silver (Yagci 2008a) and

Figure 10. Dependence of the yield of silver nanoparticles on concentration of the precursor. The graf is based on data (2010, Vescovo).

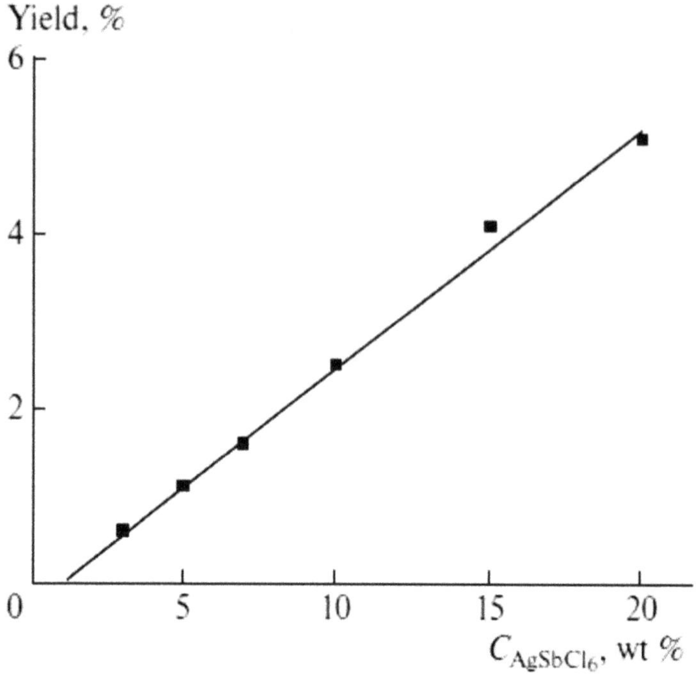

gold (Yagci, 2008b), but 2,3-bornadione (camphorquinone) was used as a source of radicals; in the case of gold, the precursor was $HAuCl_4$.

The silver nanoparticles in situ via the reduction of $AgNO_3$ in the epoxy resin by Triton-100, which simultaneously functioned as a stabilizer of nanoparticles were synthesized (Lu, 2008). A cycloaliphatic epoxy resin, hexahydro-4-methylphthalic anhydride as a curing agent, a reducing agent, and a precursor were dissolved in acetonitrile and exposed to UV radiation. After completion of the process, the solvent was removed at a reduced pressure.

It is supposed (Lu, 2006) that in this case the role of irradiation is insignificant; it is more probable that reduction occurs via the following mechanism

$$-OH + Ag^{+} \rightarrow -COOH + Ag(0),$$

where -OH is the end group of the reducing agent.

A complex of silver acetate and 2-ethyl-4methylimidazole was synthesized in an epoxy resin and during its curing Ag^{+} was reduced to Ag(0) as a result of thermal decomposition of the complex (Gao, 2011). In such a manner, silver nanoparticles were generated in situ. The imidazole product of complex decomposition served as a curing agent.

The epoxy resin ED-20 with triethylamine in the presence of silver myristate was cured (Bogdanova, 2012, 2014, 2015). The reduction of the latter and the formation of silver nanoparticles occurred simultaneously during polymerization. Reduction agents were both amine and the epoxy group. Carboxylate groups compatible with the medium functioned as stabilizers of particles.

The kinetic results of this section are summarized in Table. 4.

CONCLUSION

It has been shown that, regardless of whether a filler is introduced in the reaction system or is formed in situ during the process of matrix formation, its structure changes to a greater or lesser extent compared with the unfilled cured epoxy binder. In addition, the matrix influences the character of distribution of nanoparticles over volume. This effect is especially important in the case of graphene and MMT, when exfoliation is the case in point. The matrix governs the size and shape of the resulting nanoparticles. They interact with the epoxy resin affords interfacial layers. There is no doubt that all these factors affect the properties of the epoxy nanocomposites.

The application areas of composites are defined by both the physicomechanical parameters of the epoxy matrix, its strength, thermomechanical stability, and adhesion ability and the unique properties of nanoparticles.

Table 4. Kinetic features of curing epoxy oligomers in the presence of metallic nanoparticles forming in situ

Precursor	Concentration, wt %	Method of Experiments	Reaction results	T_g	References
CH_3COOAg	10 - 50	DSC	Increase of polymerization rate	Decrease	Gao, 2011
$AgSbCl_6$	5, 10, 15	-	-	Decrease	Kortaberria, 2012
$AgSbCl_6$	1, 3, 5	FTIR	Decrease of polymerization rate	Increase	Sangermano, 2007
$AgSbCl_6$	5, 10, 20	FTIR	No effect	Decrease	Vescovo, 2010
$AgSbCl_6$	0.3, 0.5, 1	FTIR	No effect	Increase	Yagci, 2008a
$HAuCl_4$	0.5, 1	FTIR	No effect	Increase	Yagci, 2008b

Nanoparticles of gold, silver, copper, TiO_2, ZnO, fullerene, and CNTs demonstrate effective antibacterial properties; therefore, composites containing these nanoparticles may be used for the microbiological control and purification of water, disinfection of surfaces, and creation of germicidal coatings, protective films, etc. Silver shows anti-inflammatory behavior and features antiviral and antifungal abilities. Its application in the form of nanoparticles (compared with the ionic form) decreases the cellular toxicity rather than the antibacterial efficiency (Kortaberria, 2012, Gu, 2012).

Dielectric and magnetic epoxy nanocomposites have found wide use in such fields as Fourier spectroscopy, NMR, data storage, and absorption of electromagnetic radiation from other objects. The role of nanoparticles shows itself as improvement of electric strength and stress durability, suppression of space charge, and increase in the stability of dielectric discharge. For example, in the case of built-in planar capacitors, the insertion of dielectric films between copper sheets makes it possible to efficiently reduce the number of assembly devices. This not only leads to the miniaturization of circuit boards and electric wiring but also improves the characteristics of devices (e.g., promotes reduction in electromagnetic interference and switching noises) (Huang, 2014, Novikov, 2017a, 2017b, Gu, 2012).

Epoxy resins are often employed in antiwear applications. The use of such fillers as graphene oxide (Shen, 2013) or complexes MMT + SiO_2 (Ni, 2015) and cloisite + TiO_2 (Thind, 2015), even at their very low content, decreases the rate of material wear by almost an order of magnitude.

The epoxy binders modified with carboxylated carbon nanotubes are more resistant to the action of aging factors. The presence of aggregates of carboxylated carbon nanotubes in the epoxy matrix positively influences the preservation of physicomechanical properties of the composite subjected to heat and humidity aging. Microscopic examination revealed structural features of the epoxy nanocomposite and their effect on the resistance of the composite to the heat and humidity aging (Kondrashov, 2017).

As for the use of epoxy nanocomposites in aerospace science and technology, this important problem, in author view, should be devoted to a special work. In short, it can be formulated as follows:

Structural nanocomposites, which are reinforcement structures based on carbon or glass fibers embedded in a polymer matrix modified with nanofillers, are the most important application of nanocomposites in the aerospace field (Dinca, 2012, Rathod, 2017), laminates and sandwich structures. In addition, they can be used as anti-lightning (Katunin, 2017, Raimondo, 2018), anti-radar (Da Silva 2016, Ates, 2017) protection devices, flame retardant (Toldy, 2011, Govindaraj, 2018) and heat resistant paints (Meenakshi, 2016), ameliorating anti-corrosion performances (Monetta, 2015).

REFERENCES

Abdalla, M., Dean, D., Robinson, P., & Nyairo, E. (2008). Cure behavior of epoxy/MWCNT nanocomposites: The effect of nanotube surface modification. *Polymer*, *49*(16), 3310–3317. doi:10.1016/j.polymer.2008.05.016

Acocella, M. R., Corcione, C. E., Giuri, A., Maggio, M., Maffezzoli, A., & Guerra, G. (2016). Graphene oxide as a catalyst for ring opening reactions in amine crosslinking of epoxy resins. *RSC Advances*, *6*(28), 23858–23866. doi:10.1039/C6RA00485G

Agareva, N. A., Aleksandrov, A. P., Smirnova, L. A., & Bityurin, N. M. (2009). Synthesis of block polymethylmethacrylate containing precursor for photoinduced formation of gold nanoparticles, *Perspektivnye Materialy*, *1*, 5-12.

Alzina, C., Mija, A., Vincent, L., & Sbirrazzuoli, N. (2012). Effects of incorporation of organically modified montmorillonite on the reaction mechanism of epoxy/amine cure. *The Journal of Physical Chemistry B*, *116*(19), 5786–5794. doi:10.1021/jp3021722

Alzina, C., Sbirrazzuoli, N., & Mija, A. (2010). Hybrid nanocomposites: Advanced nonlinear method for calculating key kinetic parameters of complex cure kinetics. *The Journal of Physical Chemistry B*, *114*(39), 12480–12487. doi:10.1021/jp1040629

Alzina, C., Sbirrazzuoli, N., & Mija, A. (2011). Epoxy-amine based nanocomposites reinforced by silica nanoparticles. Relationships between morphologic aspects, cure kinetics, and thermal properties. *The Journal of Physical Chemistry C*, *115*(46), 22789–22795. doi:10.1021/jp2066259

Ates, M., Eker, A. A., & Eker, B. (2017). Carbon nanotube-based nanocomposites and their applications. *Journal of Adhesion Science and Technology*, *31*(18), 1977–1997. doi:10.1080/01694243.2017.1295625

Baller, J., Becker, N., Ziehmer, M., Thomassey, M., Zielinski, B., Müller, U., & Sanctuary, R. (2009). Interactions between silica nanoparticles and an epoxy resin before and during network formation. *Polymer*, *50*(14), 3211–3219. doi:10.1016/j.polymer.2009.05.020

Baller, J., Thomassey, M., Ziehmer, M., & Sanctuary, R. (2011). The catalytic influence of alumina nanoparticles on epoxy curing. *Thermochimica Acta*, *517*(1–2), 34–39. doi:10.1016/j.tca.2011.01.029

Bardaji, M., Uznanski, P., Amiens, C., Chaudret, B., & Laguna, A. (2002). Aurophilic complexes as gold atom sources in organic media. *Chemical Communications*, (6): 598–599.

Becker, O., Simon, P. G., & Dusek, K. (2005). Epoxy layered silicate nanocomposites. *Advances in Polymer Science*, *179*, 29–82. doi:10.1007/b107204

Bogdanova, L., Kuzub, L., Dzhavadjan, E., Rabenok, E., Novikov, G., & Pomogailo, A. (2012). In situ synthesis and properties of epoxy nanocomposites. *Macromolecular Symposia*, *317–318*(1), 117–122. doi:10.1002/masy.201100099

Bogdanova, L. M., Kuzub, L. I., Dzhavadyan, E. A., Torbov, V. I., Dremova, N. N., & Pomogailo, A. D. (2014). Mechanical properties of epoxy composites based on silver nanoparticles formed in situ, *Journal of Polymer Science. Ser. A*, *56*(3), 304–310.

Bogdanova, L. M., Kuzub, L. I., Spirin, M. G., Dzhardimalieva, G. I., & Irzhak, V. I. (2015). Features of the synthesis of silver nanoparticles in a curing epoxy matrix. *Vestnik KGTU*, *18*(16), 10–15.

Bogdanova, L. M., Shershnev, V. A., Spirin, M. G., Irzhak, V. I., & Dzhardimalieva, G. I. (in press). Evolution of silver nanoparticles synthesized in situ in a polycondensation epoxy matrix, *Russian. The Journal of Physical Chemistry A*.

Britz, D. A., & Khlobystov, A. N. (2006). Noncovalent interactions of molecules with single walled carbon nanotubes. *Chemical Society Reviews*, *35*(7), 637–659. doi:10.1039/b507451g

Cai, Z.-Q., Movva, S., Chiou, N.-R., Guerra, D., Hioe, Y., Castro, J. M., & Lee, L. J. (2010). Effect of polyaniline surface modification of carbon nanofibers on cure kinetics of epoxy resin. *Journal of Applied Polymer Science*, *118*(4), 2328–2335.

Charan, S., Singh, N., Khanna, P. K., & Patil, K. R. (2006). Direct synthesis of nanocrystalline silver from the reaction between silver carboxylates and n-trioctylphosphine. *Journal of Nanoscience and Nanotechnology*, *6*(7), 2095–2102. doi:10.1166/jnn.2006.339

Chernin, I. Z., Smekhov, F. M., & Zherdev, Yu. V. (1982). *Epoxy polymers and compositions*. Moscow: Khimiya. (in Russian)

Choi, W. J., Powell, R. L., & Kim, D. S. (2009). Curing behavior and properties of epoxy nanocomposites with amine functionalized multiwall carbon nanotubes. *Polymer Composites*, *30*(4), 415–421. doi:10.1002/pc.20571

Chua, C. K., & Pumera, M. (2015). Carbocatalysis: The state of "Metal-Free" catalysis. *Chemistry (Weinheim an der Bergstrasse, Germany)*, *21*(36), 12550–12562. doi:10.1002/chem.201501383

Cividanes, L. S., Brunelli, D. D., Antunes, E. F., Corat, E. J., Sakane, K. K., & Thim, G. P. (2013). Cure study of epoxy resin reinforced with multiwalled carbon nanotubes by Raman and luminescence spectroscopy. *Journal of Applied Polymer Science*, *127*(1), 544–553. doi:10.1002/app.37815

Cividanes, L. S. N., Simonetti, E. A., Moraes, M. B., Fernandes, F. W., & Thim, G. P. (2014). Influence of carbon nanotubes on epoxy resin cure reaction using different techniques: A comprehensive review. *Polymer Engineering and Science*, *54*(11), 2461–2469. doi:10.1002/pen.23775

Clary, D. R., & Mills, G. (2011). Photochemical generation of nanometer-sized Cu particles in octane. *The Journal of Physical Chemistry C*, *115*(30), 14656–14663. doi:10.1021/jp2040136

Clary, D. R., Nabil, M., Sedeh, M. M., El-Hasadi, Y., & Mills, G. (2012). Photochemical generation of Ag, Pd, and Pt particles in octane. *The Journal of Physical Chemistry C*, *116*(16), 9243–9250. doi:10.1021/jp210051j

Cortés, P., Fraga, I., Calventus, Y., Román, F., Hutchinson, J. M., & Ferrando, F. (2014). A new epoxy-based layered silicate nanocomposite using a hyperbranched polymer: Study of the curing reaction and nanostructure development. *Materials (Basel)*, *7*(3), 1830–1849. doi:10.3390/ma7031830

D'yachkov, P. N. (2006). *Carbon nanotubes. Structure, properties, application*. Moscow: Binom. (in Russian)

Da Silva, L. V., Pezzin, S. H., Rezende, M. C., & Amico, S. C. (2016). Glass fiber/carbon nanotubes/epoxy three component composites as radar absorbing materials. *Polymer Composites*, *37*(8), 2277–2284. doi:10.1002/pc.23405

Dinca, I., Ban, C., Stefan, A., & Pelin, G. (2012). Nanocomposites as advanced materials for aerospace industry. *INCAS Bulletin*, *4*(4), 73–83.

Esmizadeh, E., Yousefi, A., & Naderi, G. (2015). Effect of type and aspect ratio of different carbon nanotubes on cure behavior of epoxy-based nanocomposite. *Iranian Polymer Journal*, *24*(1), 1–12. doi:10.100713726-014-0281-4

Faraji, M., Yamini, Y., & Rezaee, M. (2010). Magnetic nanoparticles: Synthesis, stabilization, functionalization, characterization, and applications. *Journal of the Indian Chemical Society*, *7*(1), 1–37.

Ferdosian, F., Ebrahimi, M., & Jannesari, A. (2013). Curing kinetics of solid epoxy/DDM/nanoclay: Isoconversional models versus fitting model. *Thermochimica Acta*, *568*, 67–73. doi:10.1016/j.tca.2013.06.001

Finney, E. E., & Finke, R. G. (2008). Nanocluster nucleation and growth kinetic and mechanistic studies: A review emphasizing transition-metal nanoclusters. *Journal of Colloid and Interface Science, 317*(2), 351–374. doi:10.1016/j.jcis.2007.05.092

Frenkel', Ya. I. (1975). *Kineticheskaya teoriya zhidkostei [Kinetic Theory of Liquids]*. Leningrad: Nauka.

Galpaya, D. G. D., Fernando, J. F. S., Rintoul, L., Motta, N., Waclawik, E. R., Yan, C., & George, G. A. (2015). The effect of graphene oxide and its oxidized debris on the cure chemistry and interphase structure of epoxy nanocomposites. *Polymer, 71*, 122–134. doi:10.1016/j.polymer.2015.06.054

Gao, H., Liu, L., Luo, Y., & Jia, D. (2011). In-situ preparation of epoxy/silver nanocomposites by thermal decomposition of silver–imidazole complex. *Materials Letters, 65*(23–24), 3529–3532. doi:10.1016/j.matlet.2011.07.086

Ghaemy, M., Amini Nasab, S. M., & Barghamadi, M. (2007). Nonisothermal cure kinetics of diglycidylether of bisphenol-A/amine system reinforced with nanosilica particles. *Journal of Applied Polymer Science, 104*(6), 3855–3863. doi:10.1002/app.25994

Ghaffari, M., Ehsani, M., Vandalvand, M., Avazverdi, E., Askari, A., & Goudarzi, A. (2015). Studying the effect of micro- and nano-sized ZnO particles on the curing kinetic of epoxy/polyaminoamide system. *Progress in Organic Coatings, 89*, 277–283. doi:10.1016/j.porgcoat.2015.08.016

Govindaraj, B., & Sarojadevi, M. (2018). Microwave-assisted synthesis of nanocomposites from polyimides chemically cross-linked with functionalized carbon nanotubes for aerospace applications. *Polymers for Advanced Technologies, 29*(6), 1718–1726. doi:10.1002/pat.4275

Grachev, V. P., Kondrashov, S. V., Akatenkov, R. V., Aleksashin, V. N., Deev, I. S., Anoshkin, I. V., ... Irzhak, V. I. (2014). Modification of epoxy polymers by small additives of multiwall carbon nanotube. *Polymer Science, Series A, 56*(3), 330–336. doi:10.1134/S0965545X14030079

Grayfer, E. D., Makotchenko, V. G., Nazarov, A. S., Kim, S. J., & Fedorov, V. E. (2011). Graphene: Chemical approaches to the synthesis and modification. *Russian Chemical Reviews, 80*(8), 751–04. doi:10.1070/RC2011v080n08ABEH004181

Gu, H., Tadakamalla, S., Huang, Y., Colorado, H. A., Luo, Z., Haldolaarachchige, N., ... Guo, Z. (2012). Polyaniline stabilized magnetite nanoparticle reinforced epoxy nanocomposites. *ACS Applied Materials & Interfaces, 4*(10), 5613–5624. doi:10.1021/am301529t

Gur'eva, L. L., Tkachuk, A. I., Kuzub, L. I., Estrina, G. A., Knerel'man, E. I., Khodos, I. I., & Rozenberg, B. A. (2013). Synthesis of silver nanoparticles with polystyrylcarboxylate ligands, *Journal of Polymer Science. Ser. B, 55*(3–4), 139–146.

Gusev, A. I. (2007). *Nanomaterials, nanostructures, nanotechnologies*. Moscow: Fizmatlit. (in Russian)

Hong, S. G., & Tsai, J. S. (2001). The adsorption and curing behavior of the epoxy/amidoamine system in the presence of metal oxides. *Journal of Thermal Analysis and Calorimetry, 63*(1), 31–46. doi:10.1023/A:1010119900035

Huang, X., Xie, L., Yang, K., Wu, C., Jiang, P., Li, S., ... Tanaka, T. (2014). Role of interface in highly filled epoxy/$BaTiO_3$ nanocomposites. Part I-correlation between nanoparticle surface chemistry and nanocomposite dielectric property. *IEEE Transactions on Dielectrics and Electrical Insulation, 21*(2), 467–479. doi:10.1109/TDEI.2013.004165

Hutchinson, J. M., Shiravand, F., & Calventus, Y. (2013). Intra- and extra-gallery reactions in tri-functional epoxy polymer layered silicate nanocomposites. *Journal of Applied Polymer Science, 128*(5), 2961–2970. doi:10.1002/app.38452

Hutchinson, J. M., Shiravand, F., Calventus, Y., Fernández-Francos, X., & Ramis, X. (2014). Highly exfoliated nanostructure in trifunctional epoxy/clay nanocomposites using boron trifluoride as initiator. *Journal of Applied Polymer Science, 131*(6), 40020–40029. doi:10.1002/app.40020

Irzhak, T.F., & Irzhak, V.I. (2017). Epoxy nanocomposites. *Polymer Sciences, A, 59*(6), 791-825.

Irzhak, V. I. (2011). Epoxide composite materials with carbon nanotubes. *Russian Chemical Reviews, 80*(8), 787–806. doi:10.1070/RC2011v080n08ABEH004196

Irzhak, V. I. (2016). The mechanisms of the formation of metal-containing nanoparticles. *Review Journal of Chemistry, 6*(4), 370–404. doi:10.1134/S2079978016040026

Irzhak, V. I. (2017). Digestive ripening of nanoparticles, *Russian. The Journal of Physical Chemistry A, 91*(8), 1502–1506.

Irzhak, V. I., Rozenberg, B. A., & Enikolopyan, N. S. (1979). *Network polymers. Synthesis, structure, properties*. Moscow: Nauka. (in Russian)

Ivankovic, M., Brnardic, I., Ivankovic, H., & Mencer, H. J. (2006). DSC study of the cure kinetics during nanocomposite formation: Epoxy/poly(oxypropylene) diamine/ organically modified montmorillonite system. *Journal of Applied Polymer Science, 99*(2), 550–557. doi:10.1002/app.22488

Ivanovskii, A. L. (2012). Graphene-based and graphene-like materials. *Russian Chemical Reviews, 81*(7), 571–605. doi:10.1070/RC2012v081n07ABEH004302

Jagtap, S. B., Rao, V. S., Barman, S., & Ratna, D. (2015). Nanocomposites based on epoxy resin and organoclay functionalized with a reactive modifier having structural similarity with the curing agent. *Polymer, 63*, 41–71. doi:10.1016/j. polymer.2015.02.038

Kablov, E. N., Kondrashov, S. V., & Yurkov, G. Yu. (2013). Prospects of using carbonaceous nanoparticles in binders for polymer composites. *Nanotechnologies in Russia, 8*(3-4), 163–185. doi:10.1134/S1995078013020080

Karasinski, E. N., Da Luz, M. G., Lepienski, C. M., & Coelho, L. A. F. (2013). Nanostructured coating based on epoxy/metal oxides: Kinetic curing and mechanical properties. *Thermochimica Acta, 569*, 167–211. doi:10.1016/j.tca.2013.07.015

Kashiwagi, Y., Yamamoto, M., & Nakamoto, M. (2006). Facile size-regulated synthesis of silver nanoparticles by controlled thermolysis of silver alkylcarboxylates in the presence of alkylamines with different chain lengths. *Journal of Colloid and Interface Science, 300*(1), 169–175. doi:10.1016/j.jcis.2006.03.041

Katunin, A., Krukiewicz, K., Turczyn, R., Sul, P., Łasica, A., & Bilewicz, M. (2017). Synthesis and characterization of the electrically conductive polymeric composite for lightning strike protection of aircraft structures. *Composite Structures, 159*, 773–783. doi:10.1016/j.compstruct.2016.10.028

Khanna, P. K., Kulkarni, D., & Beri, R. K. (2008). Synthesis and characterization of myristic acid capped silver nanoparticles. *Journal of Nanoparticle Research, 10*(6), 1059–1062. doi:10.100711051-008-9366-3

Khozin, V. G. (2004). *Reinforcement of epoxy polymers*. Kazan: Dom pechati. (in Russian)

Kondrashov, S. V., Merkulova, Yu. I., Marakhovskii, P. S., D'yachkova, T. P., Shashkeev, K. A., Popkov, O. V., ... Yurkov, G. Yu. (2017). Degradation of physicomechanical properties of epoxy nanocomposites with carbon nanotubes upon heat and humidity aging. *Russian Journal of Applied Chemistry, 90*(5), 788–796. doi:10.1134/S1070427217050202

Kong, D., & Park, C. E. (2003). Real time exfoliation behavior of clay layers in epoxy-clay nanocomposites real time exfoliation behavior of clay layers in epoxy-clay nanocomposites. *Chemistry of Materials*, *15*(2), 419–424. doi:10.1021/cm0205837

Kortaberria, G., Sangermano, M., & Mondragon, I. (2012). In situ synthetized silver/epoxy nanocomposites: Electrical characterization in terms of dielectric relaxation spectroscopy. *Macromolecular Symposia*, *321-322*(1), 112–117. doi:10.1002/masy.201251119

Kuzub, L. I., Bogdanova, L. M., Kurkin, T. S. & Buzin, P. V. (2010). Principal aspects of the synthesis of silver nanoparticles with carboxylate ligands. *Proceeding of Articles "Structure and Dynamics of Molecular Systems"*, 340-45.

Kuzub, L. I., Bogdanova, L. M., Kurkin, T. S., Torbov, V. I., Gur'eva, L. L., Rozenberg, B. A. & Buzin, P. V. (2009). Investigation of the regularities of synthesis of monodisperse silver nanoparticles with chemically bound organic ligands. *Proceeding of Articles "Structure and Dynamics of Molecular Systems",* 134-39.

Kuzub, L. I., Gur'eva, L. L., Grishchuk, A. A., Estrina, G. A., Estrin, Ya. I., & Badamshina, E. R. (2015). Regularities of the formation of silver nanoparticles with oligostyrylcarboxylate ligands, *Journal of Polymer Science. Ser. B*, *57*(6), 608–615.

LaMer, V. K., & Dinegar, R. H. (1950). Theory, production and mechanism of formation of monodispersed hydrosols. *Journal of the American Chemical Society*, *72*(11), 4847–4854. doi:10.1021/ja01167a001

Lee, H., & Neville, K. (1967). *Handbook of epoxy resins*. New York: McGraw-Hill.

Li, L., Zeng, Z., Zou, H., & Liang, M. (2015). Curing characteristics of an epoxy resin in the presence of functional graphite oxide with amine-rich surface. *Thermochimica Acta*, *614*, 76–102. doi:10.1016/j.tca.2015.06.006

Li, L., Zou, H., Liang, M., & Chen, Y. (2014). Study on the effect of poly(oxypropylene) diamine modified organic montmorillonite on curing kinetics of epoxy nanocomposites. *Thermochimica Acta*, *597*, 93–100. doi:10.1016/j.tca.2014.10.008

Liang, Z., Gou, J., Zhang, C., Wang, B., & Kramer, L. (2004). Investigation of molecular interactions between (10, 10) single-walled nanotube and Epon 862 resin/DETDA curing agent molecules. *Materials Science and Engineering A*, *365*(1–2), 228–234. doi:10.1016/j.msea.2003.09.032

Lifshits, M., & Slezov, V. V. (1959). Kinetics of diffusive decomposition of supersaturated solid solutions. *Journal of Experimental and Theoretical Physics*, *35*(2), 479–492.

Lu, J., Moon, K. S., & Wong, C. P. (2008). Silver/polymer nanocomposite as a high-k polymer matrix for dielectric composites with improved dielectric performance. *Journal of Materials Chemistry*, *18*(40), 4821–4826. doi:10.1039/b807566b

Lu, Y., & Chen, W. (2012). Sub-nanometre sized metal clusters: From synthetic challenges to the unique property discoveries. *Chemical Society Reviews*, *41*(9), 3594–3624. doi:10.1039/c2cs15325d

Mauro, M., Acocella, M. R., Corcione, C. E., Maffezzoli, A., & Guerra, G. (2014). Catalytic activity of graphite-based nanofillers on cure reaction of epoxy resins. *Polymer*, *55*(22), 5612–5615. doi:10.1016/j.polymer.2014.09.019

Meenakshi, K. S., & Sudhan, E. P. J. (2016). Development of novel TGDDM epoxy nanocomposites for aerospace and high performance applications – Study of their thermal and electrical behavior. *Arabian Journal of Chemistry*, *9*(1), 79–85. doi:10.1016/j.arabjc.2011.05.020

Monetta, T., Acquesta, A., & Bellucci, F. (2015). Graphene/epoxy coating as multifunctional material for aircraft structures. *Aerospace*, *2*(3), 423–434. doi:10.3390/aerospace2030423

Montserrat, S., Román, F. M., Hutchinson, J., & Campos, L. (2008). Analysis of the cure of epoxy based layered silicate nanocomposites: Reaction kinetics and nanostructure development. *Journal of Applied Polymer Science*, *108*(2), 923–938. doi:10.1002/app.27297

Nakamoto, M., Kashiwagi, Y., & Yamamoto, M. (2005). Synthesis and size regulation of gold nanoparticles by controlled thermolysis of ammonium gold(I) thiolate in the absence or presence of amines. *Inorganica Chimica Acta*, *358*(14), 4229–4236. doi:10.1016/j.ica.2005.03.037

Navalon, S., Dhakshinamoorthy, A., Alvaro, M., & Garcia, H. (2014). Carbocatalysis by graphene-based materials. *Chemical Reviews*, *114*(12), 6179–6212. doi:10.1021/cr4007347

Ni, Y., Chen, L., Teng, K., Shi, J., Qian, X., Xu, Z., ... Ma, M. (2015). Superior Mechanical Properties of Epoxy Composites Reinforced by 3D Interconnected Graphene Skeleton. *ACS Applied Materials & Interfaces*, *7*(21), 11583–11591. doi:10.1021/acsami.5b02552

Novikov, G. F., Rabenok, E. V., Bogdanova, L. M., & Irzhak, V. I. (2017a). Temperature dependence of direct current conductivity in Ag-ED20 nanocomposite films, *Russian. The Journal of Physical Chemistry A*, *91*(10), 1971–1975.

Novikov, G. F., Rabenok, E. V., Bogdanova, L. M., & Irzhak, V. I. (2017b). Dielectric properties of films of Ag-ED20 epoxy nanocomposite synthesized in situ. Temperature dependence of direct current conductivity. *Journal of Polymer Science: Part A, General Papers*, *59*(5), 741–750.

Okabe, T., Takehara, T., Inose, K., Hirano, N., Nishikawa, M., & Uehara, T. (2013). Curing reaction of epoxy resin composed of mixed base resin and curing agent: Experiments and molecular simulation. *Polymer*, *54*(17), 4660–4668. doi:10.1016/j.polymer.2013.06.026

Omrani, A., Rostami, A. A., Ravari, F., & Mashak, A. (2011). Curing behavior and structure of a novel nanocomposite from glycerol diglycidyl ether and 3,3-dimethylglutaric anhydride. *Thermochimica Acta*, *517*(1–2), 9–15. doi:10.1016/j.tca.2011.01.024

Paiva, L. B., de Morales, A. R., & Valenzuela Diaz, F. R. (2008). Organoclays: Properties, preparation and application. *Applied Clay Science*, *42*(1–2), 8–24. doi:10.1016/j.clay.2008.02.006

Paquin, A. M. (1958). *Epoxydverbindungen und Epoxydharze*. Berlin: Springer. doi:10.1007/978-3-642-86616-6

Park, J. K., & Kim, D. S. (2014a). Effects of an aminosilane and a tetra-functional epoxy on the physical properties of di-functional epoxy/graphene nanoplatelets nanocomposites. *Polymer Engineering and Science*, *54*(4), 969–976. doi:10.1002/pen.23624

Park, J. K., & Kim, D. S. (2014b). Preparation and physical properties of an epoxy nanocomposite with amine-functionalized graphenes. *Polymer Engineering and Science*, *54*(5), 985–991. doi:10.1002/pen.23368

Pavlidou, S., & Papaspyrides, C. D. (2008). A review on polymer–layered silicate nanocomposites. *Progress in Polymer Science*, *33*(12), 1119–1198. doi:10.1016/j.progpolymsci.2008.07.008

Polte, J., Tuaev, X., Wuithschick, M., Fischer, A., Thuenemann, A. F., Rademann, K., ... Emmerling, F. (2012). Formation mechanism of colloidal silver nanoparticles: Analogies and differences to the growth of gold nanoparticles. *ACS Nano*, *6*(7), 5791–5802. doi:10.1021/nn301724z

Pomogailo, A. D., Rozenberg, A. S., & Dzhardimalieva, G. I. (2005). *Metal-polymer nanocomposites* (L. Nicolais & G. Carotenuto, Eds.). Hoboken, NJ: Wiley.

Pustkova, P., Hutchinson, J. M., Román, F., & Montserrat, S. (2009). Homopolymerization effects in polymer layered silicate nanocomposites based upon epoxy resin: Implications for exfoliation. *Journal of Applied Polymer Science, 114*(2), 1040–1047. doi:10.1002/app.30697

Qiu, J., & Wang, S. (2010). Reaction kinetics of functionalized carbon nanotubes reinforced polymer composites. *Materials Chemistry and Physics, 121*(1–2), 295–301. doi:10.1016/j.matchemphys.2010.01.039

Qiu, S. L., Wang, C. S., Wang, Y. T., Liu, C. G., Chen, X. Y., Xie, H. F., ... Cheng, R. S. (2011). Reaction kinetics of functionalized carbon nanotubes reinforced polymer composites. *Express Polymer Letters, 5*(9), 809–818. doi:10.3144/expresspolymlett.2011.79

Rahaman, A., & Mohanty, A. (2014). Effect of carbon nanotubes on the curing and thermomechanical behavior of epoxy/carbon nanotubes composites. *Polymer Composites, 35*(29), 441–449. doi:10.1002/pc.22679

Raimondo, M., Guadagno, L., Speranza, V., Bonnaud, L., Dubois, P., & Lafdi, K. (2018). Multifunctional graphene/POSS epoxy resin tailored for aircraft lightning strike protection. *Composites. Part B, Engineering, 140*, 44–56. doi:10.1016/j.compositesb.2017.12.015

Rakov, E. G. (2006). *Nanotubes and fullerenes*. Moscow: Universitetskaya kniga. (in Russian)

Rakov, E. G. (2013). Carbon nanotubes in new materials. *Russian Chemical Reviews, 82*(1), 27–47. doi:10.1070/RC2013v082n01ABEH004227

Rathod, V. T., Kumar, J. S., & Jain, A. (2017). Polymer and ceramic nanocomposites for aerospace applications. *Applied Nanoscience, 7*(8), 519–548. doi:10.100713204-017-0592-9

Román, F., Calventus, Y., Colomer, P., & Hutchinson, J. M. (2012). Effect of carbon nanotubes on the curing and thermomechanical behavior of epoxy/carbon nanotubes composites. *Thermochimica Acta, 541*, 76–85.

Rosso, P., & Ye, L. (2007). Epoxy/silica nanocomposites: Nanoparticle-induced cure kinetics and microstructure. *Macromolecular Rapid Communications, 28*(1), 121–126. doi:10.1002/marc.200600588

Rozenberg, B. A. (1986). Kinetics, thermodynamics and mechanism of reactions of epoxy oligomers with amines. *Advances in Polymer Science, 75*, 113–165. doi:10.1007/BFb0017916

Rozenberg, B. A., & Tenne, R. (2008). Polymer-assisted fabrication of nanoparticles and nanocomposites. *Progress in Polymer Science*, *33*(1), 40–112. doi:10.1016/j.progpolymsci.2007.07.004

Ryu, S. H., Sin, J. H., & Shanmugharaj, A. M. (2014). Study on the effect of hexamethylene diamine functionalized graphene oxide on the curing kinetics of epoxy nanocomposites. *European Polymer Journal*, *52*, 88–97. doi:10.1016/j.eurpolymj.2013.12.014

Sanctuary, R., Baller, J., Zielinski, B., Becker, N., Krüger, K., Philipp, U., ... Ziehmer, M. (2009). Influence of Al_2O_3 nanoparticles on the isothermal cure of an epoxy resin. *Journal of Physics Condensed Matter*, *21*(3), 035118–035126. doi:10.1088/0953-8984/21/3/035118

Sangermano, M., Yagci, Y., & Rizza, G. (2007). In situ synthesis of silver-epoxy nanocomposites by photoinduced electron transfer and cationic polymerization processes. *Macromolecules*, *40*(25), 8827–8829. doi:10.1021/ma702051g

Schaetz, A., Zeltner, M., & Stark, W. J. (2012). Carbon modifications and surfaces for catalytic organic transformations. *ACS Catalysis*, *2*(6), 1267–1284. doi:10.1021/cs300014k

Seyhan, A. T., Sun, Z., Deitzel, J., Tanoglu, M., & Heider, D. (2009). Cure kinetics of vapor grown carbon nanofiber (VGCNF) modified epoxy resin suspensions and fracture toughness of their resulting nanocomposites. *Materials Chemistry and Physics*, *118*(1), 234–242. doi:10.1016/j.matchemphys.2009.07.045

Shanmugharaj, A. M., & Ryu, S. H. (2012). Study on the effect of aminosilane functionalized nanoclay on the curing kinetics of epoxy nanocomposites. *Thermochimica Acta*, *546*, 16–23. doi:10.1016/j.tca.2012.07.026

Shen, B., Zhai, W., Tao, M., Lu, D., & Zheng, W. (2013a). Enhanced interfacial interaction between polycarbonate and thermally reduced graphene induced by melt blending. *Composites Science and Technology*, *77*, 87–110. doi:10.1016/j.compscitech.2013.01.014

Shen, X. J., Pei, X. Q., Fu, S. Y., & Friedrich, K. (2013b). Significantly modified tribological performance of epoxy nanocomposites at very low graphene oxide content. *Polymer*, *54*(3), 1234–1242. doi:10.1016/j.polymer.2012.12.064

Shiravand, F., Fraga, I., Cortés, P., Calventus, Y., & Hutchinson, J. M. (2014b). Thermal analysis of polymer layered silicate nanocomposites. Identification of nanostructure development by DSC. *Journal of Thermal Analysis and Calorimetry*, *118*(2), 723–729. doi:10.100710973-014-3709-3

Shiravand, F., Hutchinson, J. M., & Calventus, Y. (2014a). Influence of the isothermal cure temperature on the nanostructure and thermal properties of an epoxy layered silicate nanocomposite. *Polymer Engineering and Science*, *54*(1), 51–58. doi:10.1002/pen.23540

Shiravand, F., Hutchinson, J. M., & Calventus, Y. (2015). Non-isothermal cure and exfoliation of tri-functional epoxy-clay nanocomposites. *Express Polymer Letters*, *9*(8), 695–708. doi:10.3144/expresspolymlett.2015.65

Solov'ev, M. E., Irzhak, T. F., & Irzhak, V. I. Russ. (2015). Kinetics of formation of nanoparticles from first group metal carboxylates. *The Journal of Physical Chemistry A*, *89*(3), 1642–1647.

Song, S. H., Park, K. H., Kim, B. H., Choi, Y. W., Jun, G. H., Lee, D. J., ... Jeo, S. (2013). Enhanced thermal conductivity of epoxy/graphene composites by using non-oxidized graphene flakes with non-covalent functionalization. *Advanced Materials*, *25*(5), 732–737. doi:10.1002/adma.201202736

Susin, S. B., Pistor, V., Amico, S. C., Coelho, L. A. F., Pezzin, S. H., & Zattera, A. J. (2014). Investigation of cure kinetics in epoxy/multi-walled carbon nanotube nanocomposites. *Journal of Applied Polymer Science*, *131*(3), 39857–39862. doi:10.1002/app.39857

Suzdalev, I. P. (2006). *Nanotechnology. Physics and chemistry of nanaocrystals, nanostructures and nanomaterials*. Moscow: Kom Kniga. (in Russian)

Tao, K., Yang, S., Grunlan, J. C., Kim, Y.-S., Dang, B., Deng, Y., ... Wei, X. (2006). Effects of carbon nanotube fillers on the curing processes of epoxy resin-based composites. *Journal of Applied Polymer Science*, *102*(6), 5248–5254. doi:10.1002/app.24773

Thind, K. S., Singh, J., Saini, J. S., & Bhunia, H. (2015). Mechanical and wear properties of hybrid epoxy nanocomposites. *Indian Journal of Engineering and Materials Sciences*, *22*(4), 421–428.

Toldy, A., Szolnoki, B., & Marosi, G. (2011). Flame retardancy of fibre-reinforced epoxy resin composites for aerospace applications. *Polymer Degradation & Stability*, *96*(3), 371–376. doi:10.1016/j.polymdegradstab.2010.03.021

Turkevich, J., Stevenson, P. C., & Hillier, J. (1951). A study of the nucleation and growth processes in the synthesis of colloidal gold. *Faraday Discussions of the Chemical Society*, *11*, 55–75. doi:10.1039/df9511100055

Valentini, L., Puglia, D., Carniato, F., Boccaleri, E., Marchese, L., & Kenny, J. M. (2008). Use of plasma fluorinated single-walled carbon nanotubes for the preparation of nanocomposites with epoxy matrix. *Composites Science and Technology*, *68*(3–4), 1008–1014. doi:10.1016/j.compscitech.2007.07.011

Vescovo, L., Sangermano, M., Scarazzin, R., & Kortaberria, G. (2010). In-situ-synthetized silver/epoxy nanocomposites: Electrical characterization by means of dielectric spectroscopy. *Macromolecular Chemistry and Physics*, *211*(17), 1933–1939. doi:10.1002/macp.201000138

Visco, A., Calabrese, L., & Milone, C. (2009). Cure rate and mechanical properties of a DGEBF epoxy resin modified with carbon nanotubes. *Journal of Reinforced Plastics and Composites*, *28*(8), 937–949. doi:10.1177/0731684407087560

Völkle, C. M., Gebauer, D., & Cölfen, H. (2015). High-resolution insights into the early stages of silver nucleation and growth. *Faraday Discussions*, *179*, 59–69. doi:10.1039/C4FD00269E

Volmer, M. (1939). *Kinetik der Phasenbildung*. Dresden: Steinkopff.

Vyazovkin, S., & Sbirrzaauoli, N. (1996). Mechanism and kinetics of epoxy-amine cure studied by differential scanning calorimetry. *Macromolecules*, *29*(6), 1867–1873. doi:10.1021/ma951162w

Wagner, Z. (1961). Theorie der Alterung von Niderschlagen durch Umlösen (Ostwald Reifung). *Zeitschrift für Elektrochemie*, *65*, 581–591.

Wang, F., Richards, V. N., Shields, S. P., & Buhro, W. E. (2014). Kinetics and mechanisms of aggregative nanocrystal growth. *Chemistry of Materials*, *26*(1), 5–21. doi:10.1021/cm402139r

Watzky, M. A., & Finke, R. G. (1997). Nanocluster size-control and "magic number" investigations. Experimental tests of the "living-metal polymer" concept and of mechanism-based size-control reductions leading to the syntheses of Ir(0) nanoclusters centering about four sequential magic numbers. *Chemistry of Materials*, *9*(12), 3083–3095. doi:10.1021/cm9704387

Wilcoxon, J. P., & Abrams, B. L. (2006). Synthesis, structure and properties of metal nanoclusters. *Chemical Society Reviews*, *35*(4), 1162–1194. doi:10.1039/b517312b

Wuithschick, M., Paul, B., Bienert, R., Sarfraz, A., Vainio, U., Sztucki, M., ... Polte, J. (2013). Size-controlled synthesis of colloidal silver nanoparticles based on mechanistic understanding. *Chemistry of Materials*, *25*(24), 4679–4689. doi:10.1021/cm401851g

Xie, H., Liu, B., Yuan, Z., Shen, J., & Cheng, R. (2004). Cure kinetics of carbon nanotube/tetrafunctional epoxy nanocomposites by isothermal differential scanning calorimeter. *Journal of Polymer Science. Part A, Polymer Chemistry*, *42*(20), 3701–3712. doi:10.1002/polb.20220

Xin, H. L. L., & Zheng, H. M. (2012). In situ observation of oscillatory growth of bismuth nanoparticles. *Nano Letters*, *12*(3), 1470–1474. doi:10.1021/nl2041854

Xue, B., Zhu, J., Liu, N., & Li, Y. (2015). Facile functionalization of graphene oxide with ethylenediamine as a solid base catalyst for Knoevenagel condensation reaction. *Catalysis Communications*, *64*, 105–109. doi:10.1016/j.catcom.2015.02.003

Yagci, Y., Sahin, O., Ozturk, T., Marchi, S., Grassini, S., & Sangermano, M. (2011). Synthesis of silver/epoxy nanocomposites by visible light sensitization using highly conjugated thiophene derivatives. *Reactive & Functional Polymers*, *71*(7), 857–862. doi:10.1016/j.reactfunctpolym.2011.05.012

Yagci, Y., Sangermano, M., & Rizza, G. (2008a). A visible light photochemical route to silver–epoxy nanocomposites by simultaneous polymerization–reduction approach. *Polymer*, *49*(24), 5195–5198. doi:10.1016/j.polymer.2008.09.068

Yagci, Y., Sangermano, M., & Rizza, G. (2008b). Synthesis and characterization of gold-epoxy nanocomposites by visible light photoinduced electron transfer and cationic polymerization processes. *Macromolecules*, *41*(20), 7268–7270. doi:10.1021/ma801776y

Yamamoto, M., Kashiwagi, Y., & Nakamoto, M. (2006). Size-controlled synthesis of monodispersed silver nanoparticles capped by long-chain alkyl carboxylates from silver carboxylate and tertiary amine. *Langmuir*, *22*(20), 8581–8586. doi:10.1021/la0600245

Yamamoto, M., & Nakamoto, M. (2003). Novel preparation of monodispersed silver nanoparticles via amine adducts derived from insoluble silver myristate in tertiary alkylamine. *Journal of Materials Chemistry*, *13*(9), 2064–2065. doi:10.1039/b307092a

Yan, S., Wu, Z., Yu, H., Gong, Y., Tan, Y., Du, R., ... Sun, D. (2014). Time-resolved small-angle x-ray scattering study on the growth behavior of silver nanoparticles. *Journal of Physical Chemistry*, *118*(21), 11454–11463.

Yang, K., Gu, M., Jin, Y., Mu, G., & Pan, X. (2008). Influence of surface treated multi-walled carbon nanotubes on cure behavior of epoxy nanocomposites. *Composites. Part A, Applied Science and Manufacturing*, *39*(10), 1670–1678. doi:10.1016/j.compositesa.2008.07.011

Zabihi, O., Hooshafza, A., Moztarzadeh, F., Payravand, H., Afshar, A., & Alizadeh, R. (2012). Isothermal curing behavior and thermo-physical properties of epoxy-based thermoset nanocomposites reinforced with Fe_2O_3 nanoparticles. *Thermochimica Acta*, *527*, 190–198. doi:10.1016/j.tca.2011.10.026

Zabihi, O., Mostafavi, S. M., Ravari, F., Khodabandeh, A., Hooshafza, A., Zare, K., & Shahizadeh, M. (2011). The effect of zinc oxide nanoparticles on thermo-physical properties of diglycidyl ether of bisphenol A/2,2-Diamino-1,1-binaphthalene nanocomposites of surface treated multi-walled carbon nanotubes on cure behavior of epoxy nanocomposite. *Thermochimica Acta*, *521*(1–2), 49–58. doi:10.1016/j.tca.2011.04.003

Zhou, T., Wang, X., & Wang, T. (2009). Cure reaction of multi-walled carbon nanotubes/diglycidyl ether of bisphenol A/2-ethyl-4-methylimidazole (MWCNTs/DGEBA/EMI-2,4) nanocomposites: Effect of carboxylic functionalization of WCNTs. *Polymer International*, *58*(4), 445–452. doi:10.1002/pi.2558

Chapter 3
Aeronautical Impact of Epoxy/Carbon Nanotube Nanocomposite

Ayesha Kausar
National University of Science and Technology, Pakistan

ABSTRACT

Epoxy resin has been employed as an important matrix for aerospace composite and nanocomposite. In this chapter, latent and essential features of carbon nanotube (CNT) nanofiller have been considered with reference to aeronautical application. Consequently, epoxy/carbon nanotube nanocomposite are conversed here for space competency. Inclusion of CNT in epoxy resin affected the prerequisite features of space nanocomposite. Dispersion of nanotube has been altered using suitable processing technique. Uniform nanotube network formation affects mechanical, electrical, and other physical properties of nanocomposite. Key application areas in this regard are flame and thermal stability, strength, lightning strike resistance, and radiation shielding of space vehicles. Further investigations to optimize structure and properties of multiscale epoxy/CNT nanocomposite are needed for future success in the field. Hence, towards the end, challenges and future prospects of epoxy/CNT nanocomposite have also been deliberated for the improvement of nanomaterial properties for aerospace relevance.

INTRODUCTION

Nanotechnology advancements have led to several developments for technical applications. Design and properties of polymeric nanocomposite for aerospace structures have significant effect on aerospace applications. Thus, aerospace design has been made safer, faster, and cheaper compared with the conventional

DOI: 10.4018/978-1-5225-7921-2.ch003

space transportation. Polymer nanocomposite have at least one dimension in range of 1-100 nm, i.e. offered by the reinforced nanoparticles (Bogue, 2011; Ogasawara, Moon, Inoue, & Shimamura, 2011). Pristine polymers have been used in aerospace applications, however thermal and mechanical design loads were found least compatible (Rao, 2003). Employment of polymer-based nanocomposite have been focused in space vehicles owing to superior characteristics (F. Hussain, Hojjati, Okamoto, & Gorga, 2006; Liu, Du, Liu, & Leng, 2014). Carbon nanotube (CNT) is an allotrope of carbon. It is a cylindrical carbon molecule having several useful properties for electronics, optics and other nanotechnology fields (R. Saito, Dresselhaus, & Dresselhaus, 1998). Various polymers have been reinforced with nanotube including polyaniline, polyamide, polyethylene, polystyrene, polyurethane, polycarbonates, epoxies, and several others (S. T. Hussain, Abbas, Kausar, & Khan, 2013). Solicitation of polymer/CNT nanocomposite has gained remarkable research and commercial curiosity (Kausar, Rafique, & Muhammad, 2016). Consequently, CNT-based polymeric nanocomposite have been employed in organic solar cell, light-emitting diodes, field emission devices, liquid crystal, sensors, electronics and biomedical field. Moreover, nanotube-strengthened polymeric nanocomposite have been used in armor, civil engineering, automobile, sporting goods and weight-sensitive aerospace industry (Kausar & Hussain, 2013; Khan, Kausar, & Ullah, 2016). Epoxy resin is a polyepoxide. It is a class of reactive prepolymer containing epoxide groups. Polyepoxide reacts with functional diamine or diol hardener to form a thermosetting polymer. Epoxy possesses superior mechanical, thermal, and chemical resistance. Epoxy has range of applications in adhesives, coatings, electrical components, electrical insulators, strengthened materials, etc.(Ellis, 1993; Levchik & Weil, 2004). Epoxy/carbon nanotube nanocomposite possess exceptionally high strength, fatigue resistance, conductivity, thermal stability, non-flammability, adhesion, and other desired perspectives for commercial application in aerospace industry (Rafique, Kausar, Anwar, & Muhammad, 2016). In this chapter, current potential of polymeric nanocomposite especially epoxy/carbon nanotube nanocomposite have been explored in aerospace industry. In this regard, essential aspects of polymer and epoxy-based nanocomposite have been deliberated. Applications of epoxy/carbon nanotube nanocomposite in various fields of aerospace technology have been systematically conversed (Lan et al., 2009; Pandey, Ahn, Lee, Mohanty, & Misra, 2010). Consequently, this chapter reviews essential features of polymeric nanocomposites for potential aerospace applications. Towards the end, challenges and potential of epoxy/carbon nanotube nanocomposite in aerospace have also been discussed considering future advancement.

BACKGROUND

Aeronautical Composite

In aerospace industry, composites are lightweight materials having high specific strength, stiffness, and impact characteristics. Particularly, nanofiber reinforced polymers possess excellent toughness, and fatigue and corrosion resistance. Furthermore, the high performance composite materials may reveal optimum mechanical properties by orientating nanofiber direction with the load transfer paths (Gay & Hoa, 2007; Shanyi, 2007). Polymeric systems with organic as well as inorganic nanofiller have gained attention for aerospace constituents and systems on commercial scale (Ahmed, Ikram, Kanchi, & Bisetty, 2018). Nanoparticles usually have high surface area per unit volume and unique structural features (Haque, Hossain, Dean, & Shamsuzzoha, 2002). Among organic fillers carbon fiber has been progressively employed in space craft and air craft industry. Carbon fiber-based composites are well known for light weight, low density, high strength, modulus, stiffness, and fatigue life. These nanocomposites have major advantages of low manufacturing cost, weight-less ness, and outstanding mechanical and physical features to substitute metals such as aluminum alloys in aerospace applications (Hirsch & Al-Samman, 2013). Initially, the aerospace composite have been employed in passenger aircraft Boeing 747 and others. To further enhance the properties of polymer/carbon fiber composite for aerospace, nanoparticle coated carbon fiber (carbon nanotube, clay, graphene, ferrite powders) have also been focused (Andrei, Dima, & Andrei, 2006; Murr et al., 2012). Fuel consumption and maintenance cost of aerospace vehicles have also been reduced using composite materials (Tong, 2016). However, processability techniques and long and sustainable life of these composite still need research attention (Wojciechowski, 2000). Consequently, composite aerospace structures need to be replaced by potential nanocomposite materials having exciting properties and uses (Fig. 1). Extremely high non-flammability, corrosion suppression, and fatigue resistance can be achieved using new nanocomposite designs in aeronautical structures (Ganguli, 2013; Leng & Lau, 2010). Future aerospace systems pursue to further improve the impact of multifunctional nanocomposite by the integration of different type of functional nanoparticles.

Carbon Nanotube as Nanofiller

Carbon nanotube (CNT) is a nanoallotrope of carbon having cylindrical nanostructure. It is made up of sp^2 hybridized carbon atom nanosheet i.e. rolled to form a nanotube. There are many forms of single-walled carbon nanotube (SWCNT) and multi-walled carbon nanotube (MWCNT) (Fig. 2). CNT is a minute hollow cylinder having

Figure 1. Aerospace structure with composite/nanocomposite application

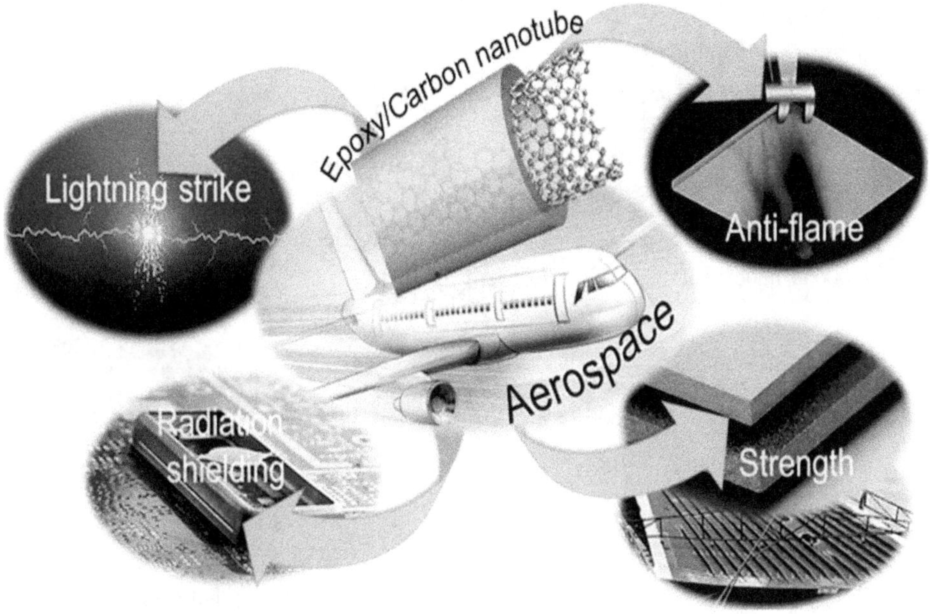

diameter of ~2.5 nm (Jorio, Kauppinen, & Hassanien, 2007). Its length may vary from few μm to 100 μm (Tans, Verschueren, & Dekker, 1998). MWCNT is made up of two or more concentric cylinders. Nanotubes possess high aspect ratio, and exceptional strength, electrical conductivity, elasticity, durability, and toughness characteristics. The remarkable features of CNT offer fine potential for electronics, energy devices, sporting goods, and biomedical applications. Moreover, high performance CNT-based nanocomposite has found relevance in aerospace and aircraft industry. Strengthened polymer/CNT nanocomposite has been obtained due to fine matrix-nanofiller compatibility. Particularly, electromagnetic interference shielding, strength, non-flammability, and lightning strike protection demands the use of high performance polymer/CNT in aerospace. The future potential of polymer/CNT-based nanocomposite has also opened doors to several advance technical applications for aerospace industry (De Volder, Tawfick, Baughman, & Hart, 2013).

Polymer/Carbon Nanotube-Based Nanomaterial

The CNT-based polymer nanocomposites have gained interest owing to superior mechanical, thermal, optical, and other useful properties. Exceptional features of carbon nanotube render it as a vital nanofiller material for nanocomposite. Several polymer/CNT nanocomposites have been synthesized by adding single-walled carbon

Figure 2. Carbon nanotube

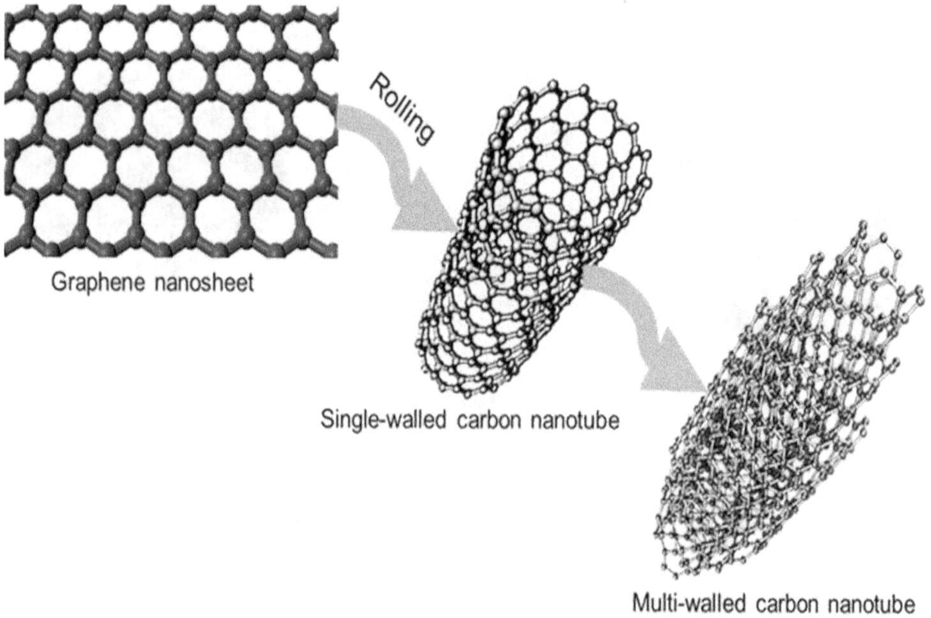

nanotube (SWCNT) or multi-walled carbon nanotube (MWCNT) in the polymer matrix. Several techniques have been developed to form high performance polymer/CNT without the destruction of CNT during processing (Cha et al.). Polymer/CNT nanocomposites have been categorized as functional composites and structural composites. In structural polymer/CNT materials, characteristic mechanical properties of CNT (tensile stress, elastic modulus, strain, compression strength) have been used to form high performance material (Cha et al., 2018; Rocha, Vale, Rocha, & Santos). On the other hand, functional polymer/CNT nanocomposite own exciting features such as high thermal, conductivity, and other functional properties (Kozlov & Dolbin, 2018). Consequently, nanocomposites of polymer and carbon nanotube have been focused owing to their remarkable properties (Shokrieh & Rafiee, 2010). Especially, thermal, electrical, and mechanical characteristics of polymer/carbon nanotube nanocomposite have been considered for space applications (Dang et al., 2012; Ma, Siddiqui, Marom, & Kim, 2010). Prospective application span of polymer/carbon nanotube nanocomposite is quite broad and covers energy systems, sensing, field emission devices, biological relevance, etc. In aerospace engineering, benefits of nanotube-based material have been observed in the development of Space Shuttle (Chang, Kisliuk, Rhodes, Brittain, & Sokolov, 2006). One of the major challenges of aerospace industry is the weight and cost reduction of spacecraft (Gohardani, Doulgeris, & Singh, 2011). In addition to the structural strength, polymer/carbon nanotube nanocomposite render advantages of low cost and light weight to the commercial aircraft and space vehicles. Polymer/SWCNT nanocomposite, for example polystyrene/SWCNT has shown increase in Young's modulus and flexural modulus with the nanofiller content. Thermal stability of polystyrene/CNT nanocomposite was also enhanced with the reinforcing nanofiller (Chang et al., 2006). Polymers bonded to covalently functional CNT also act as intumescent flame retardant materials suitable for aerospace (Manchado, Valentini, Biagiotti, & Kenny, 2005). Polypropylene wrapped CNT has shown superior flame retardancy and mechanical properties compared with the neat matrix. Functional CNT also revealed fine dispersion in polymer matrix owing to *in-situ* compatibilization. Thus, carbon nanotube has been used to improve strength, electrical conductivity, and non-flammability properties of nanocomposite (Table 1).

Polymer Matrices, Processing, and Application of Polymer/CNT Nanocomposite

Range of polymer matrices have been used for carbon nanotube reinforcement. Polystyrene and polystyrene copolymers such as hydroxylated styrene-butadiene-styrene tri-block copolymer, poly(styrene-co-p-(4-(4-vinylphenyl)-3-oxobutanol)), styrene-maleic anhydride copolymer, etc. have been employed (Bhattacharyya et al., 2007; Kaseem, Hamad, & Ko, 2016; Wu et al., 2010; Zhi, Liu, Zhang, Hong, & Yu, 2018). Both SWCNT and MWCNT have been reinforced in these matrices. Similarly, MWCNT and modified nanotubes were incorporated into poly(vinyl chloride) matrix (Alizadeh, Nayeri, & Mirzaee, 2018; Namsaeng, Punyodom, & Worajittiphon, 2018). Mechanical testing showed enhancement in tensile and impact

Table 1. Properties of polymer/carbon nanotube nanocomposite

Material	Young's Modulus	Strength	Electrical Conductivity	Reference
Polystyrene/SWCNT (0.3wt.%)	1GPa	-	Percolation threshold 0.3 wt.%	(Chang et al., 2006)
PP/SWCNT (0.75%)	1187 MPa	31.0 MPa	-	(Manchado et al., 2005)
Polyamide/MWCNT (12 wt.%)	4180 MPa	-	3.57×10^{-2} S/cm	(Meincke et al., 2004)
PS/CNT (5 wt.%)	1 TPa	-	-	(Thostenson & Chou, 2002)
PS/SWCNT (8.5 wt %)	-	-	1.34×10^{-5} S/m	(Ramasubramaniam, Chen, & Liu, 2003)

strength of the nanocomposite. Attempts on the grafting of poly(vinyl alcohol) on SWCNT and MWCNT have been observed (Dahoud et al., 2018; Purwanto & Salim, 2018). Polydimethylsiloxane has been filled with treated MWCNT to form homogeneous and viscous nanocomposite (Ghanbari & Ehsani, 2018). Silane functional CNT has been dispersed in epoxy through interfacial interaction (Kadhim et al., 2018; Zhang et al., 2018). Correspondingly, several other matrices have been explored with carbon nanotube.

Different techniques have been used for processing of polymer/CNT nanocomposites such as solution blending, melt mixing, and *in-situ* polymerization. Other techniques include dry powder mixing and surfactant-assisted mixing. The advantage of solution blending is vigorous mixing of polymer with CNT in a suitable solvent, which promotes nanotube de-aggregation and dispersion. This method mainly consists of three steps *(i)* nanotube dispersion using appropriate solvent; *(ii)* mixing of polymer and nanotube dispersion; *(iii)* and nanocomposite recovery by precipitation or film casting. Melt blending is also a valued technique for nanotube dispersion and nanocomposite formation. Melt blending has been used to form several nanocomposites such as nylon-6/MWCNT, polycarbonate/MWCNT, polyimide/SWCNT, etc. (Lee, Cruz, & Son, 2018; Otaegi, Aramburu, Müller, & Guerrica-Echevarría, 2018; Šehić et al., 2018). Although, melt blending needs high shear force and high temperature, which may deteriorate nanotube and nanocomposite properties. *In-situ* polymerization has also been used for the preparation of polymer/nanotube nanocomposite material. The main advantage of this technique is to form better physical/chemical links between the matrix-nanofiller. Nanocomposites with high nanotube loading can be easily developed using this route.

Polymer/carbon nanotube nanocomposite are much attractive for increasing number of industries. These are low cost, light weight, and high performance materials used for several engineering applications (Duc, Nguyen, Cuong, Van Sy, & Khoa, 2018). These materials have also been successfully employed for electromagnetic interference (EMI) shielding (Alegaonkar & Alegaonkar, 2018). In electronic appliances, the polymer/carbon nanotube nanocomposites are popular due to high thermal and mechanical resistance (Kong, Luo, Wang, & Wang, 2018). Additionally, these nanocomposites found success in coatings and adhesives industries (Augustin, 2018). Potential applications of polymer/CNT have been identified in sensors, energy storage and field emission systems, and biomedical (Fu, Ramos, Al-Jumaily, Meshkinzar, & Huang; Mihajlovic, Mihajlovic, Dankers, Masereeuw, & Sijbesma, 2018; Senokos). Due to extraordinary properties, integration of CNT in polymers have led to advanced materials for aviation, petroleum, and aerospace production systems (S. Chen, 2018; Smith Jr et al., 2004).

Epoxy/Carbon Nanotube Nanocomposite

Epoxy is an important type of thermosetting polymer having epoxide groups in the backbone. Epoxy is a pre-polymer having low molecular weight. Curing of epoxy using short chain dihydroxy and diamine compounds may lead to the formation of hydroxyl and secondary amine groups in the backbone (Park & Kim, 2001; Yang, Fu, & Yang, 2007). Cured epoxy also has high molecular weight. This polymer has exceptional mechanical properties, thermal stability, and adhesiveness in the cured form. Incorporation of carbon nanotube in epoxy before or while curing may lead to excellent heat resistance, chemical stability, and thermal constancy (Allaoui, Bai, Cheng, & Bai, 2002). Young's modulus, strength, and electrical properties of epoxy/CNT nanocomposite have been significantly enhanced using very small nanofiller content (Bekyarova et al., 2007). Table 2 shows some of the epoxy/CNT nanocomposite with improved desired properties at particular nanofiller loading.

Role of Carbon Nanotube in Improving the Properties of Aerospace Composite

For aerospace constituents, polymeric systems have been reinforced with carbon nanotube and carbon fibers as filler materials (Arena, Vertuccio, Barra, Viscardi, & Guadagno, 2018). Several polymeric nanocomposites have been commercially fabricated. Change in size from microparticle to nanoparticle may cause significant changes in the physical properties of composites. Scanning electron microscopy, transmission electron microscopy, scanning probe microscopy, and scanning tunneling microscopy have been employed to study the dispersion of carbon nanotube and fibers in polymers. Fibrous structured fillers have significantly diverse features than that of

Table 2. Essential properties of epoxy/carbon nanotube nanocomposite

Epoxy/CNT	Young's Modulus	Electrical Properties	Reference
Epoxy/C NT (4 wt.%)	465 MPa	1×10^{-2} S/cm	(Allaoui et al., 2002)
Epoxy/CNT (0.1 vol.%)	65 MPa	10^{-2} S/cm	(J Sandler et al., 1999)
Epoxy/SWCNT (0.25 vol.%)	500MPa	0.049 S/cm	(Bekyarova et al., 2007)
Epoxy/SWCNT (0.25 vol.%)	500MPa	0.089 S/cm	(Bekyarova et al., 2007)
Epoxy/MWCNT (0.1 wt.%)	2800MPa	-	(Gojny, Wichmann, Fiedler, & Schulte, 2005)
Epoxy/CNT (4 wt.%)	1500 MPa	1×10^{-2} S/cm	(Guadagno et al., 2011)

the nanoparticles. Carbon fiber-based composites have been used in space craft and air craft industry owing to high strength, toughness, fatigue life, modulus, corrosion resistance, and light weight compared with metals. For commercial military aircraft and satellite constituents, research have been directed towards the incorporation of CNT along with carbon fiber to further enhance the mechanical properties of composites. Especially, the properties of epoxy/fiber composite have been enhanced using carbon nanotube for low density, low manufacturing cost, and outstanding mechanical and physical features (Lonjon, Demont, Dantras, & Lacabanne, 2012; Lubineau & Rahaman, 2012). Carbon fiber coated with carbon nanotube has also been used to achieve advanced light weight aerospace composites. Properties such as light weight, high electrical conductivity, thermal stability, thermal conductivity, and electro-magnetic shielding efficiency have been attained using epoxy/carbon fiber and CNT composites. Future aerospace systems pursue to further improve the multifunctional features of composite using carbon fibers and nanotube together.

MAIN FOCUS

Aeronautical Impact of Epoxy/CNT Nanocomposite

Shielding of Space Vehicle From Radiations

Both the SWCNT and MWCNT have been studied for electromagnetic interference (EMI) shielding applications in aerospace. Electrical and mechanical properties of advanced epoxy/CNT engineering materials have been used for EMI shielding of radio frequency (JKWea Sandler, Kirk, Kinloch, Shaffer, & Windle, 2003). Intrinsic electrical conductivity, aspect ratio, and dielectric properties of nanotube are essential for EMI shielding (Chung, 2001). Epoxy/CNT nanocomposite have shown EMI shielding of -20 to 30 dB in X-band (8.2-12.4 GHz). The shielding was observed with very low nanofiller loading of 0.01 to 15 wt.%. Moreover, nanotube loading as well as thickness of nanocomposite laminates are essential to enhance the electromagnetic features (Chung, 2001). Interaction between EM radiations and nanofiller nanoparticle is illustrated in Fig. 3. EMI shielding effectiveness of ~20 dB is desired for commercial applications of aerospace structures. Inclusion of 10-15 wt.% SWCNT has shown shielding of ~20 dB in X-band. Influence of functional nanofiller ratio on EMI shielding effectiveness has also been observed. Inclusion of magnetic nanoparticles such as ferrite nanofiller caused significant effect on EMI shielding effectiveness of nanocomposite.

Figure 3. Interaction between electromagnetic radiation and nanofiller in epoxy/ CNT

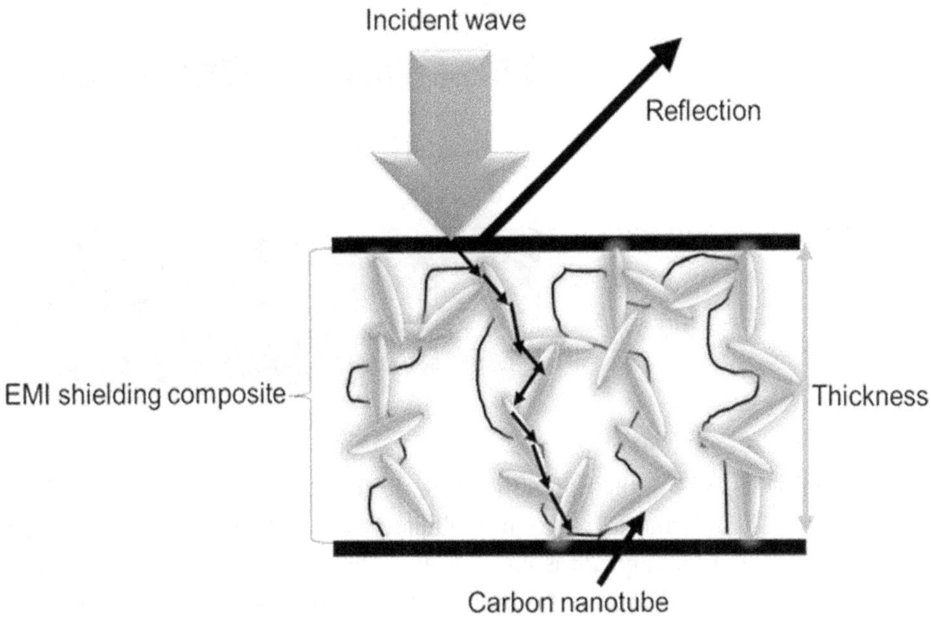

Lightning Strike Performance

Lightning strikes may cause several damages to epoxy-based aerospace structures. Studies have shown increase in the electrical conductivity of epoxy matrix with the incorporation of carbon nanotube nanofiller (Domingues, Logakis, & Skordos, 2012; Logakis & Skordos, 2012; Thostenson, Gangloff Jr, Li, & Byun, 2009). Inclusion of CNT in epoxy/fiber reinforced polymer composite has also enhanced the electrical conductivity of nanocomposite. However, dielectric nature of matrix is highly useful to cure the outer aerospace framework and enhance the lightning strike protection. However, several manufacturing complexities need to be overcome to attain high lightning strike performance. Ultrasonic C-scanning has been used to study the damage produced in the aerospace structures while lightning. Both the nanotube reinforced and non-reinforced samples have been used to study the influence of nanaofiller in lightning strike protection. Polymer alone has less capability for defense against lightning strike. Investigations on multi-wall carbon nanotube (MWCNT) doped carbon fiber epoxy composite has also found success in this case. In this regard, panels of epoxy/carbon fiber and epoxy/carbon fiber with MWCNT layer were produced. Infusion set-up to form cured laminates are given in Fig. 4.

Figure 4. Infusion set up for epoxy/CNT fabrication

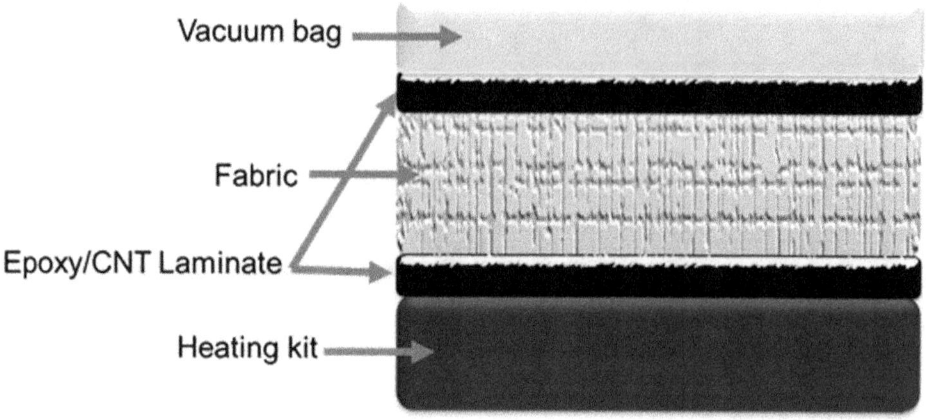

Influence of MWCNT on the lightning strike performance of epoxy/carbon fiber composite was significant. Damage in laminates without MWCNT was observed using visual observations and ultrasonic C-scans. The epoxy/carbon fiber/MWCNT panels have been found useful in primary aerospace structures (Thostenson et al., 2009; Wichmann et al., 2006).

Non-Flammability Execution

Epoxy/CNT nanocomposite has fascinated extensive research interest owing to flame resistant features. Inclusion of carbon nanotube in epoxy has found to enhance the flame retardancy and thermal features of epoxy nanocomposite (Song & Youn, 2005). Flame retardancy capability of epoxy/CNT nanocomposite has been analyzed using limiting oxygen index (LOI), Bunsen burner test, UL-94 test, and cone calorimetry. It has been observed that the efficiency of non-flammable nanocomposite has been enhanced using carbon nanotube additives (Kuan et al., 2010). LOI values of CNT loaded nanocomposites were found to enhance compared with pure epoxy. LOI results for epoxy and functional nanotube has shown more efficient flame retardancy compared with the epoxy/non-modified CNT. Strong interfacial interaction exist between epoxy and functional CNT. The interfacial bonding between matrix-nanofiller may decrease the thermal conduction in epoxy matrix, so reducing the flammability (Schartel, Bartholmai, & Knoll, 2006). Thus, ease of flammability was found to reduce even with very low nanofiller addition in epoxy/CNT nanocomposite structures. According to thermogravimetric analysis and LOI tests, 0.1-0.5 wt.% nanotube loading has enhanced both the LOI and thermal stability of aerospace structures.

Mechanical Performance

Improved mechanical performance of polymeric nanocomposite is essential to apply in aerospace application. Various methodologies have been established to reinforce the aerospace composite structures in thickness, in interface, and through various planes. The laminated plies in aerospace structures have been filled with epoxy/CNT nanocomposite to study the in-plane mechanical properties (Wicks, de Villoria, & Wardle, 2010). However, in plane application may cause volume loss, low mechanical properties, and structural damage during the edging process. Thus, inclusion of nanocomposite at ply interface without reducing the in-plane properties is challenging. Sometimes, carbon nanotube has been used as a second nanofiller in addition to the carbon fiber. In this regard, vertically aligned carbon nanotube (VACNT) has shown improved inter-laminar toughness when loaded at the interface. Accordingly, laminate engineering strength tests have been performed to analyze laminate failure (Guzmán de Villoria, Hart, & Wardle, 2011). Compared with the traditional aerospace composite, laminate in-plane strength has been found to improve with carbon nanotube reinforcement. Effect of nanofiller addition on the curing time and resulting strength properties of aerospace structural component is also important. Various facets of mechanical property improvement of different polymeric systems using CNT have been explained (Thostenson, Ren, & Chou, 2001). The CNT type, growth method, functionalization, polymer type, and processing method directly affect the mechanical strength of nanocomposite. To transfer the mechanical properties of nanotube to polymer, a chemical link is essential to develop in between. The interfacial bonding is developed by the functionalization of nanotube surface, so improvement in strength may occur (Roy, Petrova, & Mitra, 2018). The effect of functional nanotube content on the mechanical properties of nanocomposite has been studied (X. Chen et al., 2008). Inclusion of 1.5 wt. % amine-treated MWCNT enhanced the tensile strength by 51% and impact strength by 93% (Fig. 5 & Fig. 6). The amine-treated MWCNT improved the mechanical properties compared with the oxidized MWCNT. The covalent bond formation between polymer and nanotube led to effective stress transfer between matrix-nanofiller. Moreover, melt blending produce high performance nanocomposite relative to the conventional epoxy/CNT mixed materials (Ogasawara et al., 2011).

CHALLENGES, FUTURE SCOPE, AND SUMMARY

Nanocomposites have been utilized in wide range of technical applications ranging from electronics——to——biomedical——to——automobile/aerospace. Carbon nanotube is one of the most important type of nanocarbon, owing to distinctive

Figure 5. Effect of o-MWCNT (oxidized) and a-MWCNT (amine) content on tensile strength of epoxy nanocomposite (X. Chen et al., 2008)

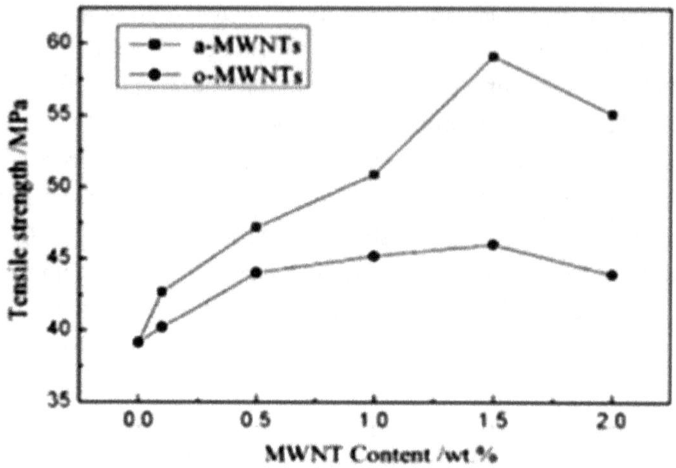

Figure 6. Effect of o-MWCNT (oxidized) and a-MWCNT (amine) content on impact strength of epoxy nanocomposite (X. Chen et al., 2008)

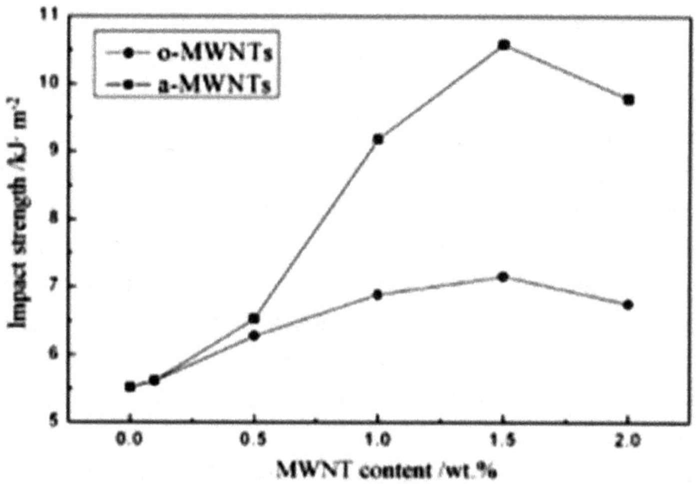

properties, employed in aerospace nanocomposite. Space applications of polymer/CNT nanocomposite seem to demand superior thermal stability, mechanical strength, and electrical features. In aerospace and aeronautics, epoxy is a well-known matrix material used (Y. Saito & Uemura, 2000). Incorporation of nanotube in epoxy has been deliberated for upgrading the strength, heat constancy, non-flammability, EMI shielding, and electrical properties of final nanocomposite. Lightning strike prevention has also been observed in aircrafts with polymer/CNT in the outer structural frame. Consequently, the role of epoxy has been modified through amalgamation with nanofiller in aerospace industry. Initially, aircrafts structures as well as space shuttles have used epoxy/CNT-based materials (Heimann, Wirts-Ruetters, Boehme, & Wolter, 2008). Further progress in aerospace and aeronautical fields has been observed through the modification of nanotube nanostructure. Functionalization of nanotube has amplified the nanocomposite application several folds. Functionalization also overcomes one of the major challenge in the implementation of nanotube in aerospace i.e. dispersion. The reliability, durability, strength, and stability properties of aeronautics can be enhanced significantly through uniform CNT dispersion in nanocomposite structure. Thus, number of key technologies have been identified for the application of epoxy/CNT in aerospace structure, missiles, rotorcraft, and next generation aircrafts.

This chapter addresses several prospects of epoxy ad carbon nanotube-based nanocomposite. Epoxy matrix incorporated with nanocarbon has been found as exceptional nanocomposite for aerospace. Even very small nanofiller content has led to substantial enhancement in the desired aerospace properties such as mechanical robustness, thermal constancy, electrical conductivity, radiation shielding, anti-flame, and anti-light strike properties. Consequently, the chapter systematically focused the potential of epoxy/carbon nanotube nanocomposite in aerospace applications.

REFERENCES

Ahmed, S., Ikram, S., Kanchi, S., & Bisetty, K. (2018). *Advancements and Potential Prospects of Polymer/Metal Oxide Nanocomposites: From Laboratory Synthesis to Commercialization. In Biocomposites* (pp. 47–84). Pan Stanford.

Alegaonkar, A. P., & Alegaonkar, P. S. (2018). Nano-Carbon/Polymer Composites for Electromagnetic Shielding, Structural Mechanical and Field Emission Applications. *Thermoset Composites: Preparation. Properties and Applications*, *38*, 128.

Alizadeh, T., Nayeri, S., & Mirzaee, S. (2018). A high performance potentiometric sensor for lactic acid determination based on molecularly imprinted polymer/MWCNTs/PVC nanocomposite film covered carbon rod electrode. *Talanta*. PMID:30348364

Allaoui, A., Bai, S., Cheng, H.-M., & Bai, J. (2002). Mechanical and electrical properties of a MWNT/epoxy composite. *Composites Science and Technology*, *62*(15), 1993–1998. doi:10.1016/S0266-3538(02)00129-X

Andrei, G., Dima, D., & Andrei, L. (2006). Lightweight magnetic composites for aircraft applications. *Journal of Optoelectronics and Advanced Materials*, *8*(2), 726.

Arena, M., Vertuccio, L., Barra, G., Viscardi, M., & Guadagno, L. (2018). Dynamic performance of self-sensing epoxy resin for aerospace structures. *AIP Conference Proceedings*. 10.1063/1.5045928

Augustin, T. (2018). *Structural health monitoring of carbon fiber reinforced polymers and carbon nanotube modified adhesive joints via electrical resistance measurement*. Technische Universität Hamburg.

Bekyarova, E., Thostenson, E., Yu, A., Kim, H., Gao, J., Tang, J., ... Haddon, R. (2007). Multiscale carbon nanotube– carbon fiber reinforcement for advanced epoxy composites. *Langmuir*, *23*(7), 3970–3974. doi:10.1021/la062743p PMID:17326671

Bhattacharyya, A. R., Bose, S., Kulkarni, A. R., Pötschke, P., Häußler, L., Fischer, D., & Jehnichen, D. (2007). Styrene maleic anhydride copolymer mediated dispersion of single wall carbon nanotubes in polyamide 12: Crystallization behavior and morphology. *Journal of Applied Polymer Science*, *106*(1), 345–353. doi:10.1002/app.26680

Bogue, R. (2011). Nanocomposites: A review of technology and applications. *Assembly Automation*, *31*(2), 106–112. doi:10.1108/01445151111117683

Cha, J., Chiashi, S., Inoue, T., Einarsson, E., Shiomi, J., & Maruyama, S. (2018). Fabrication of uniform vertically-aligned carbon nanotube–polymer composite thin films by capillary flow intrusion. *Japanese Journal of Applied Physics*, *57*(11), 115101. doi:10.7567/JJAP.57.115101

Chang, T.-E., Kisliuk, A., Rhodes, S., Brittain, W., & Sokolov, A. (2006). Conductivity and mechanical properties of well-dispersed single-wall carbon nanotube/polystyrene composite. *Polymer*, *47*(22), 7740–7746. doi:10.1016/j.polymer.2006.09.013

Chen, S. (2018). *Polymer Based Nanocomposites as Multifunctional Structure for Space Radiation Shielding: A Study of Nanomaterial Fabrications and Evaluations*. Academic Press.

Chen, X., Wang, J., Lin, M., Zhong, W., Feng, T., Chen, X., ... Xue, F. (2008). Mechanical and thermal properties of epoxy nanocomposites reinforced with amino-functionalized multi-walled carbon nanotubes. *Materials Science and Engineering A*, *492*(1-2), 236–242. doi:10.1016/j.msea.2008.04.044

Chung, D. (2001). Electromagnetic interference shielding effectiveness of carbon materials. *Carbon*, *39*(2), 279–285. doi:10.1016/S0008-6223(00)00184-6

Dahoud, M. S., Abdel-Rahem, R. A., Farrag, E. A., Ibrahim, S., Ayesh, A. S., & Alshoaibi, A. (2018). Rheological and electrical properties of multiwalled carbon nanotubes–polyvinyl alcohol nanocomposites treated at different pH conditions. *Journal of Thermoplastic Composite Materials*.

Dang, Z.-M., Yuan, J.-K., Zha, J.-W., Zhou, T., Li, S.-T., & Hu, G.-H. (2012). Fundamentals, processes and applications of high-permittivity polymer–matrix composites. *Progress in Materials Science*, *57*(4), 660–723. doi:10.1016/j.pmatsci.2011.08.001

De Volder, M. F., Tawfick, S. H., Baughman, R. H., & Hart, A. J. (2013). Carbon nanotubes: present and future commercial applications. *Science, 339*(6119), 535-539.

Domingues, D., Logakis, E., & Skordos, A. (2012). The use of an electric field in the preparation of glass fibre/epoxy composites containing carbon nanotubes. *Carbon*, *50*(7), 2493–2503. doi:10.1016/j.carbon.2012.01.072

Duc, N. D., Nguyen, P. D., Cuong, N. H., Van Sy, N., & Khoa, N. D. (2018). An analytical approach on nonlinear mechanical and thermal post-buckling of nanocomposite double-curved shallow shells reinforced by carbon nanotubes. *Proceedings of the Institution of Mechanical Engineers, Part C: Journal of Mechanical Engineering Science*.

Ellis, B. (1993). *Chemistry and technology of epoxy resins*. Springer. doi:10.1007/978-94-011-2932-9

Fu, X., Ramos, M., Al-Jumaily, A. M., Meshkinzar, A., & Huang, X. (n.d.). Stretchable strain sensor facilely fabricated based on multi-wall carbon nanotube composites with excellent performance. *Journal of Materials Science*, 1-11.

Ganguli, R. (2013). Optimal design of composite structures: A historical review. *Journal of the Indian Institute of Science*, *93*(4), 557–570.

Gay, D., & Hoa, S. V. (2007). *Composite materials: design and applications*. CRC Press. doi:10.1201/9781420045208

Ghanbari, K., & Ehsani, M. (2018). Change in surface properties of polydimethylsiloxane/polyurethane/carbon nanotubes elastomeric coatings. *Progress in Organic Coatings*, *125*, 393–402. doi:10.1016/j.porgcoat.2018.09.022

Gohardani, A. S., Doulgeris, G., & Singh, R. (2011). Challenges of future aircraft propulsion: A review of distributed propulsion technology and its potential application for the all electric commercial aircraft. *Progress in Aerospace Sciences*, *47*(5), 369–391. doi:10.1016/j.paerosci.2010.09.001

Gojny, F. H., Wichmann, M. H., Fiedler, B., & Schulte, K. (2005). Influence of different carbon nanotubes on the mechanical properties of epoxy matrix composites–a comparative study. *Composites Science and Technology*, *65*(15-16), 2300–2313. doi:10.1016/j.compscitech.2005.04.021

Guadagno, L., De Vivo, B., Di Bartolomeo, A., Lamberti, P., Sorrentino, A., Tucci, V., ... Vittoria, V. (2011). Effect of functionalization on the thermo-mechanical and electrical behavior of multi-wall carbon nanotube/epoxy composites. *Carbon*, *49*(6), 1919–1930. doi:10.1016/j.carbon.2011.01.017

Guzmán de Villoria, R., Hart, A. J., & Wardle, B. L. (2011). Continuous high-yield production of vertically aligned carbon nanotubes on 2D and 3D substrates. *ACS Nano*, *5*(6), 4850–4857. doi:10.1021/nn2008645 PMID:21591620

Haque, A., Hossain, F., Dean, D., & Shamsuzzoha, M. (2002). *s2-glass/vinyl ester polymer nanocomposites: manufacturing, structures, thermal and mechanical properties*. Paper presented at the 17th Annual Technical Conference, American Society for Composites.

Heimann, M., Wirts-Ruetters, M., Boehme, B., & Wolter, K.-J. (2008). *Investigations of carbon nanotubes epoxy composites for electronics packaging*. Paper presented at the Electronic Components and Technology Conference, 2008. ECTC 2008. 58th. 10.1109/ECTC.2008.4550214

Hirsch, J., & Al-Samman, T. (2013). Superior light metals by texture engineering: Optimized aluminum and magnesium alloys for automotive applications. *Acta Materialia*, *61*(3), 818–843. doi:10.1016/j.actamat.2012.10.044

Hussain, F., Hojjati, M., Okamoto, M., & Gorga, R. E. (2006). Polymer-matrix nanocomposites, processing, manufacturing, and application: An overview. *Journal of Composite Materials*, *40*(17), 1511–1575. doi:10.1177/0021998306067321

Hussain, S. T., Abbas, F., Kausar, A., & Khan, M. R. (2013). New polyaniline/polypyrrole/polythiophene and functionalized multiwalled carbon nanotube-based nanocomposites: Layer-by-layer in situ polymerization. *High Performance Polymers*, *25*(1), 70–78. doi:10.1177/0954008312456048

Jorio, A., Kauppinen, E., & Hassanien, A. (2007). *Carbon-nanotube metrology. In Carbon Nanotubes* (pp. 63–100). Springer.

Kadhim, N., Mei, Y., Wang, Y., Li, Y., Meng, F., Jiang, M., & Zhou, Z. (2018). Remarkable Improvement in the Mechanical Properties of Epoxy Composites Achieved by a Small Amount of Modified Helical Carbon Nanotubes. *Polymers*, *10*(10), 1103. doi:10.3390/polym10101103

Kaseem, M., Hamad, K., & Ko, Y. G. (2016). Fabrication and materials properties of polystyrene/carbon nanotube (PS/CNT) composites: A review. *European Polymer Journal*, *79*, 36–62. doi:10.1016/j.eurpolymj.2016.04.011

Kausar, A., & Hussain, S. T. (2013). Effect of multi-walled carbon nanotube reinforcement on the physical properties of poly (thiourea-azo-ether)-based nanocomposites. *Journal of Plastic Film & Sheeting*, *29*(4), 365–383. doi:10.1177/8756087913487003

Kausar, A., Rafique, I., & Muhammad, B. (2016). Review of applications of polymer/carbon nanotubes and epoxy/CNT composites. *Polymer-Plastics Technology and Engineering*, *55*(11), 1167–1191. doi:10.1080/03602559.2016.1163588

Khan, Z. U., Kausar, A., & Ullah, H. (2016). A review on composite papers of graphene oxide, carbon nanotube, polymer/GO, and polymer/CNT: Processing strategies, properties, and relevance. *Polymer-Plastics Technology and Engineering*, *55*(6), 559–581. doi:10.1080/03602559.2015.1098693

Kong, Q., Luo, Z., Wang, Y., & Wang, B. (2018). Fabrication of super-stretchable and electrical conductive membrane of spandex/multi-wall carbon nanotube/reduced graphene oxide composite. *Journal of Polymer Research*, *25*(11), 231. doi:10.100710965-018-1597-0

Kozlov, G., & Dolbin, I. (2018). Effect of a Nanofiller Structure on the Degree of Reinforcement of Polymer–Carbon Nanotube Nanocomposites with the Use of a Percolation Model. *Journal of Applied Mechanics and Technical Physics*, *59*(4), 765–769. doi:10.1134/S0021894418040259

Kuan, C.-F., Chen, W.-J., Li, Y.-L., Chen, C.-H., Kuan, H.-C., & Chiang, C.-L. (2010). Flame retardance and thermal stability of carbon nanotube epoxy composite prepared from sol–gel method. *Journal of Physics and Chemistry of Solids*, *71*(4), 539–543. doi:10.1016/j.jpcs.2009.12.031

Lan, X., Liu, Y., Lv, H., Wang, X., Leng, J., & Du, S. (2009). Fiber reinforced shape-memory polymer composite and its application in a deployable hinge. *Smart Materials and Structures*, *18*(2), 024002. doi:10.1088/0964-1726/18/2/024002

Lee, H. Y., Cruz, H., & Son, Y. (2018). Effects of incorporation of polyester on the electrical resistivity of polycarbonate/multi-walled carbon nanotube nanocomposite. *Journal of Composite Materials*.

Leng, J., & Lau, A. K.-t. (2010). *Multifunctional polymer nanocomposites*. CRC Press. doi:10.1201/b10462

Levchik, S. V., & Weil, E. D. (2004). Thermal decomposition, combustion and flame-retardancy of epoxy resins—A review of the recent literature. *Polymer International*, *53*(12), 1901–1929. doi:10.1002/pi.1473

Liu, Y., Du, H., Liu, L., & Leng, J. (2014). Shape memory polymers and their composites in aerospace applications: A review. *Smart Materials and Structures*, *23*(2), 023001. doi:10.1088/0964-1726/23/2/023001

Logakis, E., & Skordos, A. (2012). *Lightning strike performance of carbon nanotube loaded aerospace composites*. Paper presented at the 15th European Conference on Composite Materials.

Lonjon, A., Demont, P., Dantras, E., & Lacabanne, C. (2012). Electrical conductivity improvement of aeronautical carbon fiber reinforced polyepoxy composites by insertion of carbon nanotubes. *Journal of Non-Crystalline Solids*, *358*(15), 1859–1862. doi:10.1016/j.jnoncrysol.2012.05.038

Lubineau, G., & Rahaman, A. (2012). A review of strategies for improving the degradation properties of laminated continuous-fiber/epoxy composites with carbon-based nanoreinforcements. *Carbon*, *50*(7), 2377–2395. doi:10.1016/j.carbon.2012.01.059

Ma, P.-C., Siddiqui, N. A., Marom, G., & Kim, J.-K. (2010). Dispersion and functionalization of carbon nanotubes for polymer-based nanocomposites: A review. *Composites. Part A, Applied Science and Manufacturing*, *41*(10), 1345–1367. doi:10.1016/j.compositesa.2010.07.003

Manchado, M. L., Valentini, L., Biagiotti, J., & Kenny, J. (2005). Thermal and mechanical properties of single-walled carbon nanotubes–polypropylene composites prepared by melt processing. *Carbon*, *43*(7), 1499–1505. doi:10.1016/j.carbon.2005.01.031

Meincke, O., Kaempfer, D., Weickmann, H., Friedrich, C., Vathauer, M., & Warth, H. (2004). Mechanical properties and electrical conductivity of carbon-nanotube filled polyamide-6 and its blends with acrylonitrile/butadiene/styrene. *Polymer*, *45*(3), 739–748. doi:10.1016/j.polymer.2003.12.013

Mihajlovic, M., Mihajlovic, M., Dankers, P. Y., Masereeuw, R., & Sijbesma, R. P. (2018). Carbon Nanotube Reinforced Supramolecular Hydrogels for Bioapplications. *Macromolecular Bioscience*, 1800173. PMID:30085403

Murr, L. E., Gaytan, S. M., Ramirez, D. A., Martinez, E., Hernandez, J., Amato, K. N., ... Wicker, R. B. (2012). Metal fabrication by additive manufacturing using laser and electron beam melting technologies. *Journal of Materials Science and Technology*, *28*(1), 1–14. doi:10.1016/S1005-0302(12)60016-4

Namsaeng, J., Punyodom, W., & Worajittiphon, P. (2018). Synergistic effect of welding electrospun fibers and MWCNT reinforcement on strength enhancement of PAN–PVC non-woven mats for water filtration. *Chemical Engineering Science*.

Ogasawara, T., Moon, S.-Y., Inoue, Y., & Shimamura, Y. (2011). Mechanical properties of aligned multi-walled carbon nanotube/epoxy composites processed using a hot-melt prepreg method. *Composites Science and Technology*, *71*(16), 1826–1833. doi:10.1016/j.compscitech.2011.08.009

Otaegi, I., Aramburu, N., Müller, A., & Guerrica-Echevarría, G. (2018). Novel Biobased Polyamide 410/Polyamide 6/CNT Nanocomposites. *Polymers*, *10*(9), 986. doi:10.3390/polym10090986

Pandey, J. K., Ahn, S., Lee, C. S., Mohanty, A. K., & Misra, M. (2010). Recent advances in the application of natural fiber based composites. *Macromolecular Materials and Engineering*, *295*(11), 975–989. doi:10.1002/mame.201000095

Park, S. J., & Kim, H. C. (2001). Thermal stability and toughening of epoxy resin with polysulfone resin. *Journal of Polymer Science. Part B, Polymer Physics*, *39*(1), 121–128. doi:10.1002/1099-0488(20010101)39:1<121::AID-POLB110>3.0.CO;2-N

Purwanto, S., & Salim, M. (2018). *Magnetic properties α-Fe2O3Filled MWNT/PolyVinyl Alcohol Composites Film.* Paper presented at the Journal of Physics: Conference Series. 10.1088/1742-6596/1091/1/012022

Rafique, I., Kausar, A., Anwar, Z., & Muhammad, B. (2016). Exploration of epoxy resins, hardening systems, and epoxy/carbon nanotube composite designed for high performance materials: A review. *Polymer-Plastics Technology and Engineering*, *55*(3), 312–333. doi:10.1080/03602559.2015.1070874

Ramasubramaniam, R., Chen, J., & Liu, H. (2003). Homogeneous carbon nanotube/polymer composites for electrical applications. *Applied Physics Letters*, *83*(14), 2928–2930. doi:10.1063/1.1616976

Rao, M. D. (2003). Recent applications of viscoelastic damping for noise control in automobiles and commercial airplanes. *Journal of Sound and Vibration*, *262*(3), 457–474. doi:10.1016/S0022-460X(03)00106-8

Roy, S., Petrova, R. S., & Mitra, S. (2018). Effect of carbon nanotube (CNT) functionalization in epoxy-CNT composites. *Nanotechnology Reviews*, *7*(6), 475–485. doi:10.1515/ntrev-2018-0068 PMID:30637182

Saito, R., Dresselhaus, G., & Dresselhaus, M. S. (1998). *Physical properties of carbon nanotubes*. World Scientific. doi:10.1142/p080

Saito, Y., & Uemura, S. (2000). Field emission from carbon nanotubes and its application to electron sources. *Carbon*, *38*(2), 169–182. doi:10.1016/S0008-6223(99)00139-6

Sandler, J., Kirk, J., Kinloch, I., Shaffer, M., & Windle, A. (2003). Ultra-low electrical percolation threshold in carbon-nanotube-epoxy composites. *Polymer*, *44*(19), 5893–5899. doi:10.1016/S0032-3861(03)00539-1

Sandler, J., Shaffer, M., Prasse, T., Bauhofer, W., Schulte, K., & Windle, A. (1999). Development of a dispersion process for carbon nanotubes in an epoxy matrix and the resulting electrical properties. *Polymer*, *40*(21), 5967–5971. doi:10.1016/S0032-3861(99)00166-4

Schartel, B., Bartholmai, M., & Knoll, U. (2006). Some comments on the main fire retardancy mechanisms in polymer nanocomposites. *Polymers for Advanced Technologies*, *17*(9-10), 772–777. doi:10.1002/pat.792

Šehić, A., Vasiljević, J., Demšar, A., Leskovšek, M., Bukošek, V., Medved, J., ... Simončič, B. (2018). Polyamide 6 composite fibers with incorporated mixtures of melamine cyanurate, carbon nanotubes, and carbon black. *Journal of Applied Polymer Science*.

Senokos, E. (n.d.). *Development of multifunctional flexible and structural supercapacitors based on carbon nanotube fibers*. ETSI Caminos, Canales y Puertos (UPM).

Shanyi, D. (2007). Advanced composite materials and aerospace engineering. *Acta Materiae Compositae Sinica, 1*.

Shokrieh, M., & Rafiee, R. (2010). A review of the mechanical properties of isolated carbon nanotubes and carbon nanotube composites. *Mechanics of Composite Materials*, *46*(2), 155–172. doi:10.100711029-010-9135-0

Smith, J. Jr, Connell, J., Delozier, D., Lillehei, P., Watson, K., Lin, Y., ... Sun, Y.-P. (2004). Space durable polymer/carbon nanotube films for electrostatic charge mitigation. *Polymer*, *45*(3), 825–836. doi:10.1016/j.polymer.2003.11.024

Song, Y. S., & Youn, J. R. (2005). Influence of dispersion states of carbon nanotubes on physical properties of epoxy nanocomposites. *Carbon*, *43*(7), 1378–1385. doi:10.1016/j.carbon.2005.01.007

Tans, S. J., Verschueren, A. R., & Dekker, C. (1998). Room-temperature transistor based on a single carbon nanotube. *Nature*, *393*(6680), 49–52. doi:10.1038/29954

Thostenson, E. T., Gangloff, J. J. Jr, Li, C., & Byun, J.-H. (2009). Electrical anisotropy in multiscale nanotube/fiber hybrid composites. *Applied Physics Letters*, *95*(7), 073111. doi:10.1063/1.3202788

Thostenson, E. T., Ren, Z., & Chou, T.-W. (2001). Advances in the science and technology of carbon nanotubes and their composites: A review. *Composites Science and Technology*, *61*(13), 1899–1912. doi:10.1016/S0266-3538(01)00094-X

Tong, X. C. (2016). *Advanced materials and design for electromagnetic interference shielding*. CRC Press.

Wichmann, M. H., Sumfleth, J., Gojny, F. H., Quaresimin, M., Fiedler, B., & Schulte, K. (2006). Glass-fibre-reinforced composites with enhanced mechanical and electrical properties–benefits and limitations of a nanoparticle modified matrix. *Engineering Fracture Mechanics*, *73*(16), 2346–2359. doi:10.1016/j.engfracmech.2006.05.015

Wicks, S. S., de Villoria, R. G., & Wardle, B. L. (2010). Interlaminar and intralaminar reinforcement of composite laminates with aligned carbon nanotubes. *Composites Science and Technology*, *70*(1), 20–28. doi:10.1016/j.compscitech.2009.09.001

Wojciechowski, S. (2000). New trends in the development of mechanical engineering materials. *Journal of Materials Processing Technology*, *106*(1-3), 230–235. doi:10.1016/S0924-0136(00)00619-1

Wu, G., Zhou, L., Ou, E., Xie, Y., Xiong, Y., & Xu, W. (2010). Preparation and properties of hydroxylated styrene–butadiene–styrene tri-block copolymer/multi-walled carbon nanotubes nanocomposites via covalent bond. *Materials Science and Engineering A*, *527*(20), 5280–5286. doi:10.1016/j.msea.2010.04.094

Yang, G., Fu, S.-Y., & Yang, J.-P. (2007). Preparation and mechanical properties of modified epoxy resins with flexible diamines. *Polymer*, *48*(1), 302–310. doi:10.1016/j.polymer.2006.11.031

Zhang, D., Villarreal, M. G., Cabrera, E., Benatar, A., Lee, L. J., & Castro, J. M. (2018). Performance study of ultrasonic assisted processing of CNT nanopaper/solventless epoxy composite. *Composites. Part B, Engineering*.

Zhi, X., Liu, J., Zhang, H.-B., Hong, S., & Yu, Z.-Z. (2018). Simultaneous enhancements in electrical conductivity and toughness of selectively foamed polycarbonate/polystyrene/carbon nanotube microcellular foams. *Composites. Part B, Engineering*, *143*, 161–167. doi:10.1016/j.compositesb.2018.01.022

KEY TERMS AND DEFINITIONS

Aerospace Technology: Development in science, technology, and engineering to hover the atmosphere of Earth and surrounding space.

Carbon Nanotube: It is a nanoallotropic form of carbon having rolled graphene sheet like structure.

Electromagnetic Interference Shielding: It is the practice of reducing electromagnetic field in space using certain barriers.

Nanocomposite: It is a composite material having nanometer size particles.

Nanofiller: Reinforcement of matrices having size in the nanometer range.

Non-Flammability: Ability of materials to resist flame/fire.

Chapter 4
Cryogenic Treatment of Polymer/MWCNT Nano-Composites for Mechanical and Tribological Applications

Swamini Chopra
Visvesvaraya National Institute of Technology, India

Swaksha P. Halde
Visvesvaraya National Institute of Technology, India

S. Sreya
Visvesvaraya National Institute of Technology, India

Kavita A. Deshmukh
Visvesvaraya National Institute of Technology, India

Rohit V. Babhulkar
Visvesvaraya National Institute of Technology, India

D. R. Peshwe
Visvesvaraya National Institute of Technology, India

ABSTRACT

The cryogenic treatment of material has been known to motivate structural stability by rearranging its crystallographic structure in metals and by promoting intermolecular as well as intramolecular rearrangements in polymers. Additionally, in case of polymers reinforced with micro fillers, the structural changes brought about by cryogenic treatment are still largely governed by the polymer matrix itself. Thus, when investigated for their mechanical and tribological properties, the response of polymer/MWCNT nano-composites after cryogenic treatment was found to be depending on the cryo-structural modifications in the polymer matrix, followed by the MWCNT interaction to some extent. The enhancement in the mechanical properties of the polymer/MWCNT nano-composites is attributed to the increasing

DOI: 10.4018/978-1-5225-7921-2.ch004

% crystallinity, changes in crystal structure, conversion of less stable phases into more stable phases, change in the nature of bonding and strengthening of interphase between polymer and MWCNT. Thus, for the cryogenic treatment temperature of -185 °C, the optimum soaking period for PA and PA/MWCNT nano-composite was 24 hrs, whereas for PBT and PBT/MWCNT nano-composite it was 12 hrs and 16 hrs, respectively. This agrees well with the popular claim that each polymer has a specific functional group and/or structural characteristic that readily responds to the cryogenic treatments conditions (irrespective of the filler type, content and/or interaction), thereby, modifying the structure and giving superior properties, which makes cryogenic treatment a material specific process.

INTRODUCTION

On a global scenario; the material sector has witnessed momentous growth of polymers as a potential alternative for conventional metals and alloys. The major advantages of using polymers are good mechanical strength coupled with low stiffness-to-weight ratio, ease of manufacturing and excellent reproducibility. Due to the ease of tailoring the polymer properties; they are used in enormous and expanding range of products from paper clips to spaceships. The degree of crystallinity and the extent of rigidity of a polymer can also be controlled during its processing itself or by the incorporation of different types of additives such as fillers, plasticizers, stabilizers, etc. This blending and compounding technology has opened many alternatives to pursue tailoring of any polymer, thus, making a variety of choices available. As the worldwide market grows continuously; the current global production of plastics is approx. 322 million metric tonne per year (Federation of Indian Chambers of Commerce and Industry [FICCI], 2017). Such huge production propose the need of improving quality of polymers and extending their field of application, as even a small enhancement in property seems to be very important economically.

According to FICCI (2017), India is a growing market for plastics and consumes about 16 million tonne annually against a global consumption of 322 million tonne per year. The polymer consumption of India is growing at an average rate of 10% and is expected to touch 30 million tonnes by 2025. It is estimated that current low levels of per capita polymer consumption in India, as compared to developed countries, along with the increased growth in the end use industries offers a huge opportunity for the growth of polymers over long term (FICCI, 2017).

Polymer Composites in Automobile, Aerospace and Structural Applications

The automotive industry saw a revolution when polymers started replacing the conventional metallic components. Thermoplastics made their debut in automotive industry in 1950. The continuous development of advanced, high performance polymers has dramatically increased their usage since then. But, now more than ever, the automotive industry is under increasing pressure to meet higher fuel efficiency, environmental constrains and performance demands at competitive costs (American Chemistry Council – Plastic Division, 2014). All material sectors, metals as well as polymers, are working to respond to the automotive industry's changing needs. Many see weight reduction as a key to achieve the above requirements at once. Weight reduction can be achieved in three different ways; by use of light weight materials, by optimizing the design of different components and system layouts or by innovations in manufacturing processes. However, the challenge is to keep an optimum balance between cost, weight and performance, to comply with regulations as well as satisfy customers (Mubarak, 2009). One material class that promises such opportunity is high-performance polymer composites. In addition to potential innovative aerodynamic design and styling aesthetics, polymer composites with high strength-to-weight and stiffness-to-weight ratios have made them the material of choice in industries like motor sports and aerospace for years (American Chemistry Council – Plastic Division, 2014).

Nowadays, the polymers and their composites are used to make cars more energy efficient by reducing weight, together with providing durability, corrosion resistance, toughness, design flexibility, resiliency and high performance at low cost. An average vehicle uses about 150 kgs. of polymeric parts versus 1163 kgs. of metal – currently around 13-15% of total weight of car. It is estimated that every 10% reduction in vehicle weight will result in 5% to 7% fuel saving. Thus, for every kilogram of vehicle weight reduction, there is potential to reduce 20 kgs. of CO_2 emissions (Szeteiová, 2010). This makes the incorporation of lightweight, high-performance materials in automobiles as a necessity and a common need. Such applications ask for higher wear resistance, improved fatigue life, better deformation behavior and good mechanical strength.

During the enormous growth of plastics components in automotives, the advantages of using plastics have changed. Mounting costs are being met by the ability of plastics to be molded into components of complex geometries, often replacing several parts in other materials and offering integral fitments that all add up to easier assembly. Many types of polymers are used commercially in more than hundred different parts of all shapes and sizes. A quick look inside any model of the car shows that plastics are now used in exterior and interior components such as bumpers, doors, safety and windows, headlight and side view mirror housing, trunk lids, hoods, grilles and wheel covers. Upto 16 different types of polymers are used for various automobile components, which are listed in Table 1. In automotive sector, polymers have contributed to a multitude of innovations in safety, performance and fuel efficiency. However, while providing for the aim of weight reduction in the cars, these plastics components are usually made from virgin polymers, occasionally mixed with strengthening additives and impact modifiers (Szeteiová, 2010). In order to make a regular car as economically friendly as possible, the performance to cost ratio of fibre reinforced polymer composites need to be minimized for good.

While the glass fiber reinforced polymers have been in successful use for over fifty year or so, the need of composites with even superior mechanical properties gave momentum to carbon fiber research (Gohardani, Elola, & Elizetxea, 2014). The emerging challenges of crack propagation and encountered fiber defects however called for alternative solution for obtaining ultra high modulus fibers. This led to accidental discovery of carbon nanotubes (CNTs) by Iijima in 1991. Since then the CNT's have been looked at extensively by researchers in various fields such as chemistry, physics, materials science and electrical engineering. CNT's are unique nanostructured materials with remarkable physical and mechanical properties. Since carbon-carbon bonds are among the strongest bonds in nature, a structure based on a perfect arrangement of these bonds oriented along the axis of the nanotubes produces a very strong material with an extremely high strength-to-weight ratio. These properties have inspired interest in using CNT as filler in polymer composite systems to obtain ultra-light structural materials with enhanced mechanical, thermal and electrical characteristics. The prospect of obtaining advanced nanocomposites with multifunctional features has attracted the efforts of researchers in both academia and industry.

The polymer/CNT nano-composites are a radical alternative to such conventional-filled polymer composites. The high mechanical, electrical and thermal properties of the CNTs make them ideal candidates as fillers in light weight polymer composites. It is their extraordinary mechanical properties (exceptionally high tensile strength and stiffness) that has aroused particular interest and promoted research into the fabrication of polymer/CNT composite materials for high end applications in various load bearing and severe abrasive conditions. Recent examples include strong and lightweight composite turbine blades and hulls for maritime security boats that are made by using carbon fiber composite with CNT enhanced resin (Coleman, Khan, Blau, & Gun'ko, 2006 and Chou, Gao, Thostenson, Zhang, & Byun, 2010). Other multifunctional applications under investigation include lightning strike protection,

Table 1. Plastics used in a typical car (Szeteiová, 2010)

Sr. No.	Component	Main type of polymers
1	Bumpers	PS, ABS, PC, PBT
2	Seating	PU, PP, PVC, ABS, PA
3	Dashboard	PP, ABS, SMA PPE, PC
4	Fuel system	HDPE, POM, PA, PP, PBT
5	Body (including panels)	PP, PPE, PU
6	Under-bonnet components	PA, PP, PBT
7	Interior trim	PP, ABS, PET, POM, PVC
8	Electrical components	PP, PET, PBT, PA, PVC
9	Exterior trim	ABS, PA, PBT, POM, ASA, PP
10	Lighting	PC, PBT, ABS, PMMA, PU
11	Upholstery	PVC, PU, PP, PE
12	Liquid reservoirs	PP, PE, PA
PS (poly-styrene), ABS (acrylonitrile-butadiene-styrene), PC (poly-carbonate), PBT (poly-butylene-terephthalate), PU (poly-urethane), PP (poly-propylene), PVC (poly-vinyl-chloride), PA (poly-amide), SMA (styrene-maleic-anhydride), PPE (poly-phenylene-ether), HDPE (high-density-poly-ethylene), POM (poly-oxy-methylene), PET (poly-ethylene-terephthalate), ASA (Acrylonitrile-styrene-acrylate), PMMA (poly-methyl-methacrylate), PE (poly-ethylene)		

deicing and structural health monitoring for aircraft (Coleman et al., 2006 and De Volder, Tawfick, Baughman, & Hart, 2013).

Polymer and their composites are extensively used in aerospace applications also. However, if one considers the most widely used polymer composite; carbon fiber reinforced epoxy, it has limitations in terms of material failure (Thawre et al., 2015). These composites not only underperform after surpassing the yield limits, but also increase the chances of catastrophic failure of components, all because of the delamination phenomenon undergone by carbon fibers (Thawre et al., 2015). With the next generation aircraft, rotorcraft, unmanned aerial vehicles and missiles demanding stricter requirements in terms of lightweight, visual and thermal signature, increased speed, maneuverability, etc; they incite a need for advanced materials and systems that can incorporate these functionalities (Gohardani et al., 2014). Similar to that of automobile sector, the development of newer aircrafts and aerospace devices aims as weight reduction by replacing current materials, which ultimately results in the reduction of fuel consumption and directly affects the cost of operation for both commercial and military aircrafts and space vehicles. But the constant challenge that the aerospace industry faces is the introduction of a light aircraft or spacecraft, without sacrificing its structural integrity.

Polymer nano-composites are a subject of many research endeavors in different scientific fields due to their unique combination of properties. However, the dispersion of nano-particles and their adhesion at the particle–matrix interface are of crucial importance, which otherwise have adverse effects on the mechanical properties. In a study conducted by O'Donnell, Sprong, & Haltli (2004), a carbon nano-tube reinforced polymer (CNRP) was utilized airframe material on four conceptual aircraft structures, viz., Boeing 747-400, Boeing 757-200, Airbus A320, and Embraer E145. The simulation was done by replacing the conventional aluminum and its alloys with same structural volume of CNRP. The simulation showed a reduction in take-off mass as a result of replacing the airframe material. This presented approx. 14% weight reduction and 10% decrease in fuel consumption. Other studies in the field of aerospace materials and structures indicate that conductive epoxy adhesives based on MWCNTs bond on metal substrates, with a contact resistance, several order of magnitudes higher than their adhesive resistance (Rosca and Hoa, 2011). Also, for aerospace grade carbon/epoxy laminates, the modification of interfaces by using CNT forests has shown improvement in the interlaminar fracture toughness of the material (García, Wardle, & Hart, 2008). The application of vertically aligned CNTs as reinforcements in aero-structures has also been implemented by Guzman de Villoria, Ydrefors, Hallander, Ishiguro, Nordin, & Wardle (2012), where the reinforcement is attributed to a fiber-stitching mechanism.

Thus, numerous studies in the domain of automobile and aerospace structures indicates that CNT reinforced polymer composites exhibit multiple structural functions and integrated structural functions comprising mechanical properties such as fracture toughness, damping, strength and stiffness, along with non-structural functions, such as sensing and actuation, electrical and/or thermal conductivity, energy harvesting/storage, electromagnetic interference (EMI) shielding, recyclability, biodegradability and self-healing capability (Gohardani et al., 2014). However, for successful commercial application, it still requires never ending research and improvement.

Need of Cryogenic Treatment of Polymer and Its Composites

Even though polymer performances are enhanced using various micro and nano fillers, they are still prone to high wear and abrasion. Thus, to improve the wear resistance of such polymeric components, *Cryogenic Treatment* is proven to be an effective tool. Cryogenic treatment is an exciting, important and inexpensive technique that has already led to many discoveries and improvements in the field of steels and other metals and alloys, but it is still an emerging tool for polymers and its composites. Two review works (Indumathi, Bijwe, Ghosh, Fahim, & Krishnaraj, 1999a; Kalia, 2010) in this area, gives an overall outlook on the cryogenic processing of polymers and polymer composites. The cryogenic treatment has been previously attempted on various polymers, including poly-tetra-fluro-ethylene (PTFE), poly-carbonate (PC), poly-imide (PI), poly-ether-imide (PEI), poly-ether-ether-ketone (PEEK), poly-ether-sulphone (PES), poly-amide (PA6/PA6-6), poly-butylene-terephthalate (PBT), poly-urethane (PU), ultra high density poly-ethylene (UHDPE), Epoxy resin, and their various composites filled with a variety of fibres and micro-fillers, for exploring the potential of the treatment for enhancing the mechanical and wear resistance (Indumathi, Bijwe, & Ghosh, 1999b).

It is indeed established that the cryogenic treatment has potential for all the selected systems, although the extent of improvement depends on the type of polymer, the type of fillers, their geometry, matrix-filler concentration, etc. This has aroused a lot of debate in thc literature to standardize the parameters of cryogenic treatment process and also the outcome of the process for a specific material class. Therefore, it becomes difficult to recommend a single process for every material or even a single cycle for the components manufactured from the same material with different processing details (Chopra, Deshmukh, & Peshwe, 2017a, p. 11). This has led several researchers to assess the cryogenic treatment parameters for a specific material as per their convenience, knowledge and experience, thus, making cryogenic treatment a *material specific process*, which needs to be optimized for a particular material class.

OVERVIEW OF RESEARCH ON CRYOGENIC TREATMENT OF POLYMERS AND COMPOSITES

Cryogenic treatment is a capable process for treating a wide variety of materials such as metals, alloys, polymers, ceramics and their composites for enhancing their desired mechanical, thermal and structural properties. A typical cryo-treatment cycle is depicted in Figure 1 and consists of the following phases:

1. A *Ramp Down* phase, during which the temperature of the sample is lowered slowly to the specified level, at a cooling rate to avoid any thermal shock from developing in the part.
2. A *Soaking* phase, during which the metal sample is held at desired cryogenic temperature for the prescribed time duration.
3. A *Ramp Up* phase, during which the temperature of the object is brought back up to the room temperature. This phase is very important to the process. Ramping up too fast can cause problems such as cracking of the treated part (Farhani and Niaki, 2008 and ASM Handbook, Volume 4 - Heat treating, 1991, pp. 489-490).

Figure 1. Plot of temperature versus time for the cryogenic treatment process

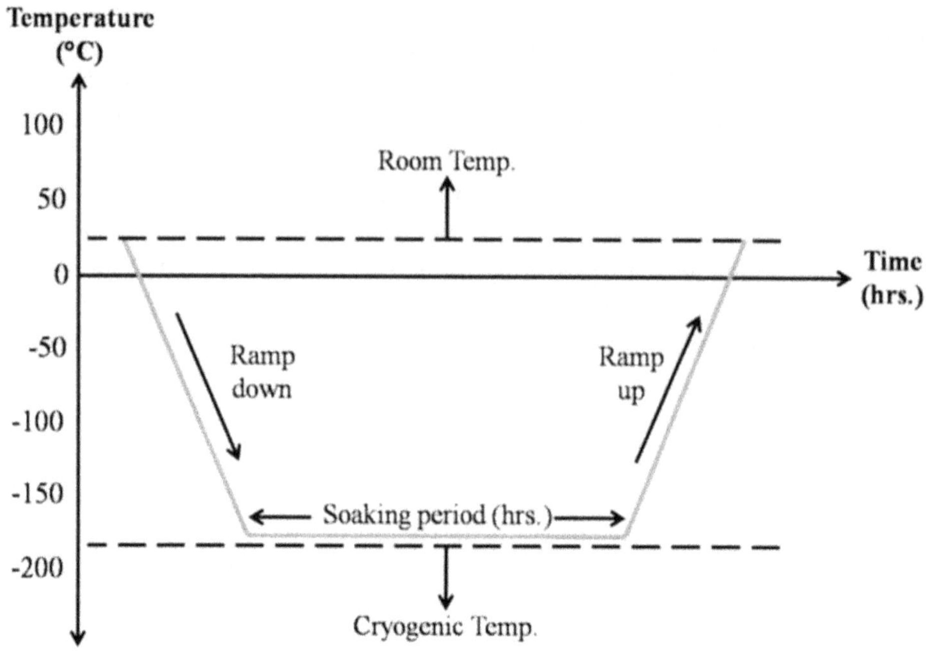

The changes brought about by the cryogenic treatment are permanent. It affects the entire volume of the material. Cryogenic processing establishes a very stable piece of material that remains distortion free due to relieving of residual stresses (Baldissera and Delprete, 2008). All the individual constituents of the material, which make up or build up the material, are rearranged into their most stable state on cryogenic treatment. These constituents are then aligned optimally with the surroundings. Also, molecular bonds are strengthened by this process. Particle alignment and grain refinement assists in relieving internal stresses, which would otherwise lead to early material failure. An extremely low temperature during cryogenic processing also slows the movement at atomic levels and promotes a pure structural balance throughout the material (Figure 2). As a result the improvements in physical and mechanical properties can be achieved in the material due to dense and uniform microstructure on cryogenic treatment. This increases the component life and ultimately reduces the frequency and cost of new production (Pande, 2012, pp. 7-8).

Cryogenic treatment, although a century older, is still considered an infant as compared to other conventional high temperature post-processing techniques. Nonetheless, it has been shown by many researchers that it can prove a potential process for enhancing the properties and ultimately the life of polymeric materials. In case of polymers and polymer composites, the cryogenic treatment brings out the

Figure 2. Schematic showing the increase in bonding energy at atomic levels with decrease in temperature

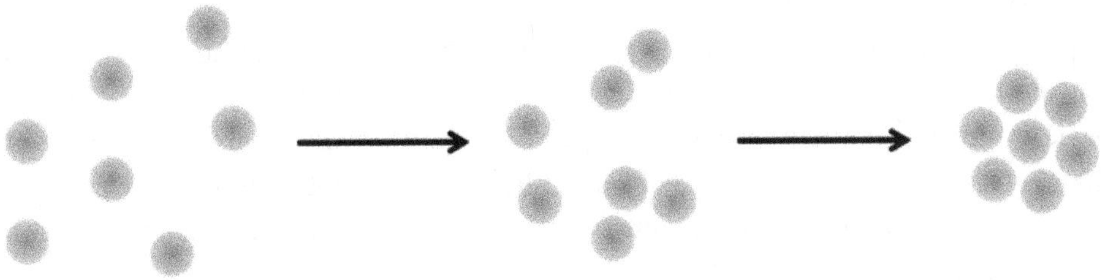

alignment of the randomly oriented chains which leads to improved physical and mechanical properties. It is said that the local intermolecular rearrangement during the cryogenic treatment results in relaxation. A dynamic mechanical relaxation occurs in polymer molecules due to the heat transfer between the intermolecular mode and the intramolecular mode (Ray, 2005a). The relaxation phenomena indicate a considerable dependence on the morphology of polymers. Polymers which are able to change their main chain bond angles seem to excel in cryogenic properties. These polymers are flexible and may undergo deformation even when their segmental motions are frozen at cryogenic temperatures. The cryogenic hardening and increase in crystallinity is reported to be one of the prominent reasons for the enhancement in wear performance of the polymers and polymer composites. The improvement in the shear value after cryogenic treatment of polymer composites is credited to the differential thermal contraction of the matrix during sudden cooling. This leads to the development of greater compressive stresses and increases the resistance to debonding and better adhesion at the interface of filler and matrix. Proper choice of polymer-filler combination and their relative proportions can affect the coefficient of contraction in cryogenic environment, thereby reducing the risks of thermal shocks. The following section summarizes the available literature related to the effect of cryogenic treatment on the mechanical properties of polymers and polymer composites.

Indumathi et al. (1999a) investigated the effect of cryogenic treatment on the wear performance of a series of thermoplastic engineering polymers such as polyimide (PI), polyetherimide (PEI), polytetrafluoroethylene (PTFE), polycarbonate (PC) and Polyurethane (PU). The selected polymers were cryogenically treated at liquid nitrogen temperature for 24 hours. The comparison of treated and untreated material revealed that this technique has the potential to increase the wear resistance of the polymers, just like metals and other cutting tool materials. The enhancement in the wear performance was observed to be maximum for PEI (by about 30%) and PTFE (by approx. 60%), however for PI and PEI copolymer it was small (i.e. by 2%) and

PC and PU showed moderate improvement (10–15%) due to cryogenic treatment. Along with the polymers, the composites were also studied by Indumathi et al. (1999a) for the cryo-treatment. It was reported here that the improvement in hardness improves the wear performance of the material due to cryo-treatment. Thus, from this work on cryogenic treatment of various polymers and their composites, it has been proved to be an effective technique for enhancing the wear performance of the material. However, it is also observed here that the potential of the treatment depended upon the filler type, geometry, orientation, quantity and their interaction with the polymers selected.

Besides this, efforts were also made to correlate the wear performance of cryo-treated polymers with other mechanical properties such as ultimate tensile strength and elongation at break. A clear correlation was observed between the mechanical properties and tribological performance by Indumathi et al. (1999b) after cryogenic treatment. In this study the influence of cryo-treatment on the mechanical and tribological properties of PTFE and composites was investigated. It was claimed here that the tribological performance improved only for those materials whose mechanical properties had either improved or did not show significant deterioration after cryogenic treatment. The composites whose mechanical properties deteriorated due to the treatment also demonstrated poor wear properties.

This additional experiment was performed in order to confirm whether some additional mechanisms were responsible for the observed changes in the mechanical behavior or not. It was observed that in case of the un-treated PTFE powder, the particles were quite separated from each other whereas, in case of cryogenically treated material, the particles were found to be closely packed. It was observed that the dense network of micro-fibrils originating from each particle was a remarkable and distinct feature of cryo-treated PTFE powder. Such micro-fibrils were rarely seen in case of un-treated powder. This typical behavior of agglomeration and excessive fibrillation of the PTFE powder is caused due to cryogenic treatment, which not only favours the formation of a dense structure but also enhances the material properties like tensile strength, hardness and wear performance.

Kumar, Sharma, & Ray (2008) reported the effect of cryogenic treatment on the interfacial interaction and debonding phenomenon of glass fibre reinforced epoxy composites. It was observed here that more amount of matrix residue adhered to the reinforced glass fibers for cryogenically treated specimens. This was credited to the increase in friction and adhesion between the fiber and the matrix due to the compression caused by contraction of the material, as the polymer contracts more than the glass fibre resulting in increased hardness. The SEM image clearly indicates that the shear value as well as the adhesion property of glass fibre reinforced epoxy composite increases after cryogenic treatment. The improvement in the shear value after cryogenic treatment was credited to the differential thermal contraction of the matrix during sudden cooling. This leads to the development of greater cryogenic compressive stresses which increases the resistance to de-bonding as a result of better adhesion by mechanical keying at the interface between fiber and the matrix (Ray, 2005b).

Pande, Peshwe, & Kumar (2012) demonstrated that the cryogenic treatment is an effective technique for enhancing the abrasive wear performance of the polyamide polymer studied herein. It was reported here that the increase in % crystallinity is the most important parameter responsible for the wear performance enhancement. The positive response of PA to the cryogenic treatment was credited to the less bulky structure of the polymer, which permitted easy chain motion at low temperatures. It was observed here that the crystallinity increases for the samples that showed minimum wear loss. The maximum improvement in the % crystallinity and wear performance was found for cryo-treatment at -185°C for 12 hours of soaking. This was claimed to be the optimum condition for the cryogenic treatment of PA. This enhancement in the wear performance of PA was attributed to the increase in % crystallinity, modification in microstructure, conversion of less stable phases into more stable phases and formation of complex phases.

Additionally, Pande and Peshwe (2015) has patented the optimum cryogenic treatment parameters for PBT composites reinforced with glass fibre and wollastonite in varying weight proportions (5, 10, 20, 30 wt. %). Based on the structural, mechanical and thermal characterization it is claimed that the cryogenic treatment shows maximum effectiveness for PBT composites at the condition of -185°C and 8 hours of soaking duration. It was reported here that the strength and the modulus of the composites increased appreciably when treated at -185°C. The theoretical modeling of the interphase also revealed a very good level of interfacial interaction, thereby indicating enhanced load transfer between the PBT matrix and filler. This patent also claims that, along with the treatment temperature, the soaking period is a decisive factor in order to determine the properties of the material. The XRD analysis indicated that structural modification of PBT crystals occur as a result of the conformational changes brought about by interfacial interaction as a result of cryogenic treatment. Similar to virgin polymers, it is reported here that the increase in crystallinity and the reduction in crystallite size after the treatments is the prime reason for increased strength and wear resistance of the cryo-treated composites. It was observed here that the ploughing and peeling-off of the composite surface suppress after the treatment. Thus, it is claimed here that on cryo-treatment, the increase in crystallinity, hardness, stiffness and especially the improved interfacial bonding, cumulatively enhances the wear performance of the composites.

Thus, it is clear that cryogenic treatment is an effective technique to improve the mechanical properties and the tribological performance of polymers and polymer composites. However, despite the advantages and commercial awareness of the treatment, the area of cryogenic treatment of polymer/MWCNT nano-composites is yet unexplored. The increasing use of these nano-composites in aerospace and automobile applications, demands for more understanding about their structure and the mechanisms involved in their enhanced properties. Though these nano-composites are used in areas where temperatures are not too high and are subjected to lesser loads than their metallic counterparts, these polymeric parts still are prone to continuous loading and wear. This gives rise to problems related to the strength of materials, performance of the components, safety and life of the part. Hence, the machine elements in such cases are expected to have better mechanical strength and abrasive resistance.

RESEARCH GAP AND EXPERIMENTAL DETAILS

It is expected that, similar to other polymer composite systems, cryogenically treating the polymer/MWCNT nano-composites will enhance their mechanical and tribological performance. The investigations of underlying mechanisms behind the property enhancement after cryogenic treatment of the polymer/MWCNT nano-composites and its difference from the micro filler reinforced polymer composites will assist in understanding the material behaviour for targeted applications. Keeping this in view, the objective of the present work is to study the mechanical and tribological properties of MWCNT reinforced PA and PBT nano-composites. This will help to generate data for polymeric systems that are actually being used in heavy duty applications demanding high strength and wear resistance. Also, it will help to understand and establish the science behind the cryogenic treatment of polymer/MWCNT nano-composites. Cryogenically treated nano-composites will be an excellent substitute for applications where reduction in weight is desirable along with sufficient strength and stiffness.

The study was planned with an aim of finding out the effect of cryogenic treatment on the polymer/MWCNT nano-composites. In accordance with the established fact that cryogenic treatment greatly influences the structure of the polymers, thus, enhancing their mechanical properties; the polymers were selected based on the applications for which they are used commercially. Polyamide-6 (PA-6) is extensively used in textile fiber for its scratch recovery nature. It is also a popular gear and bearing material and is also used for bumpers, locks and wipers in automobiles (Chopra, Deshmukh, Deshmukh, Gogte, & Peshwe, 2016). Polybutylene Terephthalate (PBT) is used in the sliding roof top assembly and door units for automobiles as well as for machine elements such as crank and gear for light duty applications. These are the areas where good abrasion resistance is desired from the component. The nano-composites were, thus, prepared by reinforcing fixed amount of MWCNT powder

[0.5 wt.% for PA and 0.3 wt.% for PBT as optimized from previous studies (Chopra et al., 2016; Chopra, Batthula, Deshmukh, & Peshwe, 2017c; Chopra, Deshmukh, Deshmukh, & Peshwe, 2018a; Chopra, Deshmukh, Deshmukh, Gogte, & Peshwe, 2018b)] via melt-compounding technique. These nanocomposites were then treated cryogenically in a controlled atmosphere using liquid nitrogen at -80 °C, -140 °C and -185 °C for varying soaking periods of 4, 8, 12, 16, 20 and 24 hours. The cryo-treated composites were then characterized for various structural and mechanical properties.

The tensile properties of the polymers and nano-composites were determined using a dog-bone shaped tensile specimen – Type I, having a gauge length of 65 mm and cross-sectional area of 13 mm × 3 mm. The testing was done at ambient conditions (Temp.: 20°C, Rel. Humidity: 55%) according to ASTM Standard D-638 using a universal testing machine (Make: INSTRON, Model: 4467) at a crosshead speed of 5 mm/min. Five test results were averaged and then reported. The wear performance of the polymers and nano-composites was evaluated at ambient temperature in a multi pass pin-on-disc configuration on a Wear and Friction monitor (Make: DUCOM, Model: TR-20LE-MI) under different load conditions of 20N, 60N and 100N. The pins with dimension 10 mm × 10 mm × 3 mm were used for experimentation purpose. The samples were first abraded on superfine SiC paper (grade 1500) for uniform contact, cleaned with acetone, dried and weighed prior to experimentation. The pin was continuously abraded for 3 minutes (i.e. multiple pass) against the SiC paper (grade 220) as counter-face at a speed of 200 RPM and track diameter of 80 mm. Three test results were averaged for each type of sample and then reported. The hardness of the polymer and nano-composites were measured against Shore D hardness scale using a Shore hardness tester (Make: STECH Engineers) according to ASTM standard D-2240 at test load of 4.5 kgs. Five hardness values were averaged and then reported.

XRD is a widely used technique for the determination of crystal structure and phase identification of materials. The X-ray diffraction patterns for MWCNT powders and composite samples were generated by using XRD machine (Make: PAN-alytical, Model: X'Pert Pro). Nickel filtered Cu Kα radiation (λ=1.5406 A°) generated at 45 kV and 40 mA is used for the angle (2θ) range from 10° to 100° using a scan step size and time per step of 0.01° and 20 sec respectively. Crystallite size measurement was done by using Scherrer's equation and the average data is reported. FTIR spectroscopy is an effective and most common spectroscopic technique for structural elucidation and compound identification. The transmittance measurement of different IR frequencies by a sample positioned in the path of an IR beam provides the information about the functional group and type of chemical bonding in a material. The changes in the intensity of a beam of infrared radiation as a function of wavenumber (cm-1) on interaction with the sample are measured using Perkin Elmer FTIR (Model: Spectrum One). The polymer and nano-composites were scanned against Zinc Selenide (ZnSe) plate as reference for wavenumber range of 500 cm^{-1} to 4000 cm^{-1} with 12 scan numbers.

The tensile fractured surfaces of the cryogenically treated and untreated polymers and nano-composites were investigated by Field Emission Scanning Electron Microscope (Make: JEOL, Model: JSM-7610F) at different magnifications. Prior to electron microscopy; the surfaces were sputter-coated with a thin platinum layer using auto fine coater (Make: JEOL, Model: JEC-3000FC) to make the surfaces conducting for obtaining images. The worn out surfaces of the cryogenically treated and untreated polymers and nano-composites were investigated by Scanning Electron Microscope (Make: JEOL, Model: 6380A) at different magnifications. Prior to this, the surfaces were sputter-coated with a thin palladium layer using auto fine coater (Make: JEOL, Model: JFC-1600) to make surface conducting.

RESULTS AND DISCUSSION

Crystal Structure of Cryo-Treated Polymers and Polymer/MWCNT Nano-Composites

It is an established fact that the low temperatures have the capacity to influence the structural properties of the polymers (Pande et al., 2012). Thus, the XRD analysis was done for cryogenically treated polymers and nano-composites. Figure 3 shows the XRD diffractograms for virgin PA cryo-treated at different temperatures for

Figure 3. XRD patterns for Virgin PA cryo-treated at (a) -80°C; (b) -140°C and (c) -185°C for varying soaking periods

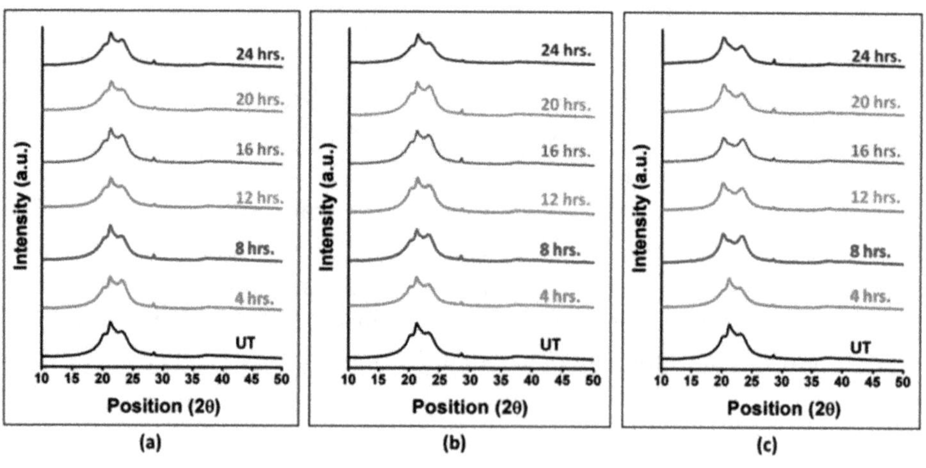

varying soaking period. The % crystallinity and the average crystallite size of all the samples are given in Table 2.

The XRD pattern of untreated virgin PA shows prominent diffraction peaks at 2θ positions of 20.13°, 23.44° and 28.52°, corresponding to *α-phase* and at 21.22° position corresponding to *γ-phase*, both monoclinic in structure. These peaks are the reflections from (200), (002,220), (210) and (100,010) planes, respectively (Mahmood, Islam, Hameed, & Saeed, 2013a; Chopra et al., 2016). The *α-phase* of PA is a more thermally stable phase consisting of sheets of hydrogen-bonded chains that are packed in an anti-parallel fashion, whereas the *γ-phase* is the less stable phase formed due to random hydrogen bonding between parallel chains (Chopra et al., 2016).

It can be seen from Figure 3 that the cryogenically treated virgin PA retains all the characteristics peaks mentioned above. For cryogenic treatment at -80°C and -140°C, there is no change in the crystal structure of PA, irrespective of the soaking period. However, the peak of *γ-phase* at 21.2° reduced when virgin PA was cryo-treated at -185°C beyond 4 hours of soaking period. The peak disappeared completely after 16 hours and the intensity of the *α-phase* peak at 20.1° increases significantly. This confirms that the virgin PA reacts to cryogenic treatment when the temperature is significantly lower and the material is exposed for longer duration. The increasing % crystallinity of PA after cryogenic treatment (as given in Table 2) also confirms the same. It can be observed form Table 2 that the crystallinity of virgin PA increases with increasing soaking periods at all treatment temperatures. There is a prominent increase of 29% in the crystallinity of cryo-treated PA as the soaking period increases form 8 hours to 12 hours at -185°C. This can be attributed

Table 2. Average crystallite size and % crystallinity of Virgin PA after cryogenic treatment

Soaking Period (hours)	-80°C		-140°C		-185°C	
	Avg. crystallite size (Å)	% crystallinity	Avg. crystallite size (Å)	% crystallinity	Avg. crystallite size (Å)	% crystallinity
UT	98	26.9	98	26.9	98	26.9
4 hrs.	122	30.7	112	32.4	99	32.8
8 hrs.	119	31.4	110	34.2	99	39.2
12 hrs.	117	33.0	110	37.3	93	50.7
16 hrs.	116	34.9	103	39.7	87	51.1
20 hrs.	116	35.5	100	40.2	81	54.1
24 hrs.	115	36.2	99	40.9	78	54.7

to the fact that during cryogenic treatment, the extremely low temperatures retards the molecular movements at an atomic level which increases the internal molecular bonding energy and hence induces a structural balance within the material (Pande, 2012, p. 81). The prolonged soaking periods at such low temperature stimulates the less stable *γ-phase* and gives sufficient time to rearrange into a more stable *α-phase*, thus changing the crystal structure of virgin PA from *partial-α/partial-γ* to complete *α-structure*.

The XRD diffractograms for PA/MWCNT nano-composites cryo-treated at different temperatures for varying soaking period is shown in Figure 4, while the % crystallinity and the average crystallite size of all the samples are given in Table 3. The comparison between the XRD patterns of untreated virgin PA (Figure 3) and untreated PA/MWCNT nano-composite (Figure 4) shows that the characteristic peaks for *α-phase* of PA were observed in case of untreated PA/MWCNT nano-composites also, but the peak for *γ-phase* disappeared after MWCNT addition. This indicates that the nano-reinforcing tendency of the MWCNTs assists in improved nucleation and crystallization of PA matrix, thereby changing the crystal structure of PA matrix from *partial-α/partial-γ* into complete *α-structure* (Chopra et al., 2016). However, it is clear from Figure 4 that the crystal structure of PA/MWCNT nano-composites remained unaffected after cryogenic treatment at different temperatures and varying soaking periods.

It can be observed from Table 3 that, unlike cryo-treated virgin PA, the % crystallinity of PA/MWCNT nano-composites reduced initially for smaller soaking periods and later increased. The crystallinity of PA/MWCNT nano-composite cryo-

Figure 4. XRD patterns for PA/MWCNT nano-composites cryo-treated at (a) -80°C; (b) -140°C and (c) -185°C for varying soaking periods

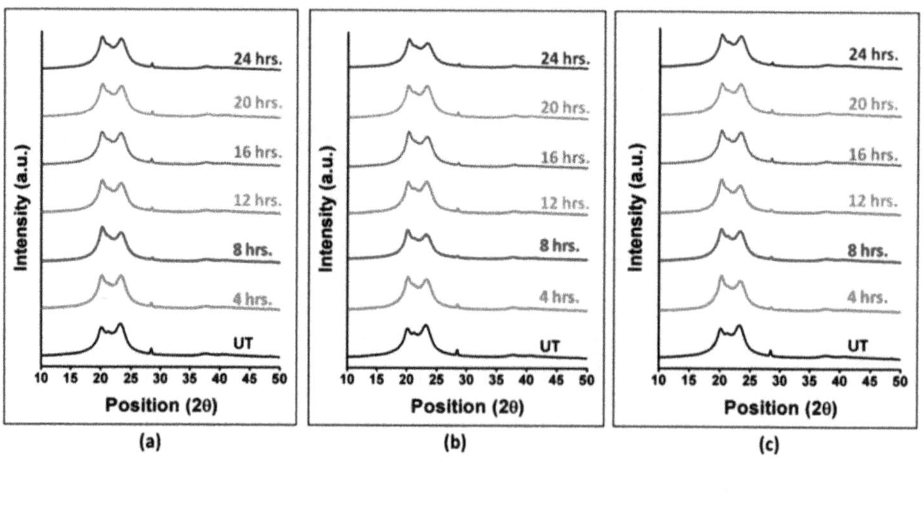

Table 3. Average crystallite size and % crystallinity of PA/MWCNT nano-composites after cryogenic treatment

Soaking Period (hours)	-80°C		-140°C		-185°C	
	Avg. crystallite size (Å)	% crystallinity	Avg. crystallite size (Å)	% crystallinity	Avg. crystallite size (Å)	% crystallinity
UT	36	54.6	36	54.6	36	54.6
4 hrs.	58	45.1	47	49.1	38	49.3
8 hrs.	57	45.5	40	51.0	38	51.4
12 hrs.	54	46.7	39	52.7	37	55.3
16 hrs.	54	47.5	37	55.3	35	57.4
20 hrs.	53	48.0	37	57.1	33	60.3
24 hrs.	52	49.1	36	58.0	33	60.4

treated at -80°C for 24 hours remained lower than the untreated material, while that for cryo-treated at -140°C, the crystallinity increased beyond the untreated material after 16 hours of soaking period. Similarly, for PA/MWCNT nano-composites cryo-treated at -185°C, the crystallinity increased after 12 hours of soaking period. The maximum increase in % crystallinity of PA/MWCNT nano-composites amounts to ~11% at -185°C for 24 hours of soaking period.

The XRD diffractograms for virgin PBT cryo-treated at different temperatures for varying soaking period is shown in Figure 5, while the % crystallinity and the average crystallite size of all the samples are given in Table 4.

Figure 5. XRD patterns for virgin PBT cryo-treated at (a) -80°C; (b) -140°C and (c) -185°C for varying soaking periods

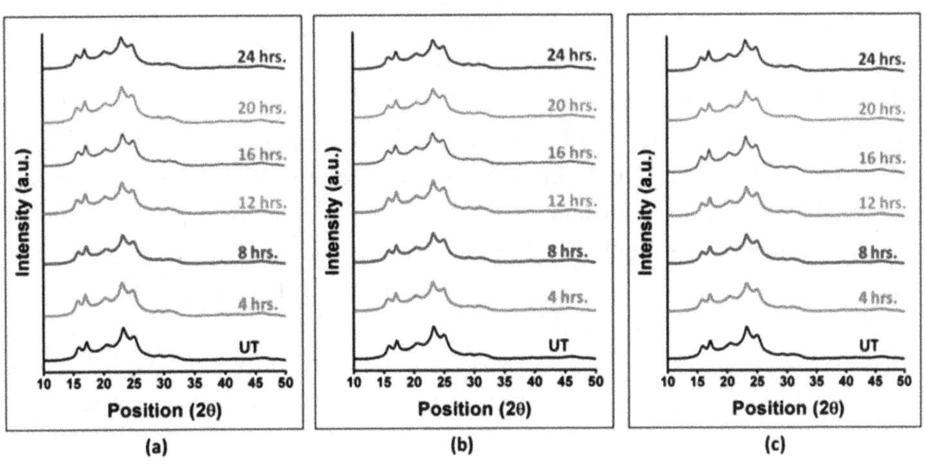

Table 4. Average crystallite size and % crystallinity of virgin PBT after cryogenic treatment

Soaking Period (hours)	-80°C		-140°C		-185°C	
	Avg. crystallite size (Å)	% crystallinity	Avg. crystallite size (Å)	% crystallinity	Avg. crystallite size (Å)	% crystallinity
UT	239	47.9	239	47.9	239	47.9
4 hrs.	258	47.9	215	49.2	196	50.9
8 hrs.	248	50.2	213	50.1	196	53.5
12 hrs.	235	50.3	209	50.5	158	54.2
16 hrs.	223	50.7	207	51.0	156	54.5
20 hrs.	208	50.7	180	51.6	169	54.1
24 hrs.	206	49.6	167	53.0	168	50.8

From the XRD pattern of virgin PBT, strong diffraction peaks are observed at 2θ positions of 15.61°, 16.97°, corresponding to the *β-crystallite* form of PBT and at 20.30°, 23.15° and 24.94°, corresponding to the *α-crystallite* form of PBT, both with a triclinic configuration. These peaks are the reflections from (011), (010), (110), (100) and (111) planes respectively (Fakirov, 2005 and Kim, 2011). It can be inferred from Figure 5 that all the characteristic peaks of untreated virgin PBT were observed after cryogenic treatment also, irrespective of the treatment temperature and soaking period. This indicates that, unlike virgin PA, the cryogenic treatment of virgin PBT does not change its crystal structure in terms of polymeric phases present in the material. However, the improvement in the % crystallinity and average crystallite size given in Table 4 establish the fact that some type of crystallographic rearrangement is taking place after cryogenic treatment.

Akin to cryo-treated virgin PA, the % crystallinity of virgin PBT increases with increasing soaking periods at all treatment temperatures. However, this increase is not as noteworthy as that for PA, specifically at -80°C and -140°C. A maximum increase of 14% in the crystallinity of virgin PBT is observed for cryo-treatment at -185°C for the soaking period of 12-20 hours. This is because virgin PA is highly susceptible to structural transformations as compared to PBT. The reason for this is the higher irregularity present in PA beforehand. A considerable amount of *γ-phase* is inherently present in untreated virgin PA, as indicated by the highest peak at 21.2° in XRD pattern of untreated PA given in Figure 3. This imparts a lower crystallinity to the material i.e. 26.9% (Table 2) under present study. As the *γ-phase* is thermally more unstable it imparts more probability for crystallographic rearrangement even for a small external stimulus, such as prolonged exposure at cryogenic temperatures. Thus, it was noted here that the response of virgin PA to cryogenic treatment is more profound than virgin PBT. However, it is imperative to note here that the improvements in the crystal structure and crystallinity of PA as well as PBT both occurs at a cryogenic treatment temperature of -185°C and for longer soaking periods. This is because of the energy required by the material to attain structural balance at such low temperatures

Figure 6 shows the XRD diffractograms for PBT/MWCNT nano-composites cryo-treated at different temperatures for varying soaking period. The % crystallinity and the average crystallite size of all the samples are given in Table 5. The XRD pattern of untreated PBT/MWCNT nano-composite indicates that all the characteristic peaks of untreated virgin PBT remains almost unchanged with the incorporation of MWCNTs. This indicates that the addition of MWCNTs in PBT matrix does not affect the crystal structure of the nano-composite. It can be observed from Figure 6 that, akin to cryo-treated virgin PBT, all the characteristic peaks of untreated PBT/MWCNT nano-composites were retained after cryogenic treatment. This indicates that the response of PBT/MWCNT nano-composites to the cryogenic treatment is same as that of virgin PBT. However, the presence of MWCNTs also plays a key role on the crystallographic arrangement in the material. Thus, the crystallinity of cryo-treated PBT/MWCNT nano-composites shows a similar trend to that of PA/MWCNT nano-composites, as evident from Table 5.

The unusual behaviour of the nano-composites can be due to the difference in the thermal expansion coefficients of the polymer matrix and the MWCNT filler. The sudden drop in temperature subjects the material to differential contraction (Pande,

Figure 6. XRD patterns for PBT/MWCNT nano-composites cryo-treated at (a) -80°C; (b) -140°C and (c) -185°C for varying soaking periods

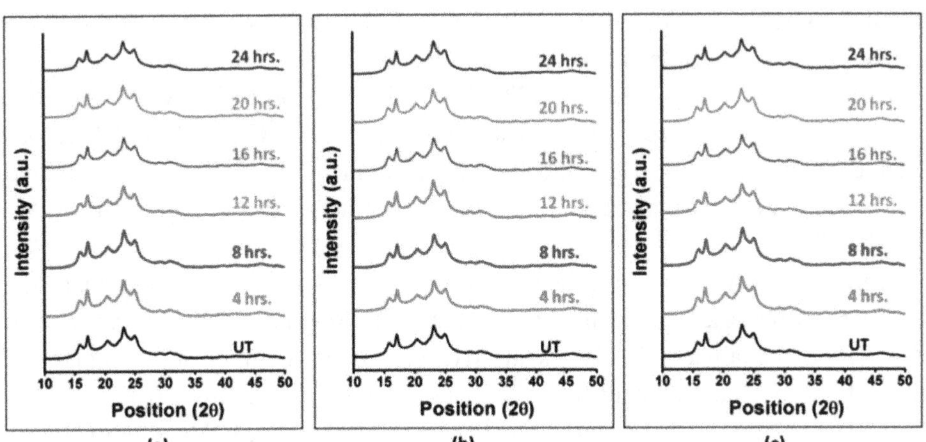

Table 5. Average crystallite size and % crystallinity of PBT/MWCNT nano-composites after cryogenic treatment

Soaking Period (hours)	-80°C		-140°C		-185°C	
	Avg. crystallite size (Å)	% crystallinity	Avg. crystallite size (Å)	% crystallinity	Avg. crystallite size (Å)	% crystallinity
UT	91	50.8	91	50.8	91	50.8
4 hrs.	98	44.4	88	48.3	69	51.5
8 hrs.	97	45.6	85	49.5	66	51.5
12 hrs.	97	47.0	84	49.7	65	52.2
16 hrs.	94	47.1	82	50.8	59	55.4
20 hrs.	87	47.2	81	51.1	71	53.9
24 hrs.	87	48.0	75	48.6	72	52.2

2012, pp. 163-164). Due to lower thermal coefficient of matrix, it will contract more as compared to the filler. This will distort the matrix structure and cause embrittlement of the sample. When held at low temperature for smaller durations, the molecular rearrangement encouraged by cryogenic treatment happens at a slower rate and does not get sufficient time to overcome the structural inhomogeneities caused by differential contraction. However, at prolonged soaking periods, there is sufficient time for the material to overcome this difference sustained by the sample as well as rearrange itself into a more stable crystal structure.

Therefore, it was observed from the XRD analysis that for both virgin polymers as well as the MWCNT nano-composites; the crystallite size increased at first and then reduced as the soaking period increases. In case of virgin PA and PA/MWCNT nano-composites, the crystallite size remained larger than the untreated material itself for all the samples cryo-treated at -80°C and -140°C, however, when cryo-treated at -185°C, the crystallite size reduced more significantly. On the other hand, the crystallite size of virgin PBT and PBT/MWCNT nano-composites was larger than untreated material when cryo-treated at -80°C. But at -140°C and -185°C there was a reduction in crystallite size, which became prominent at -185°C. This is credited to the systematic re-arrangement of polymeric segments after cryogenic treatment (Pande, 2012).

During the polymerization process and consecutive solidification of the material, not every polymeric chain will grow to the same length, resulting in an uneven distribution of chain lengths. This leads to extensive intertwining and entanglement of neighboring chain molecules. Any chain disorder or misalignment will result in an amorphous region, thereby affecting the crystallinity of the polymer (Gowariker,

Viswanathan, & Sreedhar, 1986). Secondly, the molecular chemistry also influences the ability of a polymer to crystallize. In case of linear polymers, crystallization is easily accomplished whereas, the presence of side branches or bulkier groups restricts the ordered chain alignment in polymers. Infact, excessive branching, cross linking and/or presence of much bulkier groups, are responsible for inducing amorphous nature, as it prevents the polymer chains from aligning into a crystalline structure (Pande, 2012, pp. 92-93).

On cryogenic treatment, as the material is brought to the low temperature gradually, the randomly oriented chains get aligned, resulting in increased volume fraction of crystalline regions. This is well demonstrated by the increasing crystallinity of the virgin polymers as well the MWCNT nano-composites. Given sufficient time at such low temperatures, i.e. higher soaking periods, the material not only achieves structural equilibrium but also forms smaller sized crystalline segments. Thus, on cryo-treating the material at longer soaking periods causes the crystallites to contract and get reduced in size, thereby reflecting reduced crystallite size for virgin polymers and MWCNT nano-composites cryo-treated at -185°C. Thus, in general it is observed that on cryogenic treatment there is increase in % crystallinity and reduction in the crystallite size of the virgin polymers as well the polymer/MWCNT nano-composites.

It is well known that the localized intermolecular rearrangement at low temperatures results in relaxation, which considerably depends on the morphology of the polymer (Ray, 2005a). The polymers which are able to change their main chain bond angles are reported to excel in cryogenic properties (Indumathi et al., 1999a; Pande, 2012). These polymers are flexible and may undergo deformation even when their segmental motions are frozen due to cryogenic treatment. In order to investigate this, the FTIR of cryogenically treated virgin polymers and their nano-composites was done.

Structural Analysis of Cryo-Treated Polymers and Polymer/MWCNT Nano-Composites by FTIR

The FTIR spectra of untreated and cryo-treated virgin PA are shown in Figure 7 (a)-(c). According to previous literature (Xu, jia, Wu, Han, & Meek, 2006; Motozuka et al., 2013; Vidal, 2013); the characteristic peaks in untreated virgin PA can be assigned as: amide group aliphatic at 3837 and 3751 cm^{-1}, N-H stretch at 3413 cm^{-1}, amide I C=O stretching around 1649 cm^{-1} and amide II vibration at 1557 cm^{-1} i.e. combination of N-H bend and C-N stretch.

It was observed that after cryogenic treatment, the FTIR spectra of virgin PA showed emergence of a new peak while retaining all but one of the characteristic peaks of untreated PA. These changes were same for all the cryo-treated samples, irrespective of the treatment temperature and soaking period, thus, indicating cryo-

Figure 7. FTIR patterns of virgin PA cryo-treated at (a) -80°C; (b) -140°C and (c) -185°C for varying soaking periods and (d) Schematic representation of PA chain structure

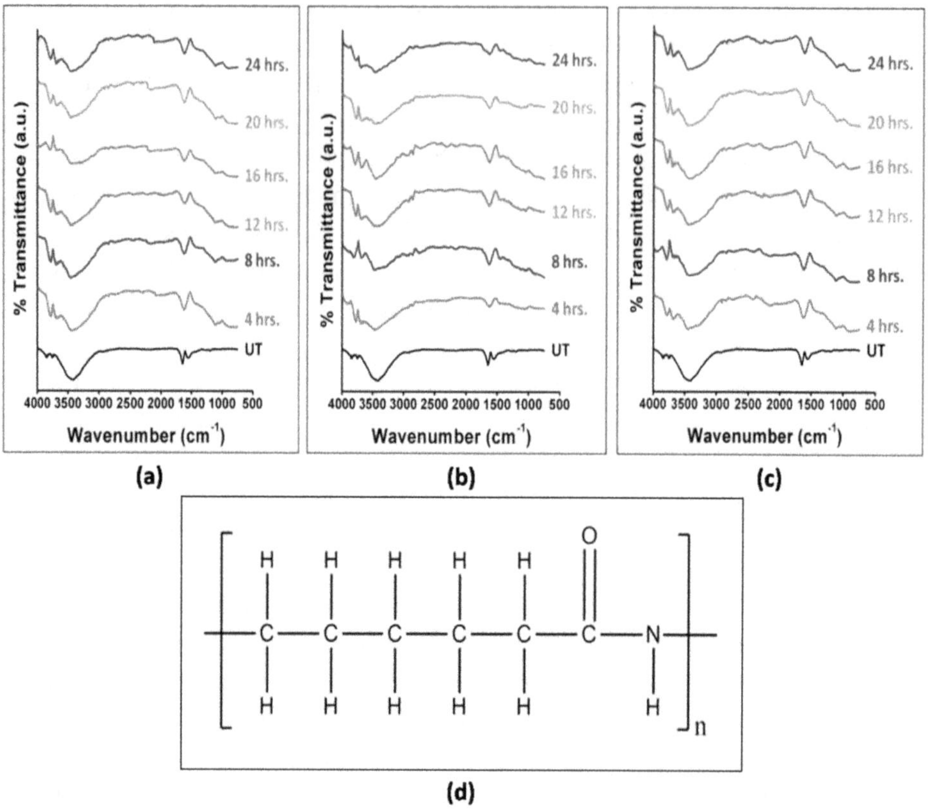

structural modifications in the chemistry of virgin PA. The significant change observed in PA after cryogenic treatment is the disappearance of the amide II vibration at 1557 cm^{-1}, along with the broadening of amide I C=O stretch peak at 1649 cm^{-1}. In addition, a new peak appears at approx. 1130 cm^{-1} corresponding to the C-CO stretch in the *α-phase*. The appearance of this peak confirms the claim that during cryogenic treatment, the volume fraction of *α-phase* crystallites increases by forming at the expense of *γ-phase* crystallites by intrasheet hydrogen bonding mechanism, also reported in previous studies (Ahmed, Gotoha, & Ohkoshia, 2005;Pande, 2012, p. 114). The low temperature during cryogenic treatment contracts the PA molecules and forms longer chains owing to weak hydrogen bonding between the consecutive sheets. As it was observed from Table 2, the cryo-treated structure of PA consists of smaller crystallites.

This indicates lesser distance between the parallel PA sheets when forming into smaller sized crystallites. The formation of longer chains arranged in a compact crystalline pattern with increased proximity imposes motion restriction on the amide group present at the chain open end in PA, the structure of which is shown in Figure 7 (d). This causes the disappearance of amide II peak, which is basically the combination vibration of N-H bend and C-N stretch. This also leads to a down-shift in the position of C=O stretch peak from 1649 cm^{-1} to 1624 cm^{-1}. Moreover, the cryo-treated virgin PA reflects an overtone of CH vibrations (i.e. CH_2 stretch, CH_2 rocking, C-H bend, etc.) within 2000 cm^{-1} to 3000 cm^{-1} wavenumber region, which results due to more regular alignment of long chains causing in-plane long chain vibrations (Pande, 2012, pp. 113-114). Thus, these FTIR outcomes support the results obtained by XRD for the disappearance of γ-*phase* after cryogenic treatment.

The FTIR patterns for untreated and cryo-treated PA/MWCNT nano-composites are shown in Figure 8 (a)-(c). With the addition of MWCNTs, new peaks can be seen for untreated PA/MWCNT nano-composite as compared to the peaks mentioned above for untreated virgin PA. It is observed from the figure that there is splitting of the tip of the broad N-H stretch peak at 3409 cm^{-1}. This split peak appears around 3330 cm^{-1} and is an effect of intermolecular bonding between PA and MWCNT. This is assigned as O···NH stretch vibration wherein it confirms the bonding of MWCNTs at amide group sites (Mahmood, Islam, Hameed, Saeed, & Khan, 2013b). The schematic representation of MWCNT bonding at the potential sites along the PA chain is shown in Figure 8 (d).

It is well known that the presence of N-H group at the chain end of PA forms a potential site for the filler to interact (Lin, Hill, Bentley, Allard, & Sun, 2003; Zhang, Shen, Phang, & Liu, 2004). The interaction of the MWCNTs with the open ends of PA chains causes physical motion restriction to amide N-H group, as reflected by the reduction in the intensity of N-H stretch peak after MWCNT addition. An additional peak at 1447 cm^{-1} corresponding to CH_2 scissoring vibration and another at 2931 cm^{-1} corresponding to asymmetric CH_2 stretch can be seen in case of the PA/MWCNT nano-composite. The apperance of these peaks indicates the intercalation of MWCNTs within adjacent PA chains (Lin et al., 2003). Two peaks at 2303 cm^{-1} for C-H stretch and 3661 cm^{-1} for hydrogen bonding between MWCNT and PA can also be seen. This points out towards the fact that the MWCNTs form hydrogen bond at the oxygen molecule in PA chain and also along the chain open ends, thus, causing the stretching of CH_2 backbone (Zhang et al., 2004). This indicates good extent of adhesion and/or reactive coupling between PA chains and MWCNT (Kodgire et al., 2006).

After cryogenic treatment, the PA/MWCNT nano-composites reflected a new peak along with the disappearance of few existing peaks in the untreated material, as clearly evident from Figure 8. Akin to cryogenically treated virgin PA, the changes in

Figure 8. FTIR patterns of PA/MWCNT nano-composites cryo-treated at (a) -80°C; (b) -140°C and (c) -185°C for varying soaking periods and (d) Schematic representation of MWCNT bonding at the potential sites along the PA chain

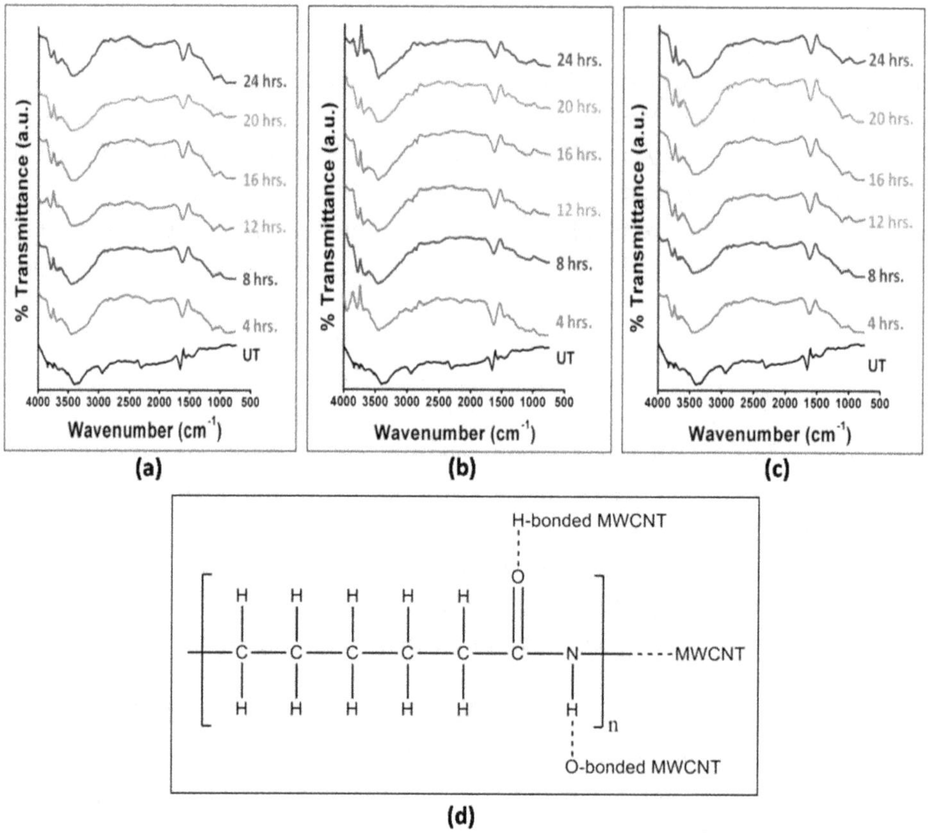

PA/MWCNT nano-composites were same for all the cryo-treatment temperature and soaking periods. Also, the major changes reflected by the cryo-treated PA/MWCNT nano-composites were very similar to that of cryo-treated virgin PA. The amide II vibration in untreated PA/MWCNT nano-composite at 1562 cm^{-1} disappeared after cryogenic treatment, along with the broadening and down shift of amide I C=O stretch peak from 1653 cm^{-1} to 1624 cm^{-1}. The peak for C-CO stretch in *α-phase* also appears in the nano-composites at approx. 1114 cm^{-1}. It was seen from XRD analysis that the crystallinity of the PA/MWCNT nano-composite increases after cryogenic treatment. This can be attributed to the structural balance attained by the material at low temperature, which in turn forms compact yet longer chains with small crystallites. This ultimately imposes motion restriction onto the amide group, analogous to that of cryo-treated virgin PA, thus, reflecting similar behavior in the vibrations of amide I and amide II groups in PA/MWCNT nano-composites. However, the down-shift of C-CO stretch from 1130 cm^{-1} for treated virgin PA to 1114 cm^{-1} in case of treated PA nano-composites indicates that the presence of MWCNTs plays a significant role in the formation of stable and compact crystal structure during cryogenic treatment as well. It was observed here that after cryogenic treatment the peak of hydrogen bonding between PA and MWCNT up-shifts from 3661 cm^{-1} to 3679 cm^{-1}. This is due to the compaction of longer PA chains during cryo-treatment. As the distance between parallel PA chains reduces during the formation of smaller crystallites, the MWCNTs attached along the PA chains, as seen in Figure 8 (d), are accommodated within the gap, thus, rotating the weaker covalent hydrogen bond between PA and MWCNT. This reflects in the form of up-shift in the peak at 3661cm^{-1} and also the down-shift of C-CO stretch peak.

Furthermore, it was observed that the peak for O···NH bond between PA and MWCNT at 3330 cm^{-1} disappeared after cryogenic treatment, while the broad hump of N-H stretch at 3409 cm^{-1} up-shifted to 3447 cm^{-1}. This is credited to motion restriction imposed on the amide group after cryogenic treatment. The compaction of material at molecular level and the systematic arrangement of PA chain in parallel configuration restrict the out-of-plane motions of the amide group, such as N-H bend in amide II vibrations. However, the physical presence of MWCNT in the vicinity of the amide group, as represented in Figure 8 (d), causes the in-plane N-H stretch vibration to overlap the weaker O···NH bond between PA and MWCNT, thereby reflecting as a cumulative vibration in the form of a broad hump at 3447 cm^{-1}. Finally, the disappearance of asymmetric CH_2 stretch at 2931 cm^{-1} and C-H stretch at 2303 cm^{-1} in PA/MWCNT nano-composites further confirms the claim that the cryogenic treatment helps in regular alignment of long chains, thereby, restricting out-of-plane vibrations.

The FTIR spectra of virgin PBT before and after cryogenic are shown in Figure 9 (a)-(c). The characteristic peaks observed for virgin PBT are: free OH at 3823 cm^{-1}, molecular OH vibrations at 3426 cm^{-1} and 3245 cm^{-1}, asymmetric CH_2 stretch at 2939 cm^{-1}, C-H stretch at 2336 cm^{-1}, C=O stretch in ester groups at 1732 cm^{-1}, C=C stretch/bending in benzene aromatic ring at 1361 cm^{-1} and O=C-O stretch in esters at 1213 cm^{-1} (Koenig, 1999; Pavia, Lampman, Kriz, & Vyvyan, 2009; Deshmukh, Peshwe, Pathak, & Ekhe, 2011). It was observed here that after cryogenic treatment a new peak appeared in virgin PBT whereas many of the existing peaks mentioned above disappeared. This confirmed some cryo-structural modifications in the chemistry of virgin PBT.

It was observed here that the peaks of free O-H and molecular O-H at 3823 cm^{-1}, 3426 cm^{-1} and 3245 cm^{-1} disappear after cryogenic treatment. In addition, the peaks for C=C stretch/bending in benzene aromatic ring at 1361 cm^{-1} and O=C-O stretch in esters at 1213 cm^{-1} also disappears as a new bond appears at 1130 cm^{-1}

Figure 9. FTIR patterns of virgin PBT cryo-treated at (a) -80°C; (b) -140°C and (c) -185°C for varying soaking periods and (d) Schematic representation of PBT chain structure

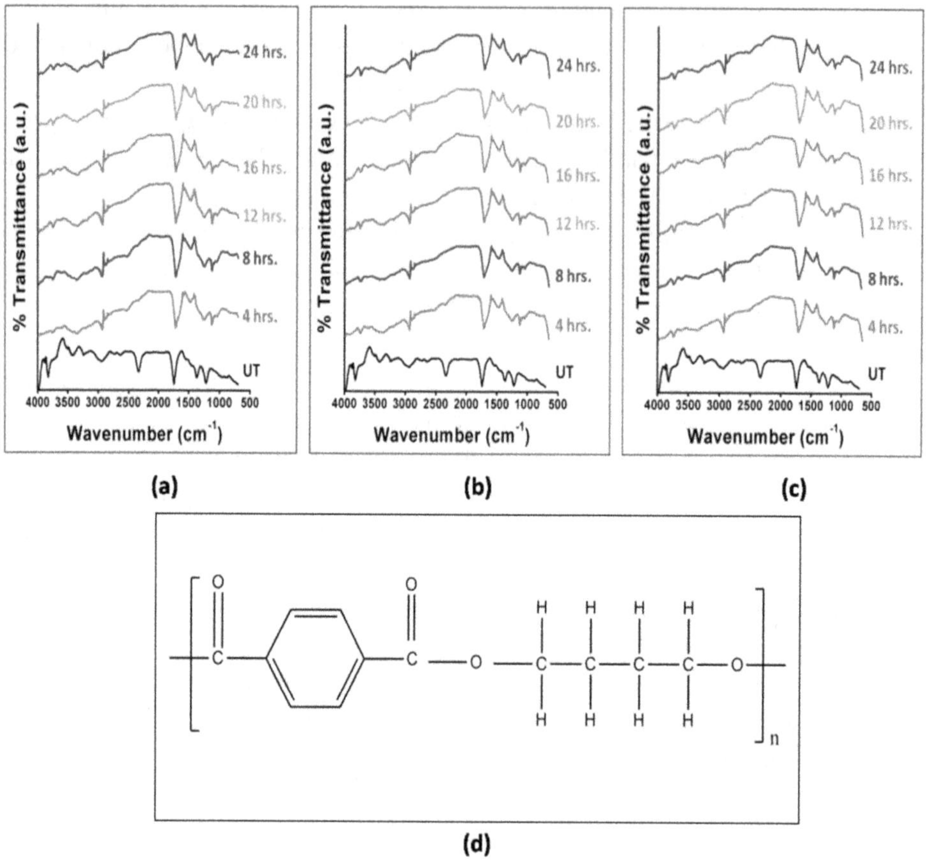

corresponding to the C-O-C stretch along the ester group on either side of benzene ring in PBT structure, as sown in Figure 9 (d). The formation of longer chains at cryogenic temperatures and consecutive contraction entails motion restriction to the benzene ring (Pande, 2012, pp. 118-119). This not only restricts the aromatic vibrations of benzene ring but also obstructs the O=C-O stretch in ester group. However, the long chain configuration and densification of PBT chain arrangement induces stress onto the ester group from the side of carbon backbone, which in turn, generates the C-O-C stretch bond along the ester group and carbon backbone. This further causes down-shift in the position of C=O stretch in ester from 1732 cm^{-1} to 1717 cm^{-1} as it is sensitive to benzene ring saturation. Furthermore, it was observed that the prominent peak of C-H stretch at 2336 cm^{-1} disappears completely for cryo-treated virgin PBT, which is credited to the decreasing distance between parallel PBT chains as the material contracts at a molecular level. Also, similar to virgin PA, cryo-treated PBT reflects CH_2 scissoring at 1469 cm^{-1} and CH_2 twisting at 1253 cm^{-1} due to the regular alignment of long chains causing the in-plane long chain vibrations after cryogenic treatment (Pande, 2012, pp. 118-119).

Figure 10 (a)-(c) shows the FTIR patterns for PBT/MWCNT nano-composites before and after cryogenic treatment. When comparing the FTIR pattern of untreated PBT/MWCNT nano-composite with that of untreated virgin PBT, it was noted that the characteristic peaks of virgin PBT reflected in PBT/MWCNT nano-composites also. Additionally, a new peak appears at 3748 cm^{-1} which signifies the hydrogen bonding between PBT and MWCNT at ester group, as depicted in Figure 10 (d).

Figure 10. FTIR patterns of PBT/MWCNT nano-composites cryo-treated at (a) -80°C; (b) -140°C and (c) -185°C for varying soaking periods and (d) Schematic representation of MWCNT bonding at the potential sites along the PBT chain

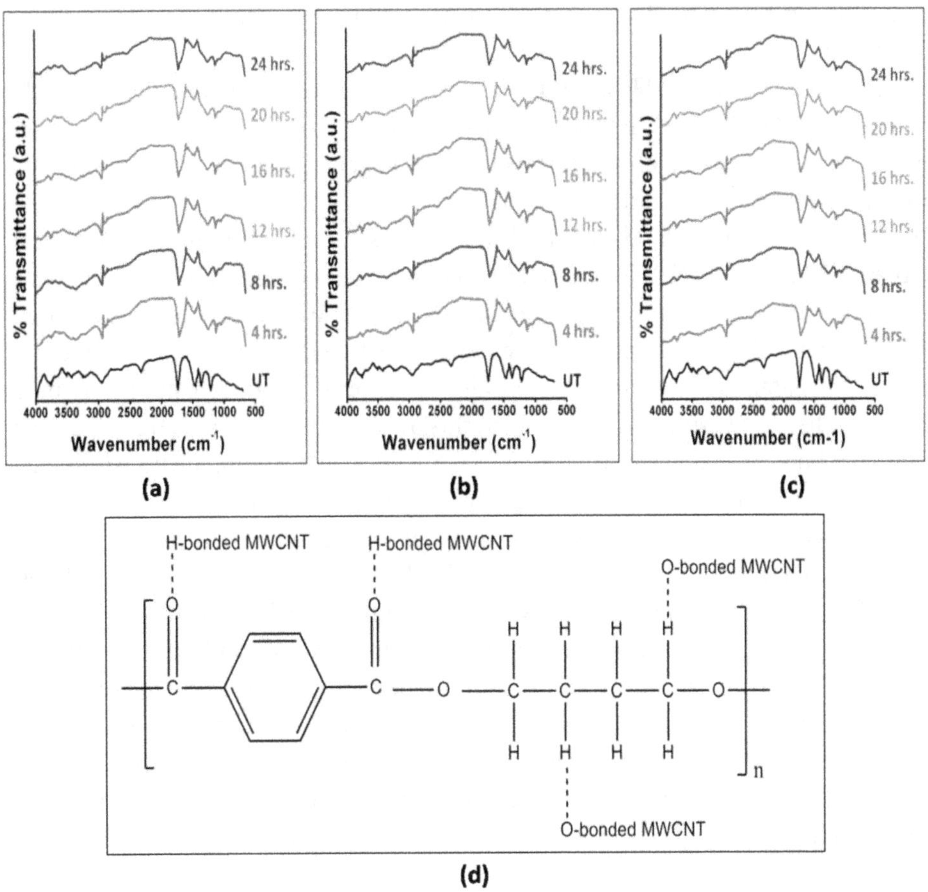

It is known that the graphitic struture of carbon atoms in MWCNTs tend to bind towards oxygen atoms during processing (Motozuka et al., 2013). Thus, the C=O ester groups present in PBT molecules create a potential site for the MWCNTs to bond via H···O=C with the PBT matrix. There is intermolecular bonding also taking between PBT and MWCNT which is reflected as a new peak at 1130 cm^{-1} and can be assigned as O···CH_2 bonding along PBT backbone. This points out towards the fact that with hydrogen bonding at ester group and MWCNT intercalation with PBT backbone, the bonding between MWCNTs and PBT improves.

Further, it was observed from Figure 10 that the changes in the FTIR spectra of cryogenically treated PBT/MWCNT nano-composites are very similar to that of virgin PBT. For PBT/MWCNT nano-composites also, the molecular O-H at 3426 cm^{-1} and 3223 cm^{-1} disappeared after the treatment. Moreover, akin to the cryo-treated virgin PBT, the peaks for C=C stretch/bending in benzene aromatic ring at 1364 cm^{-1} and O=C-O stretch in esters at 1208 cm^{-1} disappears along with the appearance of C-O-C stretch at 1124 cm^{-1} and further down-shift of C=O stretch in ester from 1732 cm^{-1} to 1715 cm^{-1}. Similarly, the C-H stretch at 2316 cm^{-1} disappears after cryogenic treatment of PBT/MWCNT nano-composites, while the CH_2 twisting appears at 1260 cm^{-1}.

Similar to cryo-treated PA/MWCNT nano-composites, the major changes in the chemistry of PBT/MWCNT nano-composites after cryogenic treatment is governed by the polymer matrix only. However, the cryogenic treatment does affect the bonding between PBT matrix and MWCNT. It was observed here that for all the PBT/MWCNT nano-composite samples, the intensity of the hydrogen bonding peak between PBT and MWCNT at 3748 cm^{-1} reduces after cryogenic treatment. This is credited to motion restriction imposed onto the aromatic benzene ring present along the PBT chain. The systematic arrangement of PBT chains and densification of the crystallites entails restriction to the bulkier benzene ring and the ester group in its vicinity. The C-O-C stretch arising due to this overlaps the vibrations from the weak hydrogen bonding between PBT and MWCNT, thereby reducing its transmittance.

It is a well known fact that cryogenic treatment motivates structural stability in the polymer by forming long chains and promoting intermolecular as well as intramolecular rearrangements (Ahmed et al., 2005). This was indeed observed from the FTIR analysis of the cryogenically treated virgin PA and PBT polymer. However, in present study the reason for cryo-structural modifications in both PA and PBT are different. In present study, virgin PA polymer is more susceptible to structural transformations as compared to PBT. The reason for this is the more irregularity present in PA beforehand. A considerable amount of *γ-phase* is inherently present in virgin PA, as indicated by the highest peak at 21.2° in XRD pattern of untreated PA given in Figure 3. This imparts a lower crystallinity to the material i.e. 26.9% (Table 2). As the *γ-phase* is thermally more unstable it imparts more probability for crystallographic rearrangement even for a small stimulus, such as low temperatures attained during cryogenic treatment. This is well reflected by the XRD analysis of the polymers and its nano-composites after the treatment.

The crystal structure of PA changes drastically, while for PBT it remains almost same. Nonetheless, the improvements in crystallinity and crystallite size of the polymers hinted towards beneficial cryo-structural modifications. The tendency of the polymer to form longer chains increases at low temperatures. This, along with the contraction of material at molecular level, results in the formation of a denser, stable and uniform structure. However, the formation of this structure differs for PA and PBT. In case of PA, the volume fraction of *α-phase* crystallites increases at the expense of *γ-phase* crystallites. This transformation occurs due to the intrasheet hydrogen bonding between adjacent PA chains. The simultaneous contraction due to low temperature and sufficient time for the formation of longer chains gives rise to a superior crystal structure in PA after cryogenic treatment. On the other hand, the PBT exhibits less randomness as compared to PA under present study. Thus, there is no significant molecular rearrangement in PBT as evident from XRD also. However, the cryo-treated PBT demonstrates enhanced crystallinity due to the physical rearrangement of the long chains in a denser configuration, unlike the structural modifications undergone by cryo-treated PA. The polymer/MWCNT nano-composites also exhibit this behavior upon cryogenic treatment.

Tensile Properties of Cryo-Treated Polymers and Polymer/MWCNT Nano-Composites

The mechanical properties of the material play a crucial role, as these properties decide the behavior of the material during actual applications. It is established above that the cryogenic treatment results in a refined, dense and uniform structure in virgin polymers and their MWCNT nano-composites under present study. In order to verify the effectiveness of cryogenic treatment parameters on the tensile properties of the material; the enhancement in the tensile strength and % elongation of the cryo-treated samples is studied and discussed in detail. Figure 11 illustrates the tensile strength and % elongation of virgin polymers PA and PBT cryo-treated at different temperatures with varying soaking period.

It can be noted from the figure that the tensile strength of both the polymers remains almost same or increases marginally when cryogenically treated at -80°C and -140°C. However, for the samples cryo-treated at -185°C, the tensile strength increases appreciably. The strength of virgin PA increases by approx. 14% for 24 hours of soaking period, while the increase in the strength of virgin PBT is 20% for 12 hours of soaking period, when both are cryo-treated at -185°C. On the other hand, it can be seen that the % elongation of PA increased significantly after cryogenic

Figure 11. (a) Tensile strength and (b) % elongation of cryo-treated virgin PA and (c) Tensile strength and (d) % elongation of cryo-treated virgin PBT

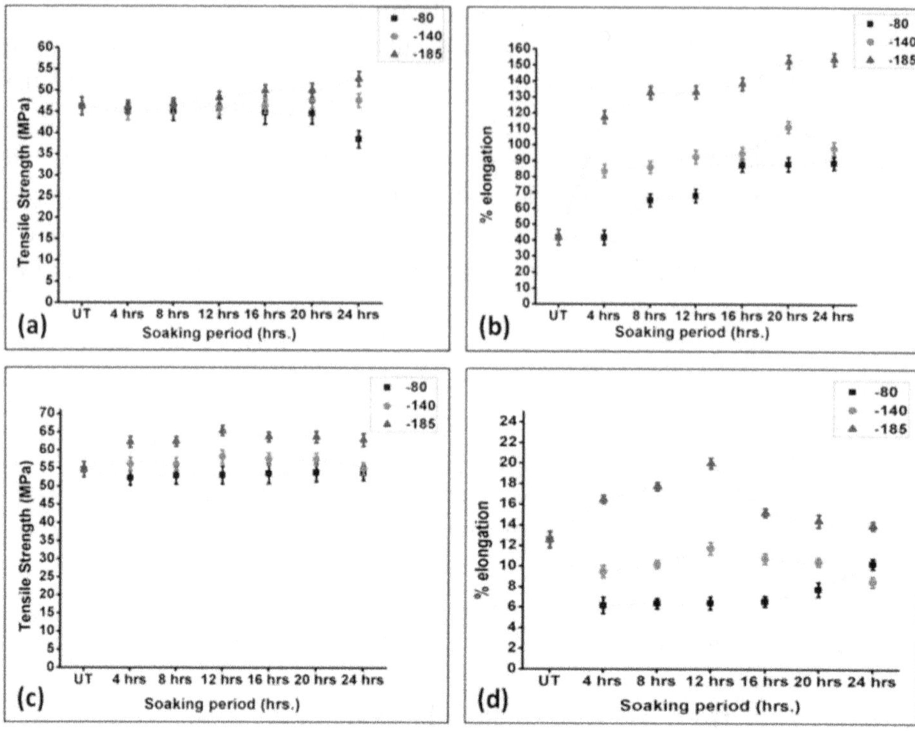

treatment, irrespective of the treatment temperature or soaking period. However, the response of % elongation of PBT was similar to that of its tensile strength, where the properties increased at -185°C of cryogenic treatment. The maximum increase in % elongation of virgin PA is by 267% for 24 hours of soaking period, while for virgin PBT it was by approx. 59% for 12 hours of soaking period, both again at -185°C. The improvement exhibited by the tensile properties of the cryo-treated polymers is analogous to the crystal structure modifications explained above. The polymer samples that demonstrate enhanced crystallinity and smaller crystallite size responds significantly to the tensile load, thereby, improving its strength and elongation. Thus, it can be inferred here that the structural refinement brought about by cryogenic treatment helps the material to sustain more loads. The relationship between the crystallite size and strength of the polymer can be correlated by the Hall-Petch phenomenon.

The Hall-Petch phenomenon demonstrates that with the reduction in the crystallite size, both the strength and the toughness of the material increases (McCombs, Golland, & Mayer, 1974). This can be explained by the resistance offered by the domain boundaries of a crystallite to plastic deformation and/or the arresting of cracks during fracture. The schematic of this phenomenon and its effect on tensile and wear properties of polymers is illustrated in Figure 12. The smaller size of crystallites means more number of crystalline segments per unit area. This means that the density of domain boundaries also increases, as shown in Figure 12. As a crack initiates within the material, numerous boundaries around and about the crack propagating path arrests the progression of crack and thus, gives sufficient time for the material to undergo more loading. This increases strength of the material until the crack finally overcomes the domain boundaries and fractures. Thus, the improvement in tensile strength of the virgin polymers after cryogenic treatment is found to be in good accordance with the crystallite size data.

Also, the remarkable increase in the % elongation of virgin polymers can be understood from the relationship between % crystallinity and yielding behavior of polymers. In polymers, the yield point increases linearly with % crystallinity (Askeland, 1996). This linear increase is due to the fact that more amount of tensile load is needed to disturb the orderly arrangement of crystallites in order to get fractured. Basically, elongation depends on the volume of the crystalline regions and not on the morphology of the crystals (Askeland, 1996). As seen from the structural analysis

Figure 12. Schematic representation of Hall-Petch phenomenon demonstrating the effect of crystallite size on tensile and wear properties of polymers

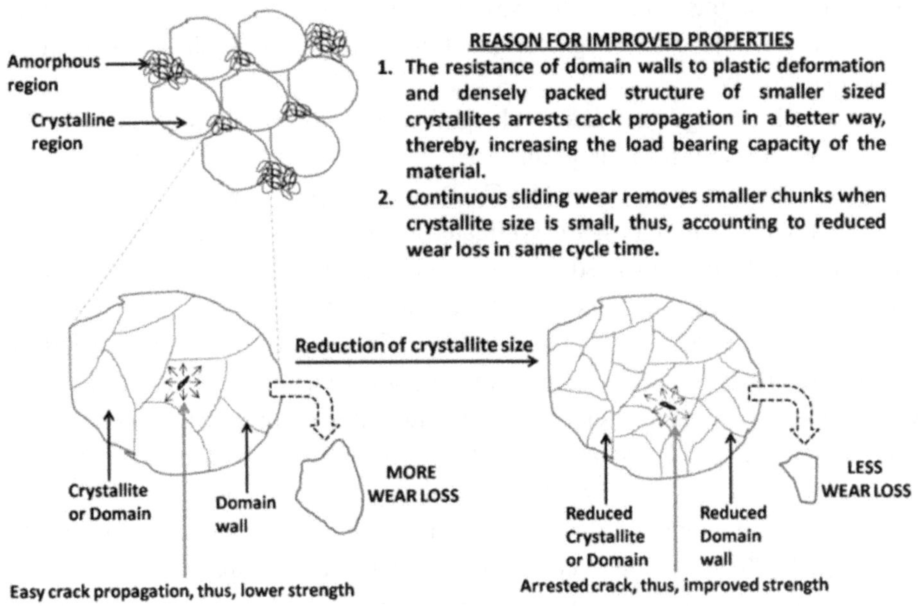

of treated polymers; the cryogenic treatment rearranges the polymeric crystalline segments into a more systematic and dense configuration. Also, the structural stability of the polymer increases as the crystallite density increases due to the presence of more number of smaller sized crystallites per unit area, as illustrated in Figure 12.

It is well known that the stretching and distortion of polymeric chains upto the yield point is elastic in nature. The chains recover almost instantaneously when the load is removed. Once the yield point of the polymer is surpassed, plastic deformation sets in. The plastic deformation of polymeric crystalline segments is the result of chain sliding, stretching, rotating and finally disentangling under load (Askeland, 1996; Callister, 2000). The initial loading causes the tangled and intertwined polymer chains in amorphous segments to untangle themselves and straighten out in the direction of applied load. A further loading separates the crystalline segments and initiates necking; where further sliding, stretching and deformation of the chains continues. The continuous necking causes the chains to move extremely closer in an almost parallel pattern (Askeland, 1996; Callister, 2000). This physical arrangement requires higher stress and more time to completely deform and finally fracture.

Thus, when the cryo-treated polymer is subject to tensile pull, more amount of load is needed in order to disturb the highly dense, compact and orderly arrangement of smaller sized crystallites to initiate necking and generate deformation, thereby, contributing towards improved elongation of the polymer after cryogenic treatment.. As the structural stability increases with improved crystallinity in cryo-treated polymers, the deformation of the material at a molecular level improves as it elongates under tensile loading. Thus, in this case, both the motions; viz. the mobility of the molecule as a whole and the segmental motion are improved, leading to an increase in chain flexibility.

The tensile strength and the % elongation of cryo-treated polymer/MWCNT nano-composites are shown in Figure 13.

The tensile strength of the PA/MWCNT nano-composites was found to be lower than the untreated material when cryo-treated at -80°C and -140°C. Even when cryo-treated at -185°C, the tensile strength of the PA nano-composites did not increase significantly. The maximum increase noted was of only 7% when the material was cryo-treated at -185°C for 24 hours of soaking. On the contrary, irrespective of the treatment temperature and soaking period, the % elongation of PA/MWCNT nano-composites showed exceptional improvements after cryogenic treatment. When cryo-treated at -185°C for 24 hours, the elongation of PA/MWCNT nano-composite increased by 184%. It is well known that due to the nucleating tendency of MWCNTs, smaller crystallites are formed in polymer matrix. Furthermore, cryogenically treating the PA/MWCNT nano-composites has again led to a marginal reduction in crystallite size, as noted earlier from the XRD analysis. In case of cryo-treated PA/MWCNT nano-composites; the nano-reinforcing tendency and load transfer capability

Figure 13. (a) Tensile strength and (b) % elongation of cryo-treated PA/MWCNT nano-composites and (c) Tensile strength and (d) % elongation of cryo-treated PBT/MWCNT nano-composites

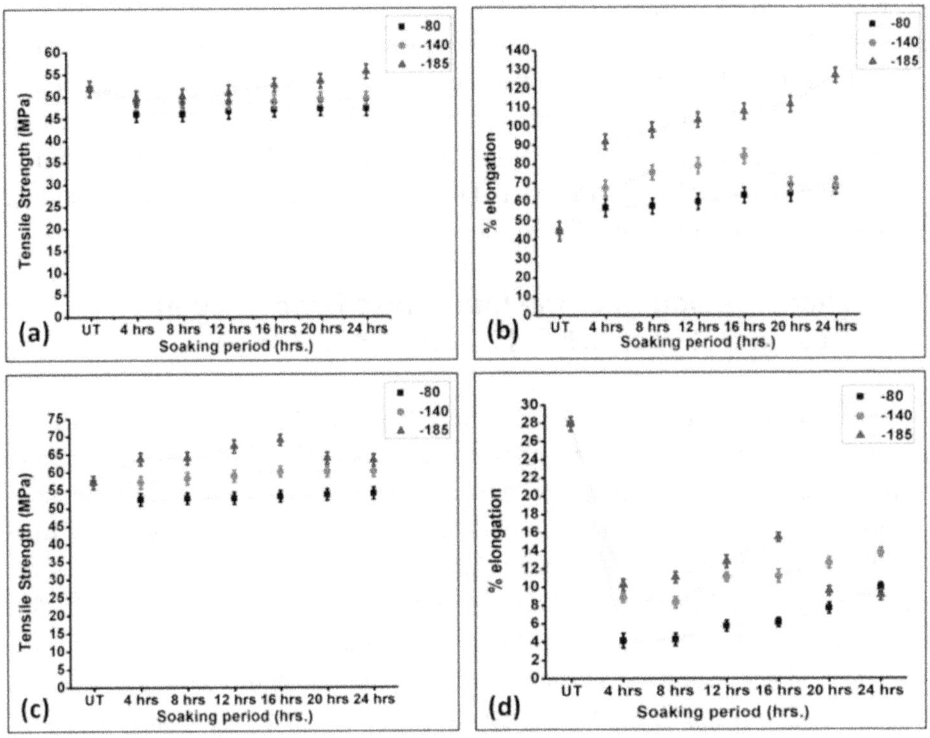

of the MWCNTs plays a fundamental role in enhancing the tensile properties, in addition to the Hall-Petch strengthening effect seen in virgin PA. Thus, during tensile loading of the cryogenically treated PA/MWCNT nano-composites, the load transfer between the filler and matrix as well as the increased number of dense and refined crystalline regions within the matrix, contributes towards increased tensile properties of the material.

It can be observed form the figure that the tensile strength of PBT/MWCNT nano-composites increased considerably only after cryogenic treatment at -185°C, similar to that of virgin PBT. This increase was by 21% when the material was subjected to 16 hours of soaking period, beyond which the strength drops marginally. However, when looking at the % elongation of PBT nano-composites, it was surprising to see that the cryogenic treatment deteriorated the elongating tendency of the material.

Thus, it can be inferred here that, akin to virgin PA, the improvement in the tensile properties of cryo-treated PA/MWCNT nano-composites is analogous to the crystal structure modifications as explained earlier. However, the same cannot be said in case of PBT/MWCNT nano-composites. Despite the enhanced crystallinity and reduced crystallite size, the low temperature during cryogenic treatment imposes rigidity to PBT chains by arresting the aromatic benzene ring and the ester group (as seen from FTIR analysis), thereby reducing the chain flexibility at a molecular level. This forms a rigid, unyielding yet a strong long chain structure that may cause embrittlement of the matrix. However, due to uniform chain rearrangement and fine structure after cryogenic treatment, the PBT nano-composite is able to sustain higher loads, thereby giving an increased tensile strength. But the densification of the structure and overall embrittlement of the bulk material do not assist in yielding and hence the elongating tendency of the sample is affected.

Theoretical Prediction of the Interphase Properties in Cryo-Treated Polymer/MWCNT Nano-Composites

In order to investigate the effect of cryogenic treatment parameters on the interfacial properties of the polymer/MWCNT nano-composites, the interphase strength (σ_i) and the thickness of interphase layer (t_i) were predicted by using mathematic modeling. The combination model of Leidnar-Woodhams and Pukanszky model was modified in earlier work for predicting the interfacial properties of polymer nano-composites reinforced with *cylindrical nanofillers* (Chopra, Deshmukh, & Peshwe, 2017b). It was established in this work that the said modified model gives a better fit as compared to other models for calculating the interphase strength (σ_i) and the thickness of interphase layer (t_i). Thus, the interfacial properties of the cryogenically treated PA and PBT nano-composites under present study are predicted using the modified combination model. The detailed derivation and modification of the said model is explained by Chopra et al. (2017b). Thus, the equations used for calculating the interphase strength (σ_i) and the thickness of interphase layer (t_i) are given below:

$$\sigma_i = \frac{\sigma_c - \sigma_m - 0.48\sigma_m \phi}{\left(1 + \frac{1}{B}\right)\phi} \tag{1}$$

$$t_i = R\left[\frac{B}{2\ln\left(\frac{\sigma_i}{\sigma_m}\right)} - \frac{1}{2}\right] \tag{2}$$

where, σ_c is the yield strength of the nano-composite, σ_m is the yield strength of the matrix, ϕ is the volume fraction of reinforced nano-filler and R is the average outer diameter of the MWCNT. The parameter B in both the above equations presents the extent of polymer-filler adhesion and is related to the load carried by the filler, given as:

$$B = \left(1 + A_c \rho_f t_i\right) \ln\left(\frac{\sigma_i}{\sigma_m}\right) \tag{3}$$

where, A_c and ρ_f are specific surface area and density of the nano-filler, respectively. The shape parameter A_c in Eq. 3 has been modified previously (Chopra et al., 2017b) to suit for *cylindrical nanofiller* reinforcement such as MWCNTs.

Using these equations, the values of the interphase strength (σ_i) and the thickness of interphase layer (t_i) were calculated for cryogenically treated polymer/MWCNT nano-composites under present study and are summarized in Table 6. It can be seen here that the σ_i and t_i values for the cryo-treated MWCNT nano-composites increases with increasing soaking period, irrespective of the temperature, analogous with the tensile strength values exhibited by the nano-composites in Figure 13. Due to the differences in the thermal coefficient of expansion between the polymer matrix and the MWCNTs, the low temperature contraction of the material during cryogenic treatment loosens the interfacial contact between the matrix and filler. When the soaking period is less, there is no sufficient time for the polymer chains to overcome the large contraction difference and form a dense interfacial region around the thermally stable MWCNTs. Further, as the polymer chains relaxes and settles in a denser configuration during ramp up stage of the treatment, the interphase remains weak and incoherent. This results in lower interphase strength of the nano-composites at relatively shorter soaking periods, as evident from Table 6. However, when the soaking period increases, the polymer chains get enough time to overcome the contraction of the matrix around the MWCNTs and form a significant interfacial region during the ramp-up stage. This region is inherently dense due to the nucleating and nano-reinforcing effect of the MWCNTs. Nonetheless, the cryo-structural modifications induced during cryogenic treatment further makes the interfacial region more dense, uniform and refined, just like the bulk matrix material. This can be established from the increasing values of σ_i and t_i values with increasing soaking period, as seen from Table 6.

The formation of a strong interphase in cryogenically treated polymer/MWCNT nano-composites is reflected in the form of improved tensile strength of the material also. A strong and dense interfacial region helps in efficient load transfer between the polymer matrix and MWCNTs. This holds true in case of cryo-treated PA and

Table 6. Interphase strength (σ_i) and Thickness of interphase layer (t_i) for cryogenically treated MWCNT nano-composites

Cryogenic treatment parameters		PA		PBT	
		σ_i (MPa)	t_i (nm)	σ_i (MPa)	t_i (nm)
UT		2069.22	154.06	5470.34	215.65
-80°C	4 hrs.	696.94	74.02	2883.01	138.09
	8 hrs.	796.68	80.81	3015.67	142.53
	12 hrs.	860.17	85.02	3480.01	157.60
	16 hrs.	928.20	89.45	3625.96	162.20
	20 hrs.	955.42	91.19	3692.30	164.27
	24 hrs.	1032.54	96.07	3944.39	172.03
-140°C	4 hrs.	950.88	90.90	4939.56	201.07
	8 hrs.	1041.61	96.64	5403.99	213.86
	12 hrs.	1268.49	110.39	5603.04	219.20
	16 hrs.	1359.25	115.67	6485.50	242.01
	20 hrs.	1518.10	124.65	6664.65	246.46
	24 hrs.	1767.76	138.13	6863.70	251.36
-185°C	4 hrs.	1676.97	133.31	7328.17	262.53
	8 hrs.	1881.25	144.02	7859.01	274.91
	12 hrs.	2307.99	165.05	8270.39	284.23
	16 hrs.	3193.36	203.94	8701.70	293.67
	20 hrs.	3629.27	221.17	7925.35	276.43
	24 hrs.	3733.71	225.14	7859.01	247.91

PBT nano-composites under study. However, the elongation in cryogenically treated MWCNT nano-composites is governed by the structural modifications in the material rather than the interphase. This can be very well correlated from the findings of XRD and FTIR of the cryo-treated nano-composites and the tensile properties exhibited by them. Thus, the cryo-structural modifications in the material have a chief impact on the mechanical properties of the polymers and their MWCNT nano-composites.

Fractographic Investigations of Cryo-Treated Polymers and Polymer/MWCNT Nano-Composites

In order to explore the tensile fracture mechanisms in the polymer/MWCNT nano-composites after cryogenic treatment, the fractured surfaces of the samples were investigated, wherein the load transfer effect between MWCNTs and the polymer matrix, role of interphase and the progression of fracture is explained in detail. Figure 14 shows the fractured samples of virgin PA and PA/MWCNT nano-composites cryo-treated at -185°C. It was observed that for both PA and its nano-composites, the necking tendency of the material increases with increasing soaking period. A similar trend was exhibited by the samples at all cryo-treatment temperatures and hence, only the samples treated at -185°C are shown here as an indicative of the fracture trend followed by the material with varying soaking period.

It is evident from Figure 14 that the cryogenically treated virgin PA and its MWCNT nano-composites undergo significant elongation before failure, which is well reflected in the % elongation values of the cryo-treated samples, as seen in Figure 11 (b) and Figure 13 (b) also. This was credited to the enhanced yielding of the polymeric crystalline segments owing to the cryo-structural modifications such as formation of smaller sized crystals and uniform rearrangement of crystalline segments, as explained previously. It was observed during the tensile loading that as the sample elongated in real time, the color of the sample surface lightened, as seen in fractured samples in Figure 14. This area is identified as the region experiencing excessive plastic deformation with continuous loading. Such behavior of the sample can be attributed to the increase in crystallinity of material (Pitchard, 1964). It is a well known fact that during tensile loading of semi-crystalline polymers, the real

Figure 14. Tensile fractured samples of (a) Virgin PA and (b) PA/MWCNT nano-composites cryo-treated at -185°C

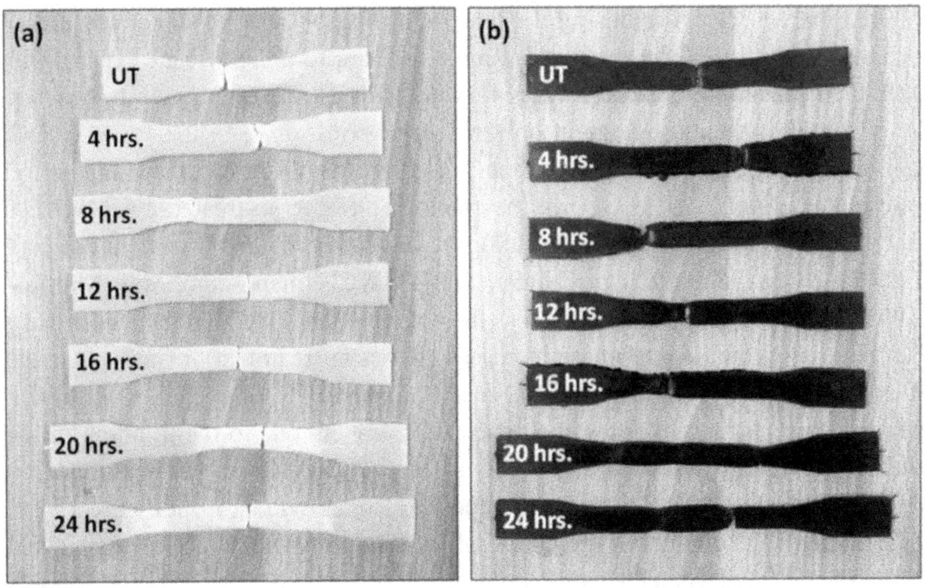

time crystallinity of the sample increases (Pitchard, 1964). As mentioned previously, the plastic deformation of polymeric crystalline segments during uniaxial tensile loading is the result of chain sliding, stretching, rotating and finally disentangling under load. Once the necking is initiated in the sample, further continuous loading causes the polymer chains to move extremely closer in an almost parallel pattern and align in an orderly manner in the direction of applied load. Because of this alignment, the crystallinity of the sample increases in the necking area. This causes further densification of the material in deformed area, which now has less number of reflection sites (i.e. defective polymeric segments) for the natural light. Because of this the color of the deformed area in the sample appears lighter and may even reach transparency if the samples were able to further elongate without fracture.

Along with the cryo-structural modifications, the improvement in the tensile properties of cryo-treated PA/MWCNT nano-composite is governed by the interfacial interactions between PA matrix and MWCNT also. In a previous study by Chopra et al. (2018b), it was reported that for PA/MWCNT nano-composites (when untreated), the *Polymer Sheathing* phenomenon was found to be the main fracture mechanism; wherein the formation of a strong interphase was credited to the wrapping of PA layer around the reinforced MWCNTs. Polymer sheathing takes place when polymer nucleates on and around the MWCNT surface and forms a thick coating around the nanotubes (Chopra et al., 2018b). As a result, the diameter of such polymer sheathed MWCNTs increases to approx. between 100 nm and 200 nm. This polymer sheathing plays a significant role in load transfer and hence advocates for the increase in tensile strength and elongation of PA/MWCNT nano-composite.

As evident from Figure 14 (b), the cryo-treated PA/MWCNT nano-composites suffered extreme necking during the tensile loading because of which the cross-sectional thickness of the fractured sample reduced to a significant extent. This made the fractographic investigations very difficult. However, in order to evaluate the effect of cryogenic treatment on the *Polymer Sheathing* in the PA/MWCNT nano-composites, the sample cryo-treated at -185°C for 24 hours of soaking period was studied. Figure 15 (a) shows the fractograph of cryo-treated PA/MWCNT nano-composite. It is evident from the figure that the cryogenic treatment does not affect the *Polymer Sheathing* mechanism in PA/MWCNT nano-composite. Multiple MWCNTs can be seen stretched across the broken PA matrix, thereby, confirming that the strength of the interphase increases after cryo-treatment. It was seen in the fractograph of untreated PA/MWCNT nano-composite (Chopra et al., 2018b) that the formation of bead-like morphology was prominent due to the thinning of the adhered polymer layer on the MWCNTs surface. This tendency was exhibited by the cryo-treated PA/MWCNT nano-composite also, as evident from Figure 15 (b). This confirms the claim that the nature of interaction between the PA matrix and MWCNT improves when the nano-composite is cryo-treated at -185°C for 24 hours

Figure 15. (a) Fractograph of PA/MWCNT nano-composite cryo-treated at -185°C for 24 hours showing multiple MWCNTs bridging across the broken PA matrix and (b) Magnified view of Polymer Sheathing on a single MWCNT showing the bead-like morphology

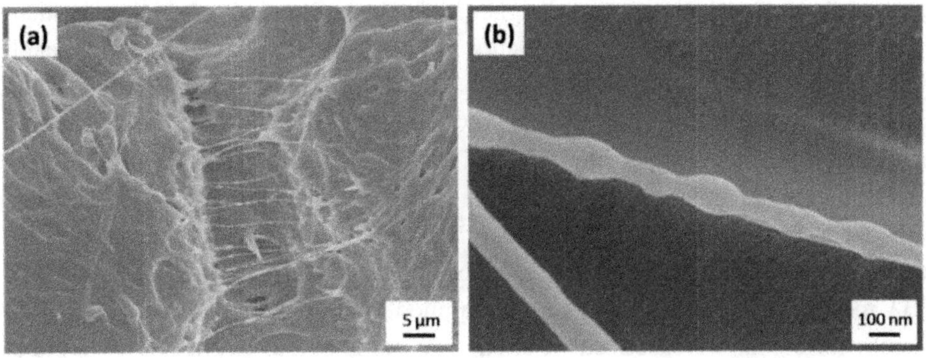

of soaking period, also indicated by the 80.4% increase in the value of interphase strength (σ_i) as observed from Table 6.

The fractured samples of untreated virgin PBT along with the samples cryo-treated at -185°C for 12 and 24 hours of soaking period are shown in Figure 16.

It can be seen from the figure that virgin PBT displays sufficient necking tendency after cryogenic treatment. During the tensile loading of cryo-treated PBT, the

Figure 16. Tensile fractured samples of Virgin PBT cryo-treated at -185°C

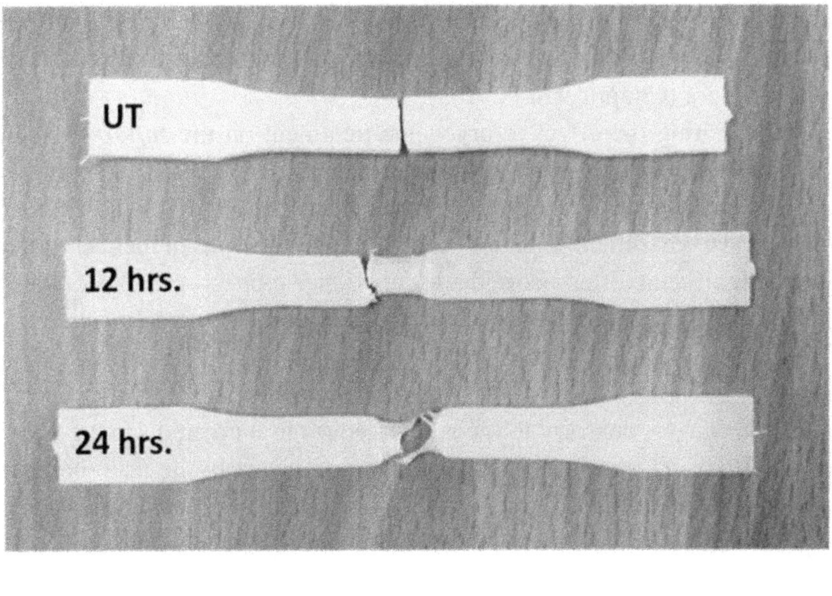

deformation of the sample was first initiated by necking and later on this deformed region elongated further due to continuous loading. It can be noted here that the elongation of cryo-treated PBT was not as high as cryo-treated PA. However, due to the structural modifications brought about by the cryogenic treatment; virgin PBT becomes strong enough to sustain higher loads than untreated material, as evident from the tensile strength shown in Figure 11 (c). A previous study by Chopra et al. (2018a) on the fractographic analysis of PBT/MWCNT nano-composites (when untreated) showed that the *Inflorescence* type morphology is a characteristic feature occurring in the nano-composite after tensile fracture; wherein the MWCNT clusters anchoring the surrounding PBT matrix is clearly visible and matrix appears to be torn out after the fracture. It was established here that during deformation, the nano-reinforcing tendency of the MWCNT clusters assists in holding down the PBT matrix from cracking apart as the applied load exceeds the matrix strength, thereby reflecting increased tensile strength as compared to virgin PBT (Chopra et al., 2018a). On closer inspection, it was observed that every MWCNT cluster and the PBT matrix surrounding it formed crater-like structures.

The distinctive fracture pattern visible in a single crater was identified as the *mirror–mist–hackle* pattern (Chopra et al., 2018a). The MWCNT cluster forms the *mirror* region, the interfacial region between the MWCNT cluster and the PBT matrix forms the *mist* zone and the rough-looking torn out polymer matrix surrounding the MWCNT cluster forms the *hackle* zone. It was concluded here that the appearance of a significant mist zone indicates better interfacial adhesion between the matrix and filler, while a rapid transition from mirror zone to hackle zone indicates insufficient time for the load transfer to take place from MWCNT cluster to relatively weaker PBT matrix. The region beyond *hackle* zone is where the initiation of matrix tearing causes severe surface roughening in the form of outward ridges. These ridges are characterized by stretched flaps and are known as the *fibrillar edges* (Chopra et al., 2018a).

For investigating the effect of cryogenic treatment on the *Inflorescence* type morphology and the *mirror–mist–hackle* pattern in PBT/MWCNT nano-composites, the fractured surfaces of the samples cryo-treated at 185°C for 16 and 24 hours of soaking period were studied. Figure 17 shows the fractographs of these cryo-treated samples. A significant difference in the *Inflorescence* type morphology of the PBT/MWCNT nano-composites can be observed after cryogenic treatment. It can be seen from Figure 17 (a) that considerable tearing of the PBT matrix when cryo-treated at -185°C for 16 hours, as signified by the sharp and torn out appearance of the *fibrillar edges*, contrary to the smooth edges in untreated sample observed in previous study. This indicates that the polymer matrix in the nano-composites becomes rigid after cryogenic treatment and its tendency to elastically recover is

Figure 17. Fractographs of PBT/MWCNT nano-composites showing Inflorescence type morphology when cryo-treated at -185°C for (a) 16 hours and (b) 24 hours; Magnified view of a single crater in PBT/MWCNT nano-composites when cryo-treated at -185°C for (c) 16 hours and (d) 24 hours and (e) Magnified view showing matrix cracking around the MWCNT cluster

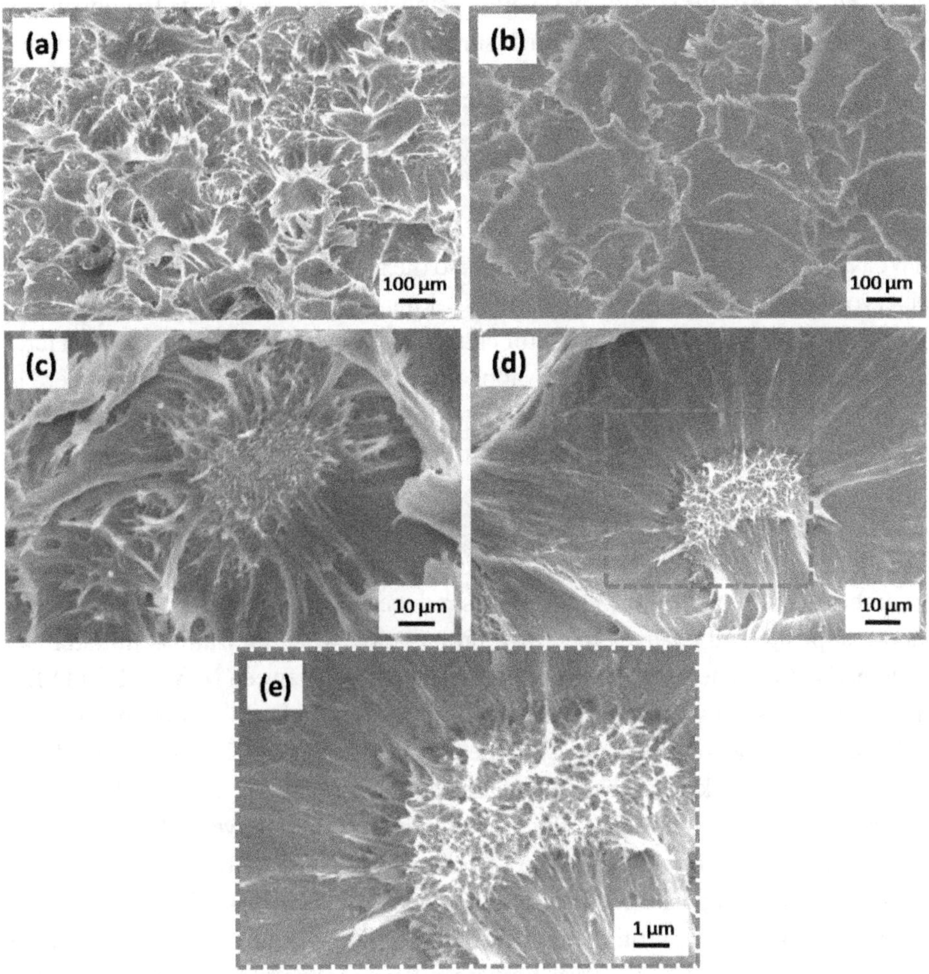

lost. As the soaking period increases to 24 hours, the fractured surface appears more brittle, as seen in Figure 17 (b).

On further inspection, the difference in the *mirror-mist-hackle* morphology in the cryo-treated PBT nano-composites was also identified. It can be seen from Figure 17 (c) that there is no significant *mist* zone around the MWCNT cluster in case of cryo-treated PBT nano-composites. Also, it can be observed from the figure that the *hackle* zone originated immediately from the vicinity of the MWCNT cluster in case of the sample cryo-treated for 16 hours. It was established by Chopra et al. (2018a) that the presence of a significant *mist* zone indicates improved interfacial adhesion between the PBT matrix and the MWCNT and its disappearance hints towards the failure of material to recover itself in real time. Nonetheless, the disappearance of *mist* zone in case of PBT/MWCNT nano-composites cryo-treated for 16 hours is not accompanied with any form of matrix surface cracking. Infact, this sample shows maximum increase in the tensile strength after cryogenic treatment, as seen earlier from Figure 13 (c). It can be seen from Figure 17 (c) that with the disappearance of the *mist* zone, the polymer matrix showed multiple markings that looked more or less like the *fibrillar edges* that were stretched due to loading and later torn out as the matrix failed. This hinted towards the efficient anchoring provided by the MWCNT clusters to the PBT matrix and also the increased load bearing capacity of the cryo-treated matrix. However, as the soaking period increased to 24 hours, the *mirror-mist-hackle* morphology again changes, as seen from Figure 17 (d). Here, it can be seen that along with the disappearance of a prominent *mist* zone, there is no *hackle* zone present and the matrix is getting torn out from the vicinity of the MWCNT cluster itself.

Figure 17 (e) shows the magnified view of the MWCNT cluster shown in Figure 17 (d), wherein the matrix cracking is clearly visible around the MWCNT cluster. This indicates that prolonged soaking period during cryogenic treatment is causing the embrittlement of the PBT matrix. This reduces the load bearing capacity of the nano-composite and changes the *craze* propagation when the flaw is initiated, as explained below. The formation of a strong interphase between MWCNT and PBT matrix (as in case of sample cryo-treated at -185°C for 16 hours of soaking period), indicates good load transfer capacity of the polymer matrix which helps to hold down the matrix from cracking as the applied load exceeds the matrix strength. Due to the presence of a strong interphase, the *craze* tends to propagate very slowly through the polymer matrix. However, due to the embrittlement of matrix in case of PBT/MWCNT nano-composite cryo-treated at -185°C for 24 hours of soaking period, the progression of *craze* promotes plastic deformation by continuous rearrangement of the polymeric network structure, with no sufficient time to offer resistance to the advancing flaw. This initiate matrix cracking as the PBT fails to take up the load transferred by the MWCNT cluster during continuous tensile loading, as seen in Figure 17 (e).

But in case of PBT/MWCNT nano-composite cryo-treated for 16 hours, due to continuous deformation and re-arrangement of the polymer chain segments, the localized recovery of the matrix is possible which reflects in the form of concentric and torn out *fibrillar edges* around the well anchored MWCNT cluster. However, it is imperative to note here that this tendency of the PBT matrix to recover in a localized area is not dominant, as indicated by the deterioration of the elongating tendency of cryo-treated PBT/MWCNT nano-composites seen from Figure 13 (d). This can be attributed to matrix embrittlement caused by the motion restriction of the benzene ring and ester groups (as inferred from FTIR analysis) when the PBT/MWCNT nano-composites are subjected to cryogenic temperatures. This reduces the chain flexibility thereby forming a rigid, unyielding yet a strong polymeric structure. The fractographs shown in Figure 17 reflects this tendency of the PBT matrix. The formation of strong interphase region when subjected to 16 hours of soaking at -185°C anchors the matrix well and produces serrated *fibrillar edges*. However, the rigidity sustained by the matrix at a molecular level and it consequent embrittlement, does not allow the material to elastically recover in a localized region, thereby, tearing off the matrix. Thus, it can be confirmed from the fractographic investigations that the cryo-structural modifications in the virgin polymers and polymer/MWCNT nano-composites affects the load bearing capacity of the material and accordingly its fracture mechanism.

Tribological Performance of Cryo-Treated Polymers and Polymer/MWCNT Nano-Composites

It is a well established fact that the cryogenic treatment enhances the wear performance of the polymers and polymer composites significantly. This has been attributed to the structural changes induced into the material by cryogenic treatment. Thus, in order to study the effect of cryogenic treatment on the wear loss of virgin polymers and polymer/MWCNT nano-composites under study, the tribological performance of the cryo-treated material is evaluated. The modes of wear and the reason for enhanced wear resistance of the polymers and nano-composites after cryogenic treatment are also studied herein with the help of microscopic investigations of the worn out surfaces.

Figure 18 shows the Shore D hardness and the wear loss values of virgin PA and PA/MWCNT nano-composite after cryogenic treatment at different temperatures and varying soaking period. It can be clearly seen from the figure that irrespective of the cryogenic treatment temperature, the hardness of both virgin PA and PA/MWCNT nano-composites increases while the wear loss reduces with increasing soaking period. The Shore D hardness of virgin PA and PA/MWCNT nano-composite increases by 21% and 15%, respectively, when both are cryo-treated at -185°C for 24 hours soaking. This can be attributed to the low temperatures during cryogenic treatment which hardens the material on exposure. The effect of this increment in the hardness values of PA and PA nano-composites after cryogenic treatment is indeed reflected in the wear performance of the material also.

Figure 18. (a) Hardness and (b) Wear loss of cryo-treated virgin PA and (c) Hardness and (d) Wear loss of cryo-treated PA/MWCNT nano-composites

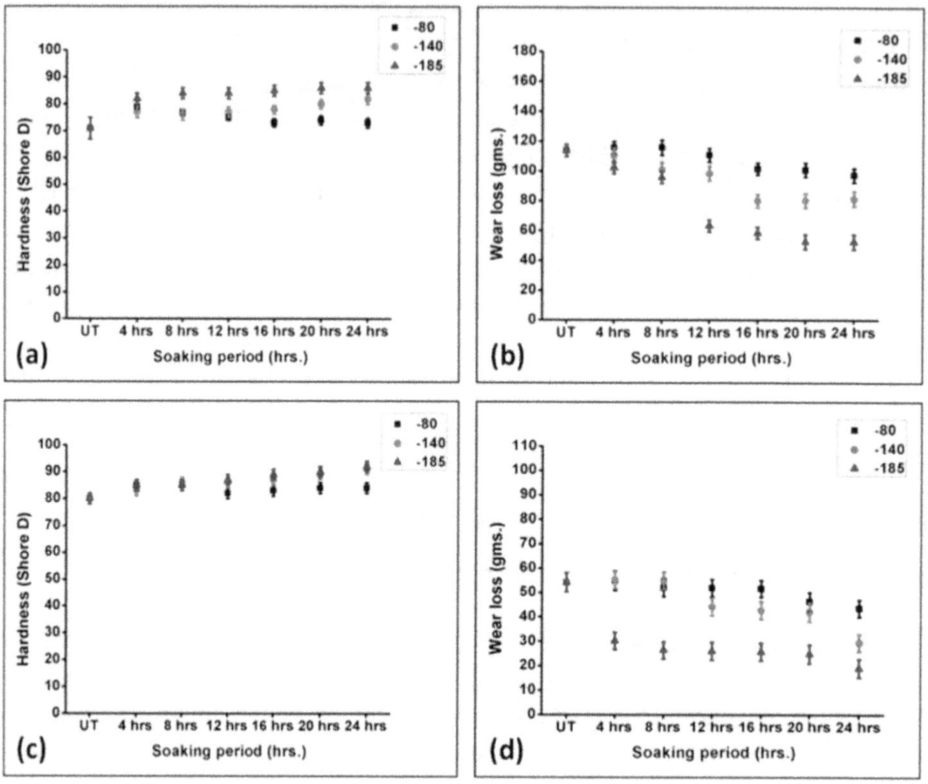

It is evident from Figure 18 (b) and Figure 18 (d) that the wear loss of virgin PA and PA/MWCNT nano-composites reduces post cryogenic treatment. This improvement is not much noteworthy in case of samples treated at -80°C for all soaking periods and -140°C for smaller durations of soaking. However, as the soaking period increases at -140°C, the wear performance of both the materials improves. Furthermore, for the samples cryo-treated at -185°C, the wear properties increase significantly. This linear reduction of the wear loss of both PA and its nano-composite with the soaking period is credited to the enough time available for the material to rearrange irregular structural segments into an aligned configuration. As more volume fraction of the bulk material undergoes cryo-structural modifications, a highly stable and uniform structure forms during the ramp-up stage. This not only enhances the polymer chain structure but also contributes towards improved properties.

It was observed here that the wear loss of virgin PA and PA/MWCNT nano-composites reduces by 54% and 65%, respectively, both at -185°C and 24 hours of soaking period. This can be attributed to the improved stiffness and crystallinity of the material after cryogenic treatment. The strong interface between the PA matrix and the MWCNT also improves the capacity of the nano-composite to sustain severe tribological conditions. Irrespective of the filler morphology and filler content, the polymers that have self-lubricating character while sliding excels in wear performance and this tendency of the polymer increases after cryogenic treatment. Similar phenomenon of wear performance enhancement of polymeric material after cryogenic treatment is reported by Indumathi et al. (1999a); Indumathi et al. (1999b) and Pande (2012) also.

The hardness and wear loss values of cryo-treated virgin PBT and PBT/MWCNT nano-composites are shown in Figure 19. The hardness of cryo-treated virgin PBT and PBT/MWCNT nano-composites increases linearly with soaking period, irrespective of the treatment temperature, similar to that of PA and its nano-composites. This increase was noted to be 19% for PBT and 14% PBT/MWCNT nano-composites, both when cryo-treated at -185°C for 24 hours. It was observed here that, similar to PA

Figure 19. (a) Hardness and (b) Wear loss of cryo-treated PBT and (c) Hardness and (d) Wear loss of cryo-treated PBT/MWCNT nano-composites

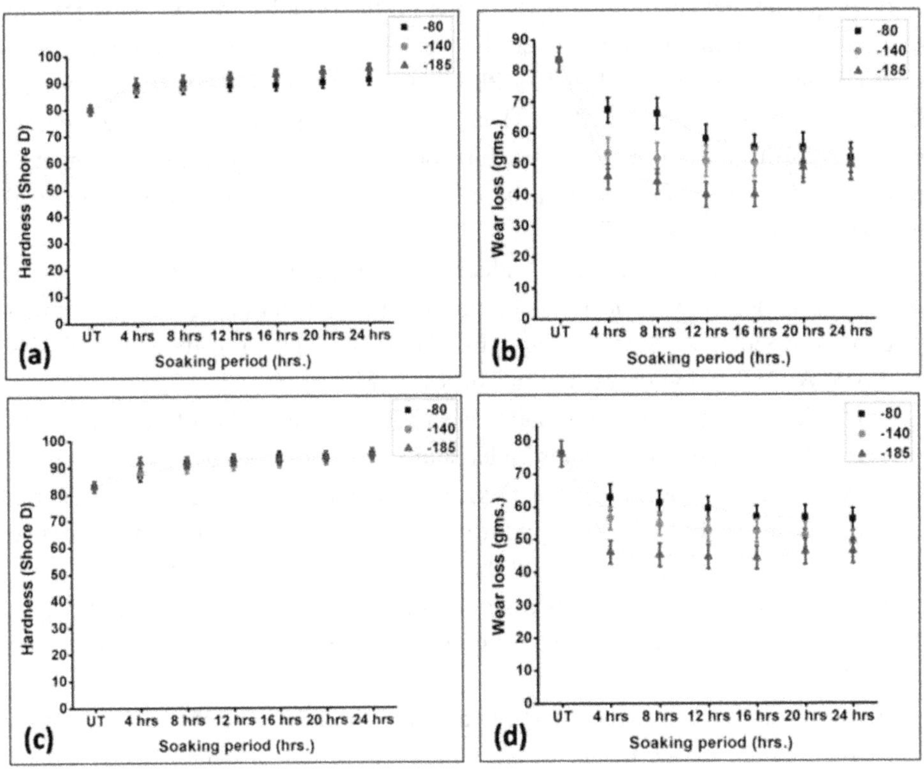

and its nano-composite, the wear loss of PBT and PBT/MWCNT nano-composites reduces as compared to untreated material for all the treatment conditions studied. However, the trend of the wear loss of PBT and its nano-composites differs at different treatment temperatures. When cryo-treated at -80°C and -140°C, the wear loss of virgin PBT and PBT/MWCNT nano-composite reduces linearly, whereas at -185°C, the wear loss first reduces and then increases. This deterioration in the wear loss values is more significant for virgin PBT than the PBT/MWCNT nano-composite, which can be attributed to the presence of MWCNTs.

As it was seen from the XRD and FTIR analysis earlier, the structure of virgin PBT becomes rigid beyond the 12 hours of soaking. The larger crystal size seen for PBT samples beyond 12 hours of soaking period at -185°C along with the rigid structure increases the wear loss in the material. However, with addition of the MWCNTs and the formation of relatively strong interfacial region, the PBT matrix is well anchored together. This, coupled with the lubricating tendency of the nanotubes, reduces the deterioration of the nano-composite surface even when the cryo-structural modification makes the material rigid. A maximum reduction of 52% in the wear loss of PBT was noted when the samples were cryo-treated at -185°C for 12 hours and it exhibited same amount of wear loss when the soaking period increased to 16 hours. On the contrary, the wear loss of PBT/MWCNT nano-composites remains more or less same for the samples treated at -185°C for 4, 8, 12 and 16 hours, beyond which the wear loss increase was marginal. Thus, the maximum reduction exhibited by PBT/MWCNT nano-composite was by 42% at both 12 and 16 hours of soaking. Thus, it can be stated here that an extremely refine, uniform and dense microstructure obtained after cryogenic treatment is responsible for an enhancement in the tribological performance of polymers and their nano-composites as well.

In order to assess the changes induced by cryogenic treatment on the wear mechanism of the polymer and nano-composites, the worn out surfaces of the samples were investigated. The SEM micrographs of untreated and cryo-treated virgin PA and PA/MWCNT nano-composites are shown in Figure 20.

It is evident that the abrasive mode of wear is dominant in case of both virgin PA and PA/MWCNT nano-composites under study. The abrasive wear mode basically occurs when a softer material (polymer) is rubbing against a relatively harder material (SiC counter-face). The arrows in Figure 20 (a) indicate that in case of untreated virgin PA, ploughing action of the SiC counter-face is the dominant abrasive wear mechanism. The *ploughing* occurs when the surface material is shifted to the sides of the wear groove created by the hard particles abrading against it (Bhushan and Gupta, 1991).

From Figure 20 (c), it is clear that, in untreated condition, the MWCNT addition reduces the extent of ploughing action suffered by the sample surface. This is credited

Figure 20. Worn out surfaces of Virgin PA (a) Untreated and (b) Cryo-treated at -185°C for 24 hours; PA/MWCNT nano-composite (c) Untreated and (d) Cryo-treated at -185°C for 24 hours

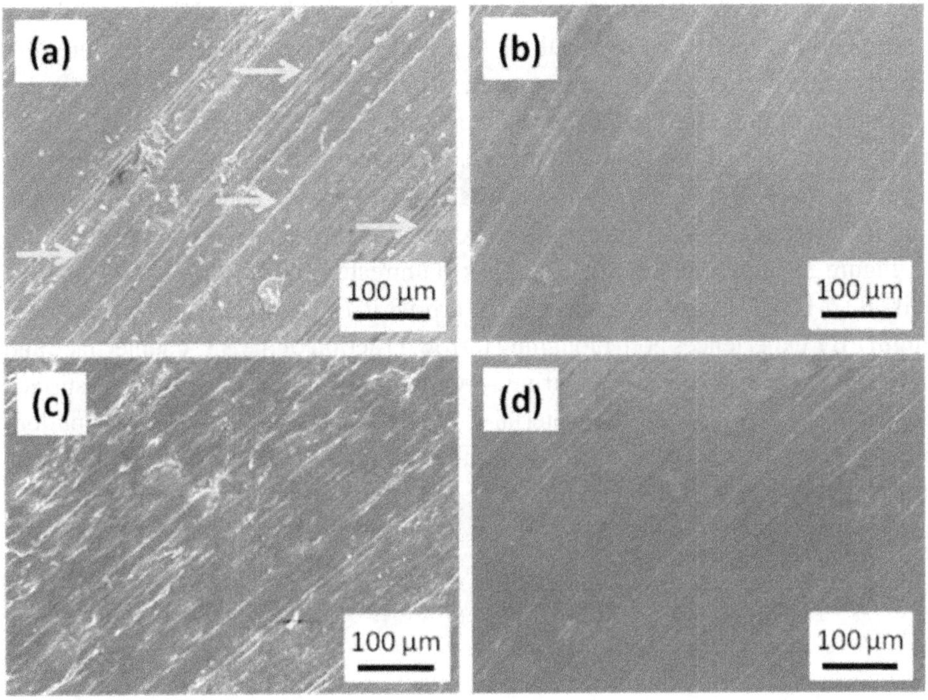

to the nano-reinforcing effect of the MWCNTs that aids in holding down the matrix material during continuous abrasion (Chopra et al., 2017b). Furthermore, as the polymer layer wears off initially, the MWCNTs exposed due to continuous sliding serve as solid lubricants and further prevent direct contact between the sliding surfaces, thereby improving the wear resistance of the nano-composites (Meng, Sui, Xie, & Yang, 2009). This confirms the claim that the well-dispersed MWCNTs help to improve the properties of PA matrix by intercalating with the polymer chains (Zhang et al., 2004) and thereby enhancing the load carrying capacity of PA against severe tribological conditions.

The worn out surface of both cryo-treated PA and PA/MWCNT nano-composite appears less damaged than the untreated sample (Figure 20). This agrees well with the enhancement of the wear performance of PA and its MWCNT nano-composites after cryogenic treatment, as evident form Figure 18 also. The ploughing phenomenon that was observed in case of untreated PA and PA/MWCNT nano-composite was greatly inhibited after cryogenic treatment, as seen clearly from Figure 20 (b) for virgin PA and Figure 20 (d) for PA/MWCNT nano-composite, both cryo-treated at

-185°C for 24 hours. This can be credited to the enhanced crystallinity of the material upon cryogenic treatment. Firstly, the dense and uniform structure after cryogenic treatment is difficult to wear and secondly, the small crystallite size further reinforces the material to a greater extent, thereby removing smaller chunks of wear debris as compared to untreated material (as depicted in Figure 20). The literature also reports same principle for the improvement in wear properties of the cryo-treated polymers and polymer composites (Indumathi et al., 1999a; Pande et al., 2012). Thus, the combination of enhanced crystal structure and increased hardness after cryogenic treatment leads to reduced wear loss (Figure 18) and relatively smooth sample surfaces (Figure 20) in case of PA and PA/MWNT nano-composites.

Figure 21 shows the SEM micrographs of untreated and cryo-treated virgin PBT and PBT/MWCNT nano-composites. The arrows in Figure 21 (a) indicate untreated virgin PBT experiences both ploughing (arrow 1) action of the SiC counter-face and the peeling (arrow 2) of surface during severe wear conditions. A polymeric surface experiences *peeling* when the material is removed in the form of fibrillar pullout and is removed from the surface (Bhushan and Gupta, 1991). From Figure 21 (c) it is clear that, in untreated condition, the ploughing and peeling of the polymeric surface was suppressed after MWCNT addition and the surface appeared relatively smoother. Thus, akin to the wear behaviour exhibited by untreated PA/MWCNT nano-composite, the improvement in the wear condition of untreated PBT/MWCNT

Figure 21. Worn out surfaces of Virgin PBT (a) Untreated, (b) Cryo-treated at -185°C for 12 hours and (c) Cryo-treated at -185°C for 24 hours; PBT/MWCNT nano-composite (d) Untreated, (e) Cryo-treated at -185°C for 16 hours and (d) Cryo-treated at -185°C for 24 hours

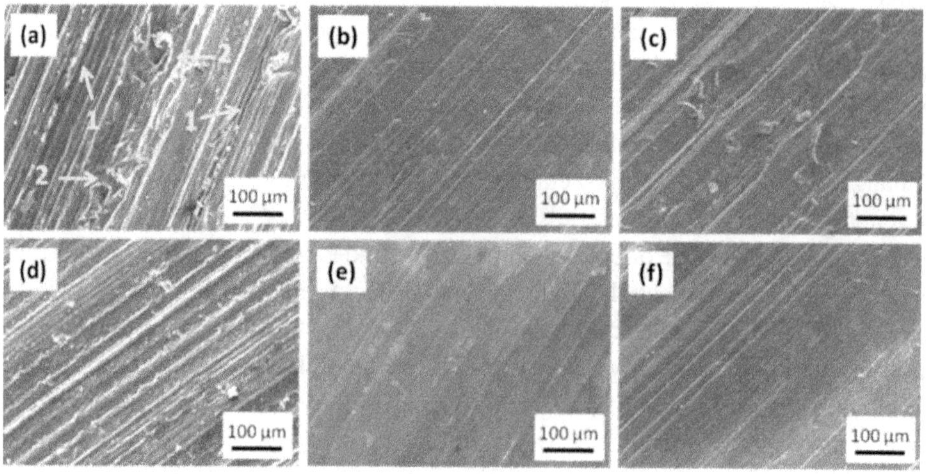

nano-composite is attributed to the self-lubricating tendency of the MWCNTs that are exposed against the SiC counter-face when the polymer matrix wears off progressively (Chopra et al., 2017c).

It is evident from the figure that the worn out surface morphology of PBT and PBT/MWCNT nano-composites after cryogenic treatment agrees well with the wear behavior of the material observed in Figure 19. It was stated earlier that the wear performance of virgin PBT and PBT/MWCNT nano-composite deteriorates after 16 hours of soaking period. This was indeed observed from the SEM micrographs shown in the figure. The cryogenic treatment of virgin PBT and PBT/MWCNT nano-composites suppresses the ploughing and peeling of the sample owing to the cryo-structural modifications brought about by the treatment.

It was noted form XRD analysis that the crystallinity of virgin PBT increases till 12 hours of soaking period (Table 4), while it increases till 16 hours of soaking period for PBT/MWCNT nano-composite (Table 5), both at -185°C. On the other hand, the hardness of both the materials increases with increasing soaking period at -185°C, as shown in Figure 19. However, Figure 19 (b) indicate that the wear loss of virgin PBT increased significantly for 20 and 24 hours of soaking at -185°C, but remained less than that of untreated material. This was indeed reflected in the SEM micrographs of cryo-treated virgin PBT. The worn out surface of PBT sample cryo-treated for 24 hours, as shown in Figure 21 (c), appeared smoother than the untreated PBT samples. However, occasional peeling, ploughing and plastic deformation was visible on the surface. Though the wear loss of cryo-treated polymer is inversely proportional to its hardness as proven by Indumathi et al. (1999a); in case of virgin PBT under present study, the increase in the hardness and the formation of larger crystallites in these samples causes embrittlement of the PBT matrix. This in turn increases the wear loss in the material and deteriorates the surface as evident from Figure 21 (c).

On the contrary, in case of PBT/MWCNT nano-composites it was observed that the wear loss decreased significantly after cryogenic treatment at -185°C with soaking period of 4 to 16 hours, beyond which the wear loss increased marginally. This was credited to the formation of strong interface between the PBT matrix and MWCNTs and the lubricating tendency of the nanotubes. Thus, the worn out surface of PBT/MWCNT nano-composites cryo-treated at -185°C looks almost similar yet smoother as compared to untreated material.

OVERALL DISCUSSION AND CONCLUSION

The results presented above clearly indicate that the *cryogenic treatment motivates structural stability in the polymer matrix* by forming long chains and promoting intermolecular as well as intramolecular rearrangements, which was confirmed from FTIR analysis. However, it was observed here that the structural changes brought about by the cryogenic treatment in polymer/MWCNT nano-composites are largely governed by the polymer matrix itself. The crystal structure and bonding changes observed in cryo-treated virgin polymers and their MWCNT nano-composites, from XRD and FTIR results, were essentially the same. Thus, it can be inferred here that *the response of polymer/MWCNT nano-composites to the cryo-structural modifications depends on the polymer matrix, followed by the MWCNT interaction* to some extent. This behavior was exhibited by macro filler reinforced polymer composites also (Indumathi et al., 1999a; Pande, 2012). Thus, the present study on the cryogenic treatment of polymer/MWCNT nano-composite agrees well with the claim by Pande and Peshwe (2015) that *each polymer has a specific functional group and/or structural characteristic that readily responds to the cryogenic treatments conditions,* thereby, modifying the structure and giving superior properties.

The reason for cryo-structural modifications in both PA and PBT differs a lot. It was established earlier that under present study PA is more susceptible to structural transformations as compared to PBT. It was indeed observed from the XRD analysis that the crystal structure of PA changes drastically after cryogenic treatment, while for PBT it remains almost same. However, the improvement in crystallinity and average crystallite size of the material hinted towards beneficial cryo-structural modifications. The formation of longer chains at low temperatures, along with the contraction of the material at molecular level results in the formation of a denser, stable and uniform structure. In case of PA, the volume fraction of *α-phase* crystallites increases at the expense of *γ-phase* crystallites due to the intrasheet hydrogen bonding between adjacent PA chains. Simultaneously, the contraction due to low temperature and the formation of longer chains gives rise to a superior crystal structure in PA after cryogenic treatment. On the other hand, no significant molecular rearrangement was seen form the XRD results of PBT. However, the cryo-treated PBT demonstrates enhanced crystallinity due to the physical rearrangement of the long chains in a denser configuration, unlike the structural modifications undergone by cryo-treated PA.

The localized intermolecular rearrangement at low temperatures of cryogenic treatment results in relaxation of the material. This promotes systematic rearrangement of the crystalline polymeric segments; however, it is largely governed by the soaking period during cryogenic treatment. Irrespective of the treatment temperature, the differential contraction sustained by the material during cryo-treatment is inevitable. When the samples are held at cryogenic temperatures, the material takes time to attain thermal equilibrium. The state of thermal equilibrium slows down the molecular movements at atomic level. This in turn increases the internal molecular bonding energy and thus motivates the material for structural balance. When the soaking period is less, this molecular rearrangement renders incomplete, whereas, with prolonged soaking periods there is enough time for the material to attain structural equilibrium. Further, during ramp up stage, the segmental motions of the polymeric chains are frozen in the rearranged configuration, thereby, retaining the cryo-structural modifications even when the material is brought back to the room temperature (Pande and Peshwe, 2015). However, depending on the type of polymer; the severity of thermal contraction during ramp down stage and the extent of stress relaxation during ramp up stage vary considerably. Therefore, it becomes necessary to optimize the cryogenic treatment parameters for every material class in order to get the best combination of properties. *This makes cryogenic treatment a material specific process.*

The mechanical properties of PA and PBT and their nano-composites studied here indeed improved after cryogenic treatment, except for the elongation of PBT/MWCNT nano-composites. Despite the XRD results of PBT nano-composites showing enhanced crystallinity and reduced crystallite size, the FTIR analysis revealed that the low temperature during cryogenic treatment imposes rigidity to PBT chains by arresting the aromatic benzene ring and the ester group, thereby reducing the chain flexibility at a molecular level. This forms a rigid, unyielding yet a strong long chain structure that causes embrittlement of the matrix. However, due to uniform chain rearrangement and fine structure after cryogenic treatment, the PBT/MWCNT nano-composites sustain higher loads, thereby giving an increased tensile strength. But the densification of the structure and overall embrittlement of the bulk material do not assist in yielding and hence the elongating tendency of the sample is affected. Thus, it can be inferred here that the cryo-structural modifications of the polymer matrix has a dominant effect over the interfacial interaction between the matrix and MWCNT, thereby governing the mechanical properties of the nano-composites.

In accordance with the literature, it was observed here that the cryogenic treatment enhances the tribological performance of the polymer/MWCNT nano-composites to a huge extent. Irrespective of the treatment temperature and time, the wear performance of PA and PBT nano-composites is significantly lower than that of untreated material. This signifies that *cryogenic treatment is superior to other post-processing techniques for achieving excellent tribological performance from polymers and its composites.* Thus, it can be said here that, akin to virgin polymers and polymer composites reinforced with micro fillers, the *cryogenic treatment is an effective technique for enhancing the mechanical properties of polymer/MWCNT nano-composites.*

Figure 22. Position of cryo-treated polymer/MWCNT nano-composites selected in present work on the materials selection chart based on the Tensile strength and Density (Ashby et al., 2009)

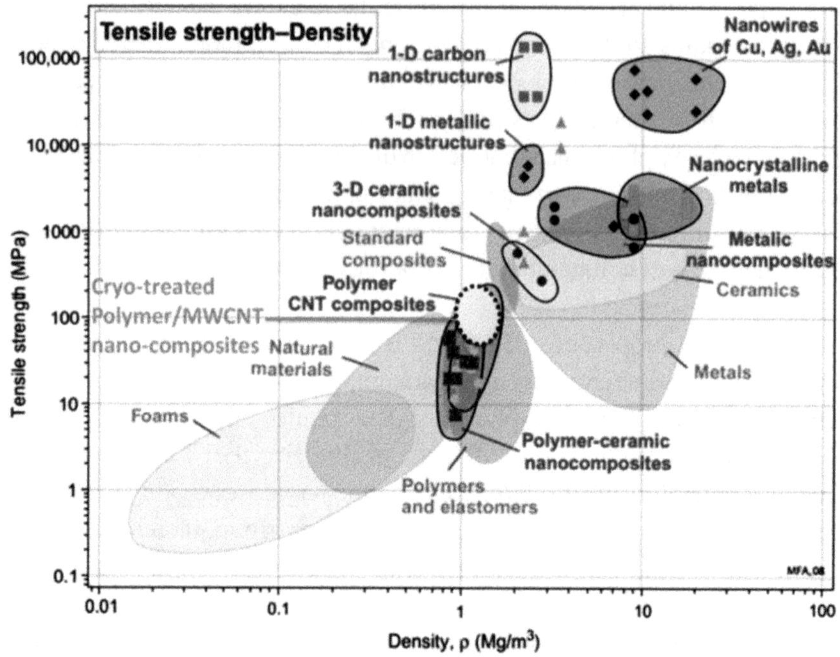

RESEARCH OUTCOME

Based on the extensive analysis of cryogenically treated polymer/MWCNT nano-composites, it is evident that there is an overall improvement in the structure and properties of the material. This introduces new possibilities for the potential applications of cryo-treated polymer/MWCNT nano-composites in near future. Thus, the data related to the Tensile strength, Young's modulus, Wear rate constant and Hardness of the studied polymer/MWCNT nano-composites are superimposed on the existing material selection Ashby diagrams (Ashby, 1992; Ashby, Ferreira, & Schodek, 2009) as shown in Figure 22 to Figure 24.

ACKNOWLEDGMENT

The authors are thankful to Dr. N. B. Dhokey, Professor, Department of Metallurgy and Material Science, College of Engineering Pune (M. S. India) for providing the

Figure 23. Position of cryo-treated polymer/MWCNT nano-composites selected in present work on the materials selection chart based on the Young's modulus and Density (Ashby et al.; 2009)

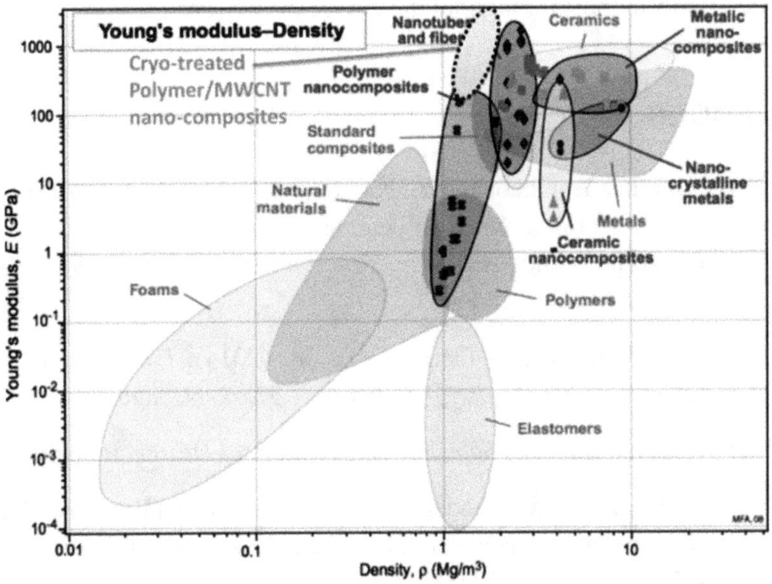

Figure 24. Position of cryo-treated polymer/MWCNT nano-composites selected in present work on the materials selection chart based on the Wear rate constant and Hardness (Ashby, 1992)

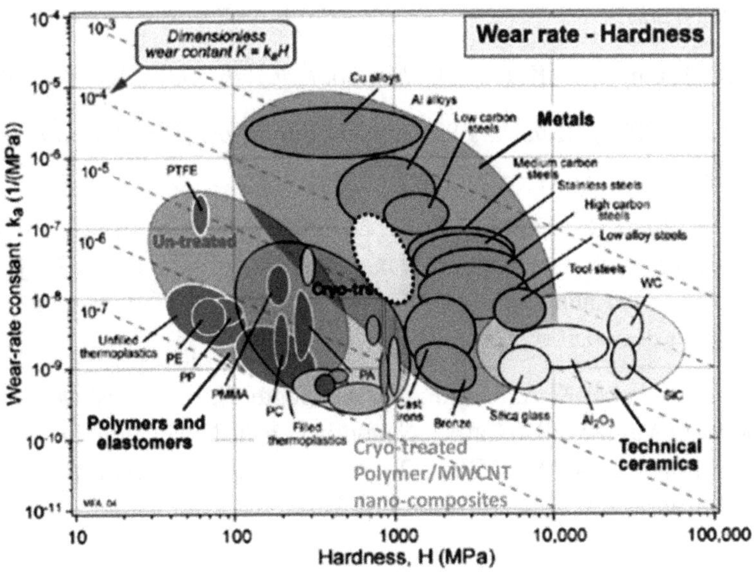

cryogenic treatment facility available at his lab. In addition, the authors are very grateful to Dr. C. L. Gogte, Professor, Department of Mechanical Engineering, M.I.T. Engineering College, Aurangabad (M. S. India) for his constant guidance, critical views and support throughout the course of this work.

REFERENCES

Ahmed, J. U., Gotoha, Y., Ohkoshia, Y., Nagura, M., Endo, R., & Hara, T. (2005). Hydration in a new semiaromatic polyamide observed by humidity-controlled dynamic viscoelastometry and X-ray diffraction. *Journal of Polymer Science*, *43*(13), 1640–1648. doi:10.1002/polb.20446

American Chemistry Council – Plastic Division. (2014). *Plastics and Polymer Composites: Technology Roadmap for Automotive Markets*. Author.

Ashby, M. F. (1992). *Materials Selection in Mechanical Design*. Pergamon Press.

Ashby, M. F., Ferreira, P. J., & Schodek, D. L. (2009). Material Property Charts and their Uses. In Nanomaterials, Nanotechnologies and Design (pp. 147-176). Elsevier. doi:10.1016/B978-0-7506-8149-0.00007-6

Askeland, D. R. (1996). *The Science and Engineering of Materials*. London: Chapman and Hall.

Baldissera, P., & Delprete, C. (2008). Deep Cryogenic Treatment: A Bibliographic Review. *The Open Mechanical Engineering Journal*, *2*(1), 1–11. doi:10.2174/1874155X00802010001

Bhushan, B., & Gupta, B. K. (1991). *Handbook of Tribology*. McGraw Hill.

Callister, W. D. (2000). *Materials Science and Engineering: An Introduction*. John Wiley and Sons.

Chopra, S., Batthula, S., Deshmukh, K., & Peshwe, D. R. (2017c). Tribological Behaviour of Multi-Walled Carbon Nanotubes (MWCNT) Filled Polybutylene Terephthalate (PBT) Nanocomposites. *Transactions of the Indian Institute of Metals*, *70*(3), 801–807. doi:10.100712666-017-1055-4

Chopra, S., Deshmukh, K. A., Deshmukh, A., Gogte, C. L., & Peshwe, D. (2016). The Influence of Multi-Walled Carbon Nanotubes on the Nucleation, Crystallization and Tensile Properties of PA-6/MWCNT Composites. *Journal of Basic and Applied Research International*, *19*(4), 247–252.

Chopra, S., Deshmukh, K. A., Deshmukh, A., Gogte, C. L., & Peshwe, D. (2018b). Prediction, evaluation and mechanism governing interphase strength in tensile fractured PA-6/MWCNT nanocomposites. *Composites. Part A, Applied Science and Manufacturing*, *112*, 255–262. doi:10.1016/j.compositesa.2018.06.012

Chopra, S., Deshmukh, K. A., Deshmukh, A., & Peshwe, D. R. (2018a). Inflorescence type morphology and mirror–mist–hackle pattern in tensile fractograph of MWCNT/PBT nano-composites. *International Journal of Materials Research*, *109*(6), 561–568. doi:10.3139/146.111630

Chopra, S., Deshmukh, K. A., & Peshwe, D. R. (2017a). *Cryogenic Treatment of Cubic Boron Nitride (CBN) Cutting Inserts*. Lambert Academic Publishing.

Chopra, S., Deshmukh, K. A., & Peshwe, D. R. (2017b). Theoretical prediction of interfacial properties of PBT/CNT nanocomposites and its experimental evaluation. *Mechanics of Materials*, *109*, 11–17. doi:10.1016/j.mechmat.2017.03.012

Chou, T. W., Gao, L., Thostenson, E. T., Zhang, Z., & Byun, J. H. (2010). An assessment of the science and technology of carbon-nanotube based fibers and composites. *Composites Science and Technology*, *70*(1), 1–19. doi:10.1016/j.compscitech.2009.10.004

Coleman, J. N., Khan, U., Blau, W. J., & Gun'ko, Y. K. (2006). Small but Strong: A review of the mechanical properties of carbon nanotube-polymer composites. *Carbon*, *44*(9), 1624–1652. doi:10.1016/j.carbon.2006.02.038

De Volder, M. F. L., Tawfick, S. H., Baughman, R. H., & Hart, A. J. (2013). Carbon Nanotubes: Present and Future Commercial Applications. *Science*, *339*(6119), 535–539. doi:10.1126cience.1222453 PMID:23372006

Deshmukh, G. S., Peshwe, D. R., Pathak, S. U., & Ekhe, J. D. (2011). Evaluation of mechanical and thermal properties of Poly(butylene terephthalate) (PBT) composites reinforced with wollastonite. *Transactions of the Indian Institute of Metals*, *64*(1-2), 127–132. doi:10.100712666-011-0025-5

Fakirov, S. (Ed.). (2005). *Handbook of Thermoplastic Polyesters: Homopolymers, Copolymers, Blends and Composites*. Wiley.

Farhani, F., & Niaki, K. S. (2008). *Cryogenic treatment of metals- a review of experimental studies, underlying mechanism and process equipment. 4th BSME-ASME International Conference on Thermal Engineering*, Dhaka, Bangladesh.

Federation of Indian Chambers of Commerce and Industry (FICCI). (2017). Plastic Industry for Infrastructure. *3rd National Conference on Sustainable Infrastructure with Plastics*.

García, E., Wardle, B., & John Hart, A. (2008). Joining prepreg composite interfaces with aligned carbon nanotubes. *Composites. Part A, Applied Science and Manufacturing*, *39*(6), 1065–1070. doi:10.1016/j.compositesa.2008.03.011

Gohardani, O., Elola, M. C., & Elizetxea, C. (2014). Potential and prospective implementation of carbon nanotubes on next generation aircraft and space vehicles: A review of current and expected applications in aerospace sciences. *Progress in Aerospace Sciences*, *70*, 42–68. doi:10.1016/j.paerosci.2014.05.002

Gowariker, V. R., Viswanathan, N. V., & Sreedhar, J. (1986). *Polymer Science*. New Age International Publishers.

Guzman de Villoria, R., Ydrefos, L., Hallander, P., Ishiguro, K., Noedin, P., & Wardle, B. (2012). Aligned carbon nanotube reinforcement of aerospace carbon fiber composites: substructural strength evaluation for aerostructure applications. In *Proceedings of the 53rd AIAA/ASME/ASCE/AHS/ASC structures, structural dynamics and materials conference*. American Institutes of Aeronautics and Astronautics. 10.2514/6.2012-1566

Indumathi, J., Bijwe, J., & Ghosh, A. K. (1999a). *Cryotreatment - an effective technique to enhance the abrasive wear performance of high performance engineering polymers and composites. International Conference on Polyimides*, Newark, NJ.

Indumathi, J., Bijwe, J., Ghosh, A. K., Fahim, M., & Krishnaraj, N. (1999b). Wear of cryo-treated engineering polymers and composites. *Wear*, *225-229*, 343–353. doi:10.1016/S0043-1648(99)00063-0

Kalia, S. (2010). Cryogenic Processing: A Study of Materials at Low Temperatures. *Journal of Low Temperature Physics*, *158*(5-6), 934–945. doi:10.100710909-009-0058-x

Kim, J. Y. (2011). Poly(butylene terephthalate) Nanocomposites Containing Carbon Nanotube. In B. Reddy (Ed.), *Advances in Nanocomposites – Synthesis, Characterization and Industrial Applications* (pp. 707–726). InTech Open Publication.

Kodgire, P. V., Bhattacharyya, A. R., Bose, S., Gupta, N., Kulkarni, A. R., & Misra, A. (2006). Control of multiwall carbon nanotubes dispersion in polyamide6 matrix: An assessment through electrical conductivity. *Chemical Physics Letters*, *432*(4-6), 480–485. doi:10.1016/j.cplett.2006.10.088

Koenig, J. L. (1999). *Spectroscopy of polymers*. Elsevier. doi:10.1016/B978-044410031-3/50005-0

Kumar, M. S., Sharma, N., & Ray, B. C. (2008). Mechanical Behavior of Glass/Epoxy Composites at Liquid Nitrogen Temperature. *Journal of Reinforced Plastics and Composites*, *27*(9), 937–944. doi:10.1177/0731684407085877

Lin, Y., Hill, D. E., Bentley, J., Allard, L. F., & Sun, Y. P. (2003). Characterization of Functionalized Single-Walled Carbon Nanotubes at Individual Nanotube-Thin Bundle Level. *The Journal of Physical Chemistry B*, *107*(38), 10453–10457. doi:10.1021/jp0306141

Mahmood, N., Islam, M., Hameed, A., & Saeed, S. (2013). Polyamide 6/Multiwalled Carbon Nanotubes Nanocomposites with Modified Morphology and Thermal Properties. *Polymers*, *5*(4), 1380–1391. doi:10.3390/polym5041380

Mahmood, N., Islam, M., Hameed, A., Saeed, S., & Khan, A. K. (2013). Polyamide-6-based composites reinforced with pristine or functionalized multi-walled carbon nanotubes produced using melt extrusion technique. *Journal of Composite Materials*, *48*(10), 1197–1207. doi:10.1177/0021998313484779

McCombs, J. B., Golland, D. I., & Mayer, G. (1974). The effects of crystallite orientation and size on the strength of polycrystals. *Materials Science and Engineering*, *15*(2-3), 275–282. doi:10.1016/0025-5416(74)90061-5

Meng, H., Sui, G. X., Xie, G. Y., & Yang, R. (2009). Friction and wear behavior of carbon nanotubes reinforced polyamide 6 composites under dry sliding and water lubricated condition. *Composites Science and Technology*, *69*(5), 606–611. doi:10.1016/j.compscitech.2008.12.004

Motozuka, S., Tagaya, M., Hotta, Y., Morinaga, M., Ikoma, T., Honma, T., ... Tanaka, J. (2013). Mechanochemical Fabrication of Carbon Fiber/Nylon-6 Composites with Interfacial Bondings. *Industrial & Engineering Chemistry Research*, *52*(5), 2182–2189. doi:10.1021/ie301333n

Mubarak, M. M. (2009). *Shedding the Extra Pounds: Weight Reduction Strategies for Fuel Efficient Cars*. A Frost and Sullivan Market Survey report.

O'Donnell, S., Sprong, K., & Haltli, B. (2004). Potential impact of carbon nanotube reinforced polymer composite on commercial aircraft performance and economics. In *AIAA fourth aviation technology, intergration and operations (ATIO) forum*. Chicago, IL: AIAA. Project no. 02MSR055-F3.

Pande, K. N. (2012). *Effect of Cryogenic Treatment on Polymers, Composites and Blends*. Nagpur: Visvesvaraya National Institute of Technology.

Pande, K. N., & Peshwe, D. R. (2015). *Effectiveness of Optimized Cryo-Treatment on Polymer Composites*. Patent no. 1724/MUM/2014 A.

Pande, K. N., Peshwe, D. R., & Kumar, A. (2012). Effect of the Cryogenic Treatment on Polyamide and Optimization of Its Parameters for the Enhancement of Wear Performance. *Transactions of the Indian Institute of Metals*, *65*(3), 313–319. doi:10.100712666-012-0135-8

Pavia, D. L., Lampman, G. M., Kriz, G. S., & Vyvyan, J. R. (2009). *Introduction to spectroscopy*. Cengage Learning.

Pritchard, R. (1964). The transparency of crystalline polymers. *Polymer Engineering and Science*, *4*(1), 66–71. doi:10.1002/pen.760040114

Ray, B. C. (2005a). Loading Rate Effects on Mechanical Properties of Polymer Composites at Ultra-low Temperatures. *Journal of Applied Polymer Science*, 95–98.

Ray, B. C. (2005b). Thermal Shock and Thermal Fatigue on Delamination of Glass-fiber-reinforced Polymeric Composites. *Journal of Reinforced Plastics and Composites*, *24*(1), 111–116. doi:10.1177/0731684405042953

Rosca, I., & Hoa, S. (2011). Method for reducing contact resistivity of carbon nanotube-containing epoxy adhesives for aerospace applications. *Composites Science and Technology*, *71*(2), 95–100. doi:10.1016/j.compscitech.2010.10.016

Szeteiová, K. (2010). *Automotive Materials: Plastics in Automotive Markets Today*. Bratislava: Slovak University of Technology.

Thawre, M. M., Pandey, K. N., Dubey, A., Verma, K. K., Pehswe, D. R., Paretkar, R. K., ... Manjunatha, C. M. (2015). Fatigue life of a carbon fiber composite T-joint under standard fighter aircraft spectrum load sequence. *Composite Structures*, *127*, 260–266. doi:10.1016/j.compstruct.2015.03.026

Vidal, B. C. (2013). Using the FT-IR linear dichroism method for molecular order determination of tendon collagen bundles and nylon 6. *Acta Histochemica, 115*(7), 686–691. doi:10.1016/j.acthis.2013.02.002 PMID:23506648

Xu, C., Jia, Z., Wu, D., Han, Q., & Meek, T. (2006). Fabrication of nylon-6/carbon nanotube composites. *Journal of Electronic Materials, 35*(5), 954–957. doi:10.1007/BF02692553

Zhang, W. D., Shen, L., Phang, I. Y., & Liu, T. (2004). Carbon Nanotubes Reinforced Nylon-6 Composite Prepared by Simple Melt-Compounding. *Macromolecules, 37*(2), 256–259. doi:10.1021/ma035594f

Section 2

Modeling and Applications of Nanotechnology in Aerospace

Chapter 5
Architect of Polymer Nanocomposites for Aerospace Applications

Sunil S. Suresh
Central Institute of Plastics Engineering and Technology, India

Kesiya George
Central Institute of Plastics Engineering and Technology, India

Smita Mohanty
Central Institute of Plastics Engineering and Technology, India

Sanjay K. Nayak
Central Institute of Plastics Engineering and Technology, India

ABSTRACT

Polymeric nanocomposites are significant engineering materials predominantly due to their enormous potential to meet a spectrum of applications, particularly in improving the strength-stiffness properties, thermal properties, optical and electrical properties. The exploitation of polymer nanocomposites in the aerospace industry is found to be attractive in recent times, since they can provide significant strength to the components with lightweight characteristics. In addition, a wide variety of polymers can be tuned with carbon and non-carbon-based nanomaterials and deployed as archetypes in the structural components of aerospace applications. Accordingly, this chapter consider the key properties of different nanomaterials in polymers as a function of nano-scale approach. Furthermore, this chapter is also dealing with the challenges that need to surmount the technological enduring of the polymer nanocomposites for advancements in the aerospace structural applications in the coming future.

DOI: 10.4018/978-1-5225-7921-2.ch005

INTRODUCTION

Aerospace manufactures are searching for the multifunctional materials which could be appropriate for its component development. In general aviation, cost of the fuel account for the 50% of the operational expenditure. Accordingly, aerospace manufacturers are looking for the suitable materials that could replace the metals in order to reduce the weight. Further, changes in the design and uses of the composite materials have been resulted in decreasing the weight by 30% of the aviation vehicles. Similarly, most of the aircrafts and spacecrafts are working in harsh atmospheric conditions, subsequently aerospace manufacturers and researchers are aiming to develop novel new materials which are durable with higher thermo-mechanical properties (Baur and Silverman; Joshi and Chatterjee, 2016; Wang et al., 2002).

The aerospace structures made up of the polymers are vulnerable to the disintegration of the polymeric chains. Therefore, researchers are attempting various techniques for toughening the polymeric matrix by introducing micro or nanomaterials into it as a filler. In a polymer matrix, reinforcing fillers is bound together by weak intermolecular forces rather than chemical interactions. In such conditions the nanomaterials are capable to serve hundred times better task than microfillers even with lower volumes like 1-5% (Joshi and Chatterjee, 2016). Moreover, due to the high aspect ratio of the nanomaterials, they can provide more surface area to the polymer matrix for the specific interactions. Similarly, if the reinforcement elements were dispersed at the nanoscale, it can attribute strong bonding with the matrix which can eventually result in the discovery of novel composite materials.

Among the different composites, polymer nanocomposites (PNCs) seek enormous attraction owing to its light weight characteristics along with multiple functionalities. Different carbon and inorganic nanomaterials are currently exploiting in the construction of different structural and non-structural components used in the aviation industries. By the proper selection of the nanomaterial as filler, the properties such as mechanical strength, thermal stability, flame retardancy, electrical and optical properties of polymer matrix can be tuned. In the current situation inorganic materials like nanoclay, SiO_2, ZnO are commonly employed for improving thermal stability, conductivity and flammability properties of various polymers (Hajibeygi et al., 2017; Liu et al., 2005; Rathod et al., 2017). Similarly, carbon based nanomaterials like carbon nanotubes (CNTs), carbon nanofibers (CNFs), graphenes and their various modified forms are also used for the generation of different PNCs (Gohardani et al., 2014; Monetta et al., 2015). Recently, such composite materials are applied in various structural parts such as fairing, interior component, primary wings, fuselage and turbine engine fan blade of the aircrafts and space craft's components (Mallick, 2007).

MATERIALS EMPLOYED IN THE AEROSPACE INDUSTRIES

Polymer Materials

Polymers and its modified forms are applied in both structural and non-structural components of aircraft and space craft's products. The excellent physico-mechanical properties, corrosion resistance, durability and cost effectiveness are the major attractive points of the polymers to be applied in the aerospace industries. Figure 1 illustrates different categories of polymers, their properties and application in aerospace components. Some of the thermoplastics are its self employed in the aircrafts structural and non structural parts. For instance, toughened and transparent polymers like acrylic plastics and polycarbonates are extensively employed in the aircrafts window shield and canopies (Vodicka, 1996). The windshield made up of polycarbonate can also resist the damage due to the hailstones and collision with birds. Additionally, high performance such as polyether ether ketones (PEEK), polyphenyl sulphide (PPS), polysufones, are employed in the aerospace applications (Vodicka, 1996; Costa et al., 2012). Furthermore, thermoplastics found as a prime candidate for landing gear strut door and access panels for F-5F aircrafts, and nose wheel door for Fokker aircrafts (Vodicka, 1996).

Similarly, low permeable polymers are commonly used as coating and laminating materials in the aerospace components. Further, epoxies, polyurethane and modified phenolics based polymers are commonly used as adhesive materials for fusing aircraft parts. Adhesives are also found applications in joining of composite structures which is employed in cabin decks, engine and transmission decks and tunnel covers. Moreover, nano-toughened adhesives are currently used in structural metal and composite bonding due to its superior mechanical and thermal properties. Recently, an epoxy based adhesive is employed in Airbus 380 fuselage for the generation of fibre-metal laminate known as glare (Jerome, 2012). Similarly, polymer based adhesives are also employed in the space craft's for joining of temperature protection tiles, solar panel and to affix ceramic tiles (Bhuvaneswari et al., 2017). Other than that, elastomers are another category of polymers generally employed in the structures which needs superior impact performance, high elongation, low stiffness and strength. Usually, elastomers are engaged in four main structural systems like airframes, landing gear systems, propulsion systems and instrument indicator and guidance systems (Bhuvaneswari et al., 2017).

Nanomaterials and Polymer Nanocomposites

Currently, both inorganic and carbon nanomaterials has been incorporated into various polymeric matrices for improving the inherent properties of the polymer.

Figure 1. Type of polymers and its application in the aviation industries

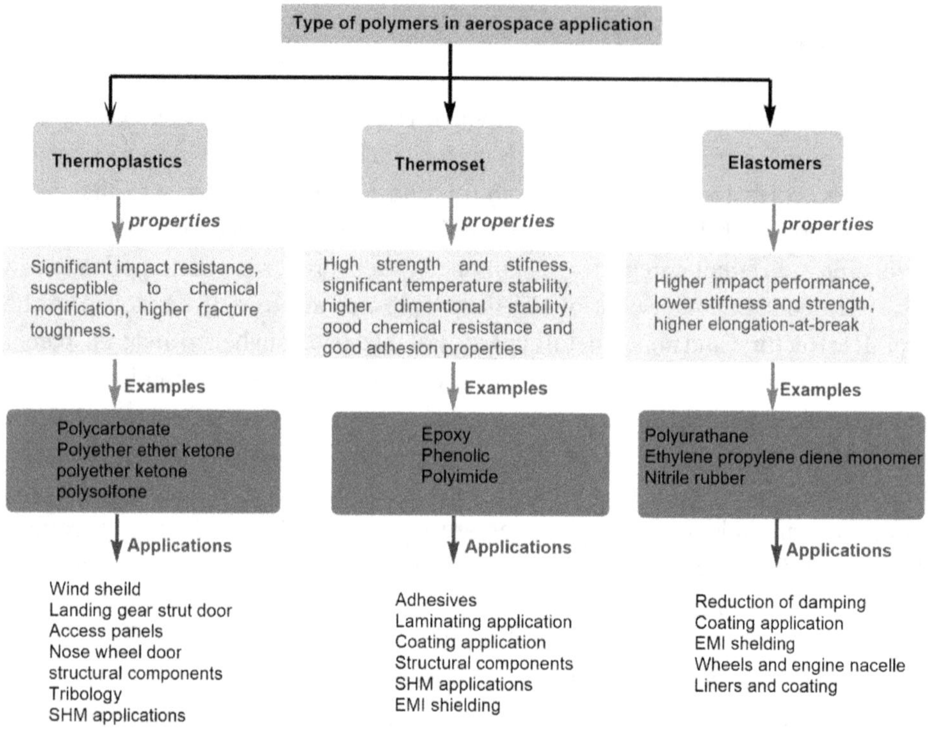

Nanomaterials like carbon nanotubes (CNT), graphenes, carbon nanofiber (CNF), nanoclays, SiO_2, TiO_2, ZrO_2, CuO, SnO_2, $BaSO_4$, Al_2O_3 and $CaSiO_3$ are commonly employed for the property enhancement of polymeric matrix. Common nanomaterials and its application in the aerospace industries are listed in table 1 (Joshi et al., 2016). Additional informations regarding the nanomaterials are detailed in the forthcoming sections.

The nanocomposites used in the aerospace applications can be majorly classified into two types based on the matrix employed. In the first case, various polymer phases are used as the base matrix and the generated composites are known as polymer nancomposites (PNC). Secondly, ceramic matrix such as alumina or a mixture of ceramics is used as a parent matrix along with various nanomaterials such as CNT, graphenes etc., such formed composites are known as a ceramic nanocomposite (Ravichandran et al., 2018). The superiority of the developed PNCs is dependent on the polymer matrix, nanomaterial and the synthesis techniques adopted. The preparation methods play a vital role in the properties of the developed PNCs. Type of PNCs employed in the aerospace industries and its primary functions are depicted

Table 1. Type of nanomaterials and its primary functions in PNCs employed in aerospace components (Joshi et al., 2016)

Type of Nanomaterial	Nanomaterial	Properties improved
Carbon based	CNT	Toughness, impact properties, electrical conductivity, heat stability, thermal stability.
	CNF	Toughness, impact properties, thermal properties.
	Graphene and graphene oxide	Gas barrier, electrical conductivity, ultraviolet stability, thermal stability.
Inorganic based	Nanoclay	Gas barrier, corrosion resistance, fire retardancy, toughness, impact properties, thermal stability.
	SiO_2 and Al_2O_3	Impact properties, toughness, scratch resistance.
	TiO_2 and ZnO	Ultraviolet stability, antimicrobial, toughness and impact resistance.
	CuO	Antimicrobial
	$BaSO_4$, CeO_2	Ultraviolet stability

in table 2 (Rathod et al., 2017). Generally, methods such as solution blending, melt compounding and *in-situ* polymerizations (figure 2) are employing for the production of PNCs (Ravichandran et al., 2018).

Application of PNCs in Aerospace Components

The design of unmanned aerial vehicles in the National Aeronautics and Space Administration (NASA) demands a long span, high elevation flights. This calls for strong, sturdy and insubstantial material for the structural parts of these aircraft

Table 2. Various PNCs in aerospace applications

Applications	PNCs
Structural applications	Epoxy/CNT, epoxy/clay, polyimide/Clay, nylon/clay, epoxy/SiC, PU/CNT, polyamide/CNT.
Coatings	PU/CNT, epoxy/CNT, Polycarbonate/CNT, epoxy/clay, PVDF/CNT, epoxy/alumina.
EMI shielding	Epoxy/CNT, epoxy/fiber/CNT.
SHM	PMMA/CNT, epoxy/carbon black, PVDF/lead zirconate tirate (PZT), epoxy/PZT,
Tribology	PPEK/TiO2

Figure 2. Different methods for the production of PNCs (a) Solution blending technique (b) melt processing technique and (c) in-situ polymerization

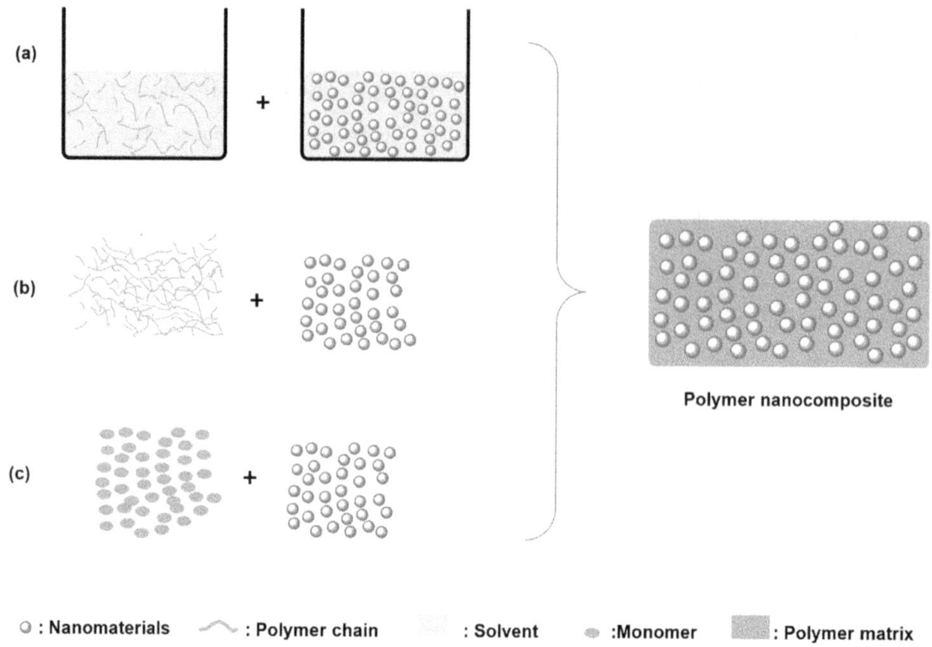

vehicles. The commercialization of polymer nanocomposites in the structural components of aerospace design makes it a safer, smoother, quicker and ultimately an economical transportation. The low density features of polymer nanocomposites accompanied with high strength and modulus values, predominantly adds to the reduced weight of the aerospace frames (Baur and Silverman; Joshi and Chatterjee, 2016; Wang et al., 2002). Other than the nanommaterials, the chosen polymer matrix is another crucial constraint for the mechanical and physical strategical performance of the aerospace components. Various polymers have been employed as the co-continuous matrix during the generation of PNCs for different components of aerospace vehicles.

Epoxy Based PNCs

The vulnerability of metals to corrosion seeks alternative materials that can replace the drastic needs attributed by the steels or the employment of superficial coatings that prevents the corrosion in stainless steel is on steady progress. Epoxy (mostly diglycidyl ether of bisphenol-A), is a thermoset polymer with low molecular weight, hardened by curing with various curing agents have been widely explored owing to the large scale utilities offered by them. The well bounded adhesive property in a variety of substrates with superior mechanical strength along with chemical as well as heat resistance appeals the application of epoxy in aerospace and most engineering materials (Jeon et al., 2013). However, nanomaterials are incorporating into the epoxy matrix in order to improve its inherent properties. For instance, Zhao et al., (2006) developed CNT/epoxy composite that further toughened with continuous carbon fibre. Epoxy with fiber reinforced systems and its PNCs have been employed in flaps, stabilizers, rudders, landing gear doors, fairings, ducts, liners, spoilers, aiierons, rudders, elevators, wing box skins, storage compartments, panels, rotor blades, tail fins, payload bay doors, filament wound case in the structural intergradient's of aircraft parts (Wright, 1991).

Polyamides Based PNCs

Earlier in the 1990s, Toyota research team has first established PNCs based on polyamides commercially (Deguchi et al., 1992). Even though, the inborn adequacy of polyamides is simulated with the incorporation of fibers such as glass fibers into the matrix (Yano et al., 1997). However, there still persists the inadequacies of uniform distribution of fiber in the matrix which can affect the potential usage of the system. Moreover, the integration of nanofillers into the matrix further escalates the applications of polyamides in the aerospace industry and engineering component parts. Polyamides and their PNCs have been hired for the cockpit and cabin parts, flare sequencer boxes, ventilation ducts, ventilator casing, brackets, engine access panels and air distribution ducting (Han et al., 2004).

Polyurethane Based PNCs

Polyurethanes are another class of polymers which serves as an effective matrix in various contributions to the aerospace industry. Properties of the polyurethanes are the resultant consequences of the molecular interactions in the polymer chains. The rigidity of the polyurethane chains can be further enhanced by nanomaterial incorporation. The nanomaterials like nanoclays can improve tensile strength and elongation-at-break of the polyurethane. Han et al. (2004) have been shown a higher endorsement in the tensile properties of the PU/montmorillonite nanocomposites prepared via solution intercalation method. Generally, PU based sandwich nanocomposites are deployed in the aircraft engine nacelle, on the protection wings of fuel tank, tailplane panels for defending the stones and pebbles while take-off timings etc. The occupancy of impermeable silicate layers in the PU matrix can reduce the water swelling that eventually adds on the increased resistance to water (Chen et al., 2003; Fletcher et al., 2010).

Elastomers Based PNCs

Elastomers are known to be good matrixes in PNCs for the development of structural components in aerospace applications. Since, viscoelastic polymers are well known to meet the damping problem offering the minimal vibration of the aircraft component parts in operation. However, the adsorption of vibrational energy by means of mechanical damping is the main concern in aerospace engineering design (Wright, 1991). There is a wide choice of fillers for rubbers to reinforce the elastomeric matrix to impart different properties to the matrix, yet the counterpart by nanofiller is quite demanding with the advancements of special instigation. For instance, Fletcher et al., (2010) indicated the possibilities of elastomeric and CNT based nanocomposites for the applications such as electromagnaetic dissipation and shielding. Similarly, Araby et al., (2014) prepared elastomer nanocomposites using solution mixing technique with enhanced thermal and electrical properties.

Other Polymer Based PNCs

Poly(p-phenylene benzbisoxazole) (PBO), polyaniline, polyarylacetylene (PAA), polyimides, polyethylene terephthalate (PET) and PEEK are the other widely utilized matrix systems for aerospace structural components. PEEK is considered as a high performing semi crystalline polymer shows outstanding mechanical, thermal and chemical resistant properties, allowing a hostile environment for different structural components. However, the advantages offered by the matrix the low temperature sustainability deprives the need for the incorporation of nanofillers into it. The mechanical properties of the PEEK can be improved with the help of nanosystems like nano-silica (Jen et al., 2005). The rigid rod polymer, PBO distinguished by its unique tensile strength and thermal stability are another class of highly deployed matrixes for aerospace. Kumar et al., (2002) developed PBO based PNCs with 10 wt% CNT which has shown enhanced tensile strength and minimal shrinkage as compared with the parent blend. Polyaniline is a conductive polymer which serves as a corrosion resistant matrix from the past decades itself (DeBerry, 1985). Liu et al., (1991) has been reported that among all the prevailing systems, double stranded polyaniline exhibits a possible substitute for the chromate systems. They have been utilized in the deicing, threat detection, surveillance sensors and in electrostatic discharges in the aerospace industry. PAAs are the steady developing polymeric matrices due to its superior heat resistance and the ablation resistance. The PAA resin matrixes with low moisture outgassing characteristics are suitable for spacecraft structures, insulative liner for rocket motor casings etc. The application of polyimide foams in aerospace industry includes thermal insulating liners for the cryogenic tanks, radomes, chairs, insulator covers, circuit substrates, wire insulation etc.

Fabrication of polyamides with nanomaterials imparts higher thermal properties to the polymers. In a study, Ogasawara et al., (2004) scrutinized the enhancement of the heat resistance attributed by the polyimide matrix with the addition of CNTs to the system. Similarly, PET based PNCs are well appropriate for the flexible laminates, balloon systems, decelerator systems, inflatable structures and pressure vessels. These flexible laminates can achieve comparatively a higher weight savings (Zeng et al., 2005).

Archetypes of Polymer Nanocomposites Modeling

Polymer nanocomposites fabrication is indeed possible with the statistical probability that encompasses the relationship between different structural as well as bulk mechanical properties in the PNC in relative to the polymer structure, nanomaterial and the polymer/nanomaterial interface. The structural properties of polymer nanocomposites and its processing are primarily ruled by the proper utilization of interatomic and intermolecular potential energy functions used in the simulation and the time span of growth imposed on the simulation. The interatomic potential function shall comprise of the vander waals systems, covalent and ionic interactions. The computer simulated modeling of PNC can be classified into molecular dynamics approach and Monte Carlo approach. Molecular modeling drafts a flow prognosis and the analogous chain conformation with regards to either the minimum energy or the abrupt drop in energy (Jin and Boyd, 1998). However, Monte Carlo method employs numbers for the discovering the particle displacement at the time of molecular simulation (Vao-soongnern et al., 2000).

For investigating the matrix fiber transfer numerous models such as Cox, Kelly and Tyson model and Lewis and Nielsen Model have been anticipated and scrutinized. The strength and stiffness properties of the PNC can be correlated with the toughness and vice versa, implies that there must be proper attention in the early stages of composite development (Njuguna and Pielichowski, 2003). In Cox model, a comprehensive study was done concerning the orientation effect of fibres on the strength and stiffness properties of the loaded matrix and other fibrous materials and had arrived a conclusion that load transport from the fiber to fiber can minimize the effective modulus, it is only possible when the fibre length is less than 100 times the fibre diameter. In addition to it, when the fibre length is approximately 10 times the diameter, the efficient modulus probably can reduce to one-half (Cox, 1952). Kelly and Tyson [36] predicted the behaviour of fibre reinforced metals from their experimental studies and their conclusions can be summarized as follows. Whenever the matrix is embedded with strong continuous fibres possessing a yield stress for plastic flow lesser than the breaking stress of fibres, the tensile stress in fibre and matrix are taken to be equal (Kelly and Tyson, 1965). Substantial consideration has been devoted to the modelling and assessing the CNTs young's modulus value. Lourie et al., (1998) conducted a study for evaluating the compressive distortion of CNTs in an epoxy matrix and the elastic modulus was founded as a result of concentric cylinder model for thermal stresses. Micro-Raman spectroscopy was utilized for monitoring these compressive deformation raised by the thermal quenching of epoxy matrix embedded CNTs. Bharadwaj (2001) invokes a tortuous based model for investigating the impacts of concentration, direction, length and degree of delamination of the layered silicates on the permeability, systematically. Distributing longer sheets in the polymeric matrix can constructively effect on the enhancement in the tortuosity and moderation in barrier property degradation. Luo et al., (2003) figures out a three phase model for explaining the partial intercalation and exfoliation present in the epoxy matrix, exfoliated clay layers and the nanolayer clusters. Employment of Mori-Tanaka method helps in the computation of the modulus of the nanocomposite with respect to parameters such as exfoliation ratio, clay layer aspect ratio, etc. (Tan et al., 2005). Luo and his group found out that with the proper experimental results, the model forecast seems to be in good agreement. Also, they have validated this model in cases, where the established mechanical models failed to outline a comprehensive measurement for the increased stiffness offered by the clay platelets in a matrix. However, the implementation of this model necessitates computational exertion; the model errand for the scrutiny of intercalation, exfoliation and unintentional dispersion of clay nanolayers and clusters in the model (Luo and Daniel, 2003).

Even though this section confers general understanding of the different types of modeling exists in the PNC fabrication, the data provided here is an outlined capsule for the basic understanding of the terminologies. The modeling of PNC reveals a path for the apparition and the examination of the processing efficiency with respect to the rate of production taking into consideration of the cost or energy consumed for the fabrication.

BREAKTHROUGH IN COMPOSITE STRUCTURE WITH NANOPARTICLES

Carbon Based Nanoparticles as Constructive Materials

The carbon based nanomaterials have been drawn attention in the field of science and technology due to its widespread properties. The carbon nanomaterials can exist in different major forms such as carbon nanotubes (CNTs), carbon nanofibers (CNFs) and graphene. Along with it, researchers also identified varied forms of carbon nanomaterials like nano-particles, cones, carbon nanoonion, scrolls, whiskers, nanoflowers, nanoporous carbon and graphite polyhedral crystals (Llobet, 2013; Yi and Shen, 2015). Even if, among the carbon nanomaterials CNT, CNF, graphene and their varied modifications are extensively considered in the aerospace industry due to its inherent nanoscale features and interesting physico-mechanical characteristics. The basic properties of carbon nanomaterials are indicated in table 3 (Al-Saleh and Sundararaj, 2011; Breuer and Sundararaj, 2004; Coleman et al., 2006; Finegan and Tibbetts, 2001; Tibbetts et al., 2007; Xie et al., 2005).

CNT Based Polymer Nanocomposites

CNTs are structurally belonged to the family of fullerenes, having cylindrical shape and the ends are either opened or closed with fullerene cap. Structurally, CNT is consisting of rolled graphene sheet with the size of nanometres. The graphene sheet can be rolled up in different orientations to produce different types of CNTs during its generation. Besides, CNTs are almost hold one-dimensional structure according to its high length to diameter ratio. CNTs have two structural adaptations like single walled carbon nanotubes (SWCNTs) and multi walled carbon nanotubes (MWCNTs). In which, SWCNTs are formed by rolling up of a single graphene layer and MWCNTs are fashioned by wrapping multiple layers of graphene (Popov, 2004; O'connell, 2006; Swain et al., 2017).

Polymer nanocomposites based on the CNTs have prime important positions in the aerospace structural and non structural applications. As compared with macro filler systems, CNTs requires only lesser quantity in order to fulfil the specific structural, physical and chemical properties. Generally, SWCNTs are found to be more attractive as compared with MWCNTs in reinforcing polymer matrixes which are usually employed for the structural applications. During the mechanical loading, the stress transferred into the surface of the incorporated SWCNTs in the polymer matrix and due to the hexagonal network of carbon atoms and high cohesive energy of C-C

Table 3. Basic properties of carbon nanomaterials

Property	CNF	SWCNT	MWCNT
Diameter	$5\times10 - 2\times10^2$	0.6–0.8	$5–5\times10$
Aspect ratio	$25\times10 -2\times10^2$	$1\times10^2 - 1\times10^4$	$1\times10^2–1\times10^4$
Density (g/m^3)	~2	~1.3	~1.75
Thermal conductivity (W/m K)	1950	$3\times10^3– 6\times10^3$	$3\times10^3– 6\times10^3$
Electrical resistivity (Ω cm)	1×10^{-4}	$1\times10^{-3}–1\times10^{-4}$	$2\times10^{-3}–1\times10^{-4}$
Tensile strength (MPa)	292×10	$5\times10^4– 5\times10^5$	$1\times10^4–6\times10^4$
Tensile modulus (MPa)	24×10^4	153×10^5	1×10^6

Figure 3. Types of carbon nanotubes (a) SWCNT (b) MWCNT

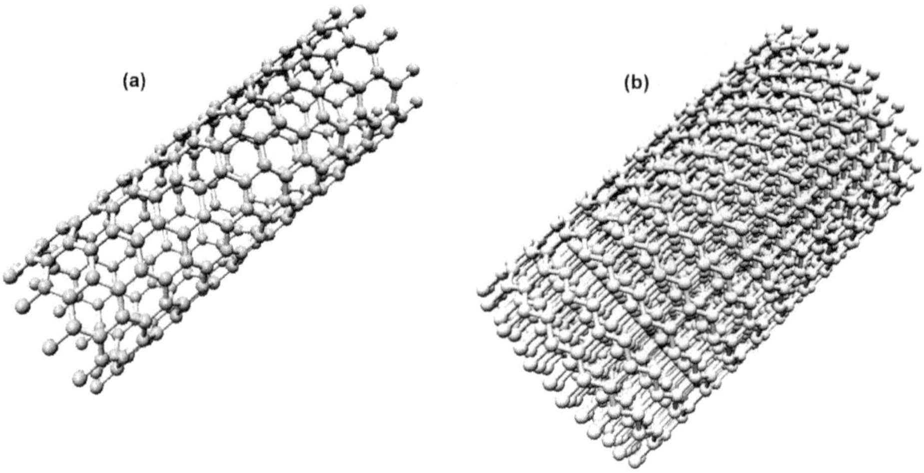

bonds they can provide good mechanical strength. While in the case of MWCNTs the inner concentric layers undergoes internal slippage during mechanical loading which arises due to comparatively weak vander walls forces present in it. Consequently, MWCNTs shows lower load bearing ability during nanocomposite formation as compared with the SWCNTs (O'connell, 2006; Moniruzzaman andWiney, 2006; Jakubinek et al., 2015; Kausar et al. 2016).

SWCNTs and chemically functionalised MWCNTs integrated polymeric composite systems improves the impact performance as well as fracture toughness. The improvement in the toughness of the nanocomposite polymeric systems is associated with the nanotube pullout mechanism and the bridging of cracks in the polymer matrix (Joshi and Chatterjee, 2016). Molecular orientation and geometry of CNTs present inside the polymer matrix is also plays important role in energy dissipation and fracture mechanism of the composite. Research over the composites is also indicated that the CNTs with 90° longitudinal orientation shows maximum toughness in the polymer nanocomposite (Joshi and Chatterjee, 2016). Due to the lower denser characteristics as compared with metals, the carbon fiber based CNT composites could be useful for the development of airframe components. Such nanocomposite systems potentially offer significant strength and corrosion resistance to the developed products. Consequently, such systems are commonly found in airframe application. The CNT reinforced epoxy materials have been used in the wingtip fairings of Lockheed Martin's F-35 fighter aircrafts (Kausar et al., 2016). Similarly, those materials are also employed in tilt-rotor aircraft like V-22 Osprey and Tomahawk missile components.

The main motive of CNTs based nanocomposite materials in aircrafts is associated to the weight reduction as compared to composite developed with CNFs and other inorganic nanomaterials. The airbus like A380 uses 61 wt% of copper and 10 wt% of steel and titanium in it structural parts, therefore, in order to reduce the weight, CNT based composites are deemed to be a suitable candidate (Ye et al., 2005). In a stimulation study, O'Donnell et al. (2004) used CNTs reinforced polymers as an airframe instead of conventional aluminium material for the theoretical aircraft structures such as Boeing 747-400, Boeing 757-200, Airbus A320 and E145. This study indicated that by the use of CNTs reinforced polymers the consumption of the fuel by the aircraft can be reduced due to loss in weight.

In addition to it, CNTs based polymer nanocomposites are also considered as prime candidates for improving the electrical conductivity. On the microscopic consideration, incorporation of CNTs in to the polymer matrix results in the generation of percolating network, which can further trigger the flow of electrons via tunnelling or hopping mechanism (Jouni et al., 2014; Min et al., 2010). In a recent study Jakubinek et al., (2015) indicated that loading of SWCNTs in lower concentrations up to 1 wt% in to epoxy system can improve the electrical conductivity while maintaining the structural bonding capacity. Such kind of fabricated systems paves a new route for developing polymer nanocomposite based conductive adhesives. Furthermore, in an interesting study Jarosz et al., (2011) indicated the possibilities of macroscopic CNT based wires and ribbons for the application of power transition due to its superior conductivity as well as lower resistivity. In the present situation, most of the cables and wires carries metallic conductors like copper for the power and information transmission (Suresh et al., 2017a; Suresh et al., 2017b). Aircrafts like Boeing 747 contains around 135 miles of copper wiring which represent a weight of 4000 lbs (Gohardani et al., 2014). By fabricating the properties such as alignment, doping, type enrichment and densification of CNTs, it could be useful in replacing metallic conductors ranging from power transmission to the energy storage and conversion devices (Lee and Ramakrishna, 2017).

Furthermore, CNT based polymer nanocomposite systems are established as a solution for managing the vibration damping and noise control in air and space crafts. Khan et al., (2011) indicated the possibilities of MWCNTs in order to improve the vibration damping properties of carbon fiber reinforced polymer. Even lower quantity (<1%) of MWCNTs could be useful for the generation of a polymer nanocomposite with good damping properties. Verdejo et al., (2008) have designed silicon nanocomposite by incorporating CNT into the polymer matrix. With the increase of nanomaterial in the polymer matrix the composite has shown improvement in the reduction of acoustic absorption. By invoking the properties of such CNT based nanocomposite it may feasible to control the vibration damping as well as noise control in the aerospace vehicles. Moreover, carbon buckypaper (aggregate of CNTs) are employed for the production of lightning protection materials (LPMs) either by incorporating into the polymer matrix or coated on the surface of the material. Han et al., (2015) synthesised carbon buckypaper and applied as a LPM material along with the epoxy adhesives for the protection of the carbon fiber reinforced plastics. Alternatively, CNT based coatings are applied onto the surface of the aircraft as coating for enhancing the electrical conductivity consequently, such coating can elevate the lightning strike efficiency. However, unlike CNT the carbon buckypaper does not need proper dispersion in the polymer matrix since they can form sheet like material so they can directly applied over the reinforced composite materials. As a result, carbon buckypaper is established as an exciting material for the lightning protection of the aerospace vehicles in the current scenario (Prabhakaran, 2011).

Other than the aforementioned properties, the materials employed in the aircrafts or space crafts required higher thermal stability as well as flame retardant properties. CNTs offer improvement in the thermal stability and flammability properties of polymeric materials. In a study, Kasshiwagi et al., (2005) demonstrated the enhancement in flame retardancy of the polymer matrix after incorporating SWCNTs into the polymer matrix. Homogeneously dispersed CNTs in the polymer matrix forms a network work structured layer over the polymeric surface such formations act as a boundary between the solid phase and the gaseous phase during combustion. Therefore, layer formed does not allows the formation of cracks and openings during the burning consequently it reduces the flammability characteristics. Schartel et al., (2005) have been evaluated the thermal and flame retardancy effect of 6 wt% MWCNT on the polyamide. This study illustrates that the percolated structure developed inside the polyamide matrix during the mixing improves thermal degradation, melt viscosity and flammability behaviour. Furthermore, several studies are indicating that functionalised CNTs can also improve the thermal and flammability of different polymers. Yu et al., (2011) investigated the effect of molybdenum-phenolic resin grafted MWCNTs for improving the characteristics such as flame retardancy and mechanical properties of epoxy resin. Interestingly, results indicate that the modified epoxy resin shows higher flame retardancy via increased char yielding as compared with unmodified counterpart. Similarly, mixtures of nanoclays, silicon and graphenes along with CNTs can impart improvement in the flammability and thermal properties of the various composites (Beyer, 2002; Kuan et al., 2010; Huang et al., 2014). These studies are pointing towards the possibilities of CNTs or modified CNTs in the production of composites for the aerospace applications since epoxy based components are frequently employed as a structural material in aerospace vehicles (Ye et al., 2005).

In addition to it, CNT nanocomposites also established its applications as advanced materials in aerospace industry such as advanced structural health monitoring systems (SHMs) and coating applications. In the present scenario, SHM sensors are either incorporated within the composites or outside surface of the structure, such sensors responding to the lower impact or strain developed due to the lower impacts (Sebastian et al., 2014). The CNTs exhibits sensitivity analogous to conventional type of materials; moreover they can easily integrate into the composites. In 2004, Fiedler et al., (2004) investigated the possibilities of CNTs in FRPs for the first time. Thereafter a number of studies have been initiated by different researchers which explore the use of CNTs for sensing the damage properties of FRPs. After the incorporation of CNTs into the polymer matrix in the FRPs they can act like a neuron sensory networks, which facilitate the detection of microcrack and its initiation. Similarly, Boger et al., (2008) demonstrated that little amount (0.3 wt%) of MWCNTs are capable of producing self sensing material with GFR epoxy laminates. In another study by Sebastian et al. (2014) indicated that CNT coated glass gibers are alternative to conventional metal foil strain gages. Even if, numbers of research articles are indicating the possibilities of CNTs as SHMs, the real industrial application is still lacking in the structural applications. Alternatively, CNTs can act as good ablative material and therefore, those could be utilized in heat shielding materials for the application in solid rocket motors (Boger et al., 2008; George et al., 2018).

In addition, studies have been indicated that CNT based nanocomposites can also be used for advanced coating technologies in the aerospace structures. Such systems could prevent UV-radiation, corrosion, thermal exposure and oxidation due to various components in the atmosphere. In the current scenario, most of the air forces and domestic aircrafts are using alloys based on aluminium (Al 2024-T3 or Al 7075-T6 for structural construction, however those are susceptible to the corrosion (Bierwagen and Tallman, 2001; Gkikas et al., 2012). Consequently, such parts are coated with polymeric components like epoxies and flexible polyurethane due to its versatile chemical and physical properties. Presently, researchers are exploring the possibilities of CNTs in coating materials since they can act as a protective barrier to Al by suppressing the corrosion rate. Gkikas et al., (2012) successfully developed epoxy/CNT nanocomposite for the coating applications. Moreover, researches over CNTs are signifying that surface modified CNTs are more effective as compared to the unmodified CNTs since former one can enhance the compatibility through interfacial bonding. Similarly, since from the early days of aviation, researchers have shown their interest over the fabrication of polymers with de-icing and anti-icing characteristics. Due to this fact that ice developed on the surface of the aircrafts appreciably blocks the air flow over the wing and alters the weight ratio. The CNT based nanocomposites are established as excellent candidates for suppressing the formation of aircraft icing during flying, since they can engender the super hydrophobic and ice-phobic capacity to the polymeric coating materials (Larson et al., 2013; Prabhakaran, 2011). The common applications of CNTs in aerospace and its major functions are listed in the table 4.

CNF Based Polymer Nanocomposites

As compared to the other carbon nanomaterials, CNFs are using as a reinforcement agent in the polymer matrix owing to its higher tensile strength and modulus. Moreover, high aspect ratio and relatively lower production cost also contributes in the remarkable consideration of CNFs in the field of polymer technology. Analogous to other nanomaterials the excellent dispersion of the CNFs in the polymer matrix is the key parameter for the overall performances of the final developed PNCs. Besides, CNFs have considerable electrochemical properties which could be used for the production of the conductive materials. Furthermore, the surface of the CNFs is possible to modify with various chemical agents. Therefore, these tailor-made

Table 4. Applications of CNTs and its functions in aerospace industry

Application of CNTs in aerospace	Function of CNTs
Structural components	CNTs can provide higher toughness and strength along with the light weight characteristics to the polymer matrix which is used for the structural components.
De-icing components	Aligned and oriented CNTs suppress the formation of the icing during flying.
Lubrication agent	CNTs act as better lubricating systems than conventional systems.
Adhesives for joints and bonding	Adhesive made up of CNTs can inhibit sparking across the joints.
LPM systems	In PNCs CNTs acts as a conductive nanoparticle thereby enhances light strike efficiency.
EMI shielding	CNTs act as EMI shielding of computer screens onboard air crafts
Flame retardant agent	Homogeneously mixed CNTs in PNCs can enhance the flame retardancy via formation of char layer during burning.
SHM systems	CNT incorporated in the PNCs can act as a neuron sensory network and which could be useful in detection of mico-cracks and voids.
Coating application	CNTs based PNC can be useful for the prevention of oxidation and corrosion in the harsh atmospheric condition.
Portable nanomesh	Applied in the space craft's for the purification of waster.
Gas and chemical detector	For the sensing and detection of the gas and chemical in the aerospace and space crafts.

properties of CNFs are utilizable in the generation of the materials that could be used in aerospace applications (Al-Saleh and Sundararaj, 2011; Swain et al., 2017; Feng et al., 2014).

The mechanical properties of the polymers can be enhanced with the incorporation of the CNFs into its matrix. The CNF present in the polymer matrix can transfer or hold a significant amount of energy applied into it. Rodriguez et al., (2011) formulated different PNCs employing aerospace grade epoxy-sized plain woven carbon layer in the presence of amine modified CNFs and non modified CNFs. Among them, amine modified CNFs have shown higher mechanical properties as compared with virgin polymer and non-modified CNF containing PNC. This is attributed to the fact that amine functionalization results in the transfer of more load to the CNFs. Similarly, Zhu et al., (2010) compared the mechanical properties of CNF and the surface treated CNF (via silanization method) in epoxy materials. Interestingly, their result demonstrated that as compared with the pristine component CNF fabricated systems have shown higher tensile properties and elongation-at-break. Among them, surface modified CNF incorporated systems have shown higher mechanical properties due to enhanced miscibility achieved with the polymer matrix. CNFs are also used to fabricate properties of other polymers such as PEEK which commonly used in aerospace components. In a study by Sandler et al., (2002) pointed out that modulus, yield stress and fracture strength of the PEEK can be improved with varying CNF concentrations in the polymer matrix.

The CNFs are also found to be applicable in electromagnetic interference shielding (EMIS) components employed in aerospace materials. The nanomaterials like CNFs can provide better shielding effect because of its higher surface area (Fu and Chung, 1996). Yang et al. (2005) formulated PNCs based on CNF and liquid crystal polymer via melt mixing technique. Further, generated PNCs have shown significant EMI shielding effects over a range of frequencies along with the enhancement in the thermal properties. The shielding mechanism mainly depends up on the reflection and multiple reflections. Yang et al., (2005) evaluated the EMIS effectiveness in the range of 8.2–12.4 GHz (X-band region) for the CNF and CNT in polystyrene matrix. Interestingly, their result indicates that with enhancing the CNF content, EMIS capacity of the polymer has been enhanced. However CNTs are more effective in offering EMIS compared to CNF in the same concentration. Alternatively, CNFs are also found to be attractive candidates for lighting protection components.

Graphene Based Polymer Nanocomposites

Graphene is consists of atomically thick single layer, planar sheet (2-dimensional) of carbon closely packed into the form of honeycomb crystal lattice (Figure 4 (a)). Each carbon atom present in the graphene are in SP^2 hybridised state and are connected to

Figure 4. (a) Graphene layer with a honey comb structure (b) Graphene oxide

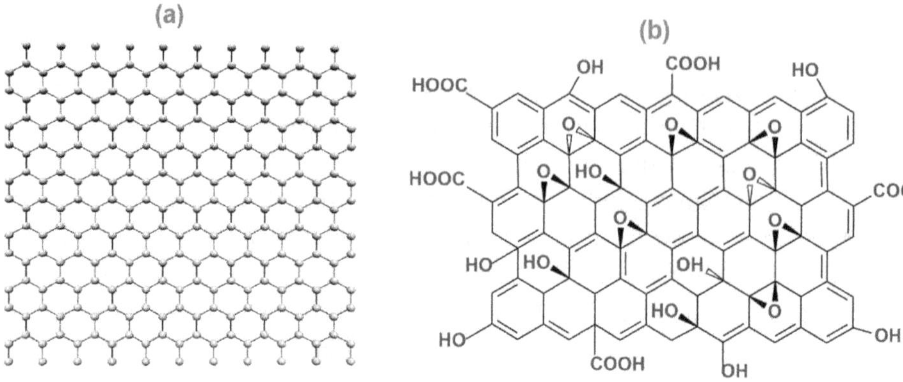

each other with σ bond having a bond length of 0.142 nm approximately. Moreover, carbon atom present in the lattice has π orbital which contributes to the delocalised network of electrons. Graphene has a large theoretical surface area of 2630 m^2g^{-1} for single layer of graphene as compared with other materials (Allen et al., 2009; Rafiee et al., 2009; Kim et al., 2010; Hussein and Kim, 2018).

The interest over the application of graphene in the aerospace industries has been come from its exceptional mechanical, electrical and thermal properties. Moreover, through the chemical modification the surface of the graphene can be activated by the incorporation of the polar functionalities, such forms are known as graphene oxides (GO) (Figure 4 (b)). As compared with the graphene, GOs have wide attractions in the nanocomposite technology, since those are readily available in great amount, easy to exfoliate and disperse in the polymeric matrix. The polar oxygen functionalities developed during the surface modification of graphene results in the use of GOs as reinforcing material in composites (Young et al., 2012). Such polar functionalities improves the interfacial interactions within the polymer matrix as a result, mechanical and thermal properties can be enriched.

Generally, graphene and GO are employed as filler in the polymeric systems inorder to improve the mechanical properties. Among them, nanocomposite composed of polymer and GO are utilized in the numerous aerospace applications due to their superior tensile strength and modulus. The GOs enhances the mechanical characteristics of epoxy materials that are commonly employed in the structural application of aircrafts due to its lightweight behaviour. NASA pointed out that epoxy nanocomposites can be employed for cryogenic storage applications (Mirabedinim and Kiamanesh, 2013). Atlan and Uysal (2016) investigated the effect of various ratios of graphene in epoxy resin and the formed composites have shown improvement in the mechanical properties such as tensile strength and stress-at-break. The improvement in the mechanical properties of nanocomposite with epoxy composite is associated with the high surface area of the GOs and accurate adhesion of GOs with the polymer matrix. The type of epoxy resins, chemical and physical nature of epoxy resin, hardener and preparation method also play crucial role in the generation of adequate mechanical properties. In a study, Gobi et al., (2017) point out that graphene quantum dots (GQDs) could be useful for the production of tougher and stronger polymer composite. Even with 2.5 wt% incorporation of GQDs into the epoxy matrix, the tensile strength, young modulus and nominal strain-at-break enhanced by 125, 153.4 and 18.1% respectively in contrast to the neat epoxy. Similarly, graphene nanoplatelets (GNPs) (short stacks of individual layer of graphite's) are also established as a reinforcing material for polymeric matrix. In a research King et al., (2015) fabricated the aerospace grade epoxy resin (EPON 862) with GNPs having various concentrations for the enhancement of mechanical properties. Furthermore, mechanical measurement indicates that GNPs have the ability to improve the mechanical properties like tensile strength and modulus in to a significant level.

In general, mechanical properties of the graphene based nanocomposites depend upon the homogenization of graphene as well as compatibility of graphene between the polymer matrixes. Graphene and GOs are able to enhance the glass transition temperature (T_g) of the polymeric matrix. The incorporation of such nanofillers affects the interchain bonding eventually the polymeric chains in the matrix become arrested. Moreover, during the mechanical stress, the stress homogeneously distributed over the graphene or GOs present in the matrix. Similarly, when the surface area of the reinforcement increases, the matrix-reinforcement is also enhanced. Subsequently, the load developed in the matrix can successfully transport in to the reinforcement in a number of routes. Consequently, graphene and GO reinforced polymeric matrix become tougher than the corresponding non-fabricated polymers (Ming et al., 2015). However, the improvement in the mechanical properties of polymer nanocomposite depends upon the 'threshold amount' of nanofillers present in it. Beyond this limit, nanofillers can adversely affect the mechanical properties of the polymer matrix due to the de-bonding and poor dispersion of nanomaterials.

Most of the polymeric substances employed in the aerospace industry are combustible due to its inherent organic nature. However, materials used in the aircrafts, skyscrapers and aerospace should not be liable to the fire. The current studies indicate that 20% of victims of airplane crashes are killed due to the exposure of fire (Ming et al., 2015). Various researches indicate that inclusion of graphene or GOs into the polymeric matrix can reduce flammability. In a study, Li et al., (2014) reported that modified GOs are able to alter the limiting oxygen index of epoxy components around 80% as compared with its virgin counterpart, which shows a transition of epoxy from flammable to non-flammable characteristics. The distinctive two-dimensional layered structure of graphene and GOs plays a vital role in reducting in flammability characteristics of incorporated matrix, such layered structures promotes dense char formation during burning. Besides, higher thermal stability of the graphene and GOs also contributes to the improved flammability characteristics of the nanocomposites. The char developed due to the graphene and GOs during the initial burning of nanocomposite can act as a physical barrier between heat and polymer substrates. Moreover, overlapped char formation further delays the elimination of pyrolysis products from the polymer substrate (Sang et al., 2016). Consequently, graphene or GO based materials become superior candidates for improving the flammability characteristics of the polymers.

Apart from it, recent researches explore the possibility of graphene nanocomposites in the development of high performance advanced materials like SHMs, LPMs, coating materials sensors, actuators, conducting application and thermal interface materials which could be useful in aircrafts and space craft applications. Nowadays, graphenes are found as a promising candidate for high performance thermal interface material due to the superior conductivity of graphene. As compared with the CNT based composites, graphene containing polymeric nanocomposites are indicating higher efficiency in the thermal conductivity (Atif et al., 2017). Besides, thermal conductivity of nanocomposites is strongly influenced by the orientation as well as the aspect ratio of graphene in the nanocomposites. Additionally, the nanocomposite preparation methods also contribute to the enhanced thermal conductivity (Atif et al., 2016).

In addition to it, graphene based polymer composite systems could be useful in the generation of SHMs particularly in aviation industries. The incorporation of functionalised graphene system into the composite system creates an electrically conductive network which is highly susceptible to the strain induced in the material (Atif et al., 2016). As compared to the graphene, modified graphenes contained in the various polymer matrixes shows enhanced SHMs properties. The polar functionalities developed in the modified graphenes are more ingrained into the polymer matrix rather than graphene. Therefore, these can offer a better conductive network than graphene in the polymer matrix. A recent study by Moriche et al. (2018) scrutinized the scope of functionalised graphene nanoplatelets for the creation of an electrically conductive network in epoxy glass fiber mutiscale composites.

To protect the structural components of aircrafts or space crafts from weathering and radiations various graphene and modified graphene are applying to it, since those can show better performance than its virgin counter parts. For instance, graphene fabricated nanocomposites have been shown superior de-icing, anti-icing capacity and anti-corrosion properties similar to the CNTs nanocomposites. Just like CNTs, graphene is also considered as an ideal candidate to restrain and improve the anti-corrosion properties. Along with the higher surface area-to-volume ratio of the graphene are also able to absorb light as well as it is showing hydrophobic behaviour. Due to these peculiar properties of graphene Monetta et al., (2015) have developed water born epoxy/ graphene nanocomposite for the coating of alloy Al 2024-T3, which is commonly employed in the aircraft structural components.

Inorganic Based Nanoparticle as Constructive Materials

Similar to the carbon based nanomaterials, inorganic (or non-carbon) nanomaterials are also engaged in the preparation of PNCs (Kumar et al., 2017a; Kumar et al., 2017b). Inorganic nanomaterials like nanoclays, nanosilicates, nanotitanium are commonly employed in polymer matrixes which are commonly applied in the aerospace industry. Among them nanoclays and nanosilicates are extensively studied due to its ability to tune for better flammability, mechanical and permeability properties inthe enhancement of mechanical properties. the polymer matrix.

Nanoclay as Constructive Material

Nanoclays are considered as naturally occurring nanoparticles segregated from the layered mineral silicates with platelets of approximately 100 nm in size. The weak vanderwals force of attraction holds the silicate layers together in nanoclays. The techniques such as melt intercalation and co-vulcanization can lead to the generation of polymer clay nanocomposites (Srinath and Gnanamoorthy, 2005; Usuki et al., 2005). Nanoclays, on lesser proportion has the capability to improve the mechanical, thermal, electrochemical and flammability properties of the polymeric matrix (Prabakaran et al., 2014a; Usuki et al., 2005; John et al, 2010). Moreover, they can be well dispersed in the matrix and can offer very high surface area at the interface of polymer matrix even without affecting the density of the developed PNCs. Nevertheless, increase in strength of polymer can follow with the improvement in stiffness and brittleness, ensuing in the damage of the PNCs derived structural components (John et al., 2008). In addition, by tuning the surface chemistry of nanoclay through chemical reactions or by means of exchange of cations, it could be possible to further enhance compatibility between polymer matrixes. Montmorillonite is the most frequently used nanoclay of all the time which belongs to the 2:1 phyllosilicates family. In addition to the high surface area, geometrical anisotropy of layered silica platelets enhances the filler-matrix interface synergy that can be manipulated for the considerable improvement in the composite properties. Among the two types of PNCs structure developed, PNCs with intercalated structure possess a well packed multilayered order that encompasses polymer/silicate layer structure in a mean distance of few nanometer scale, whereas, exfoliated PNC possesses random distribution of silicate layers in the polymer (Sengupta et al., 2007; Sadhu and Bhowmick, 2005).

Polymer foam based sandwich structured nanocomposites have been constituted in the structural applications in the aerospace materials. Highly damage tolerant hybrid syntactic foams with nanoclays can be utilized as sandwich core materials in the aerospace applications (Maharsia et al., 2006). Similarly, nanoclays could be employed for enhancing the properties of polymer matrix, especially aerospace epoxy grade materials. Jumahat et al. indicated improvement in compressive response and fracture toughness in aerospace epoxy with nanoclay content up to 5 wt%. The rigid nanoclay platelets present in the polymer matrix absorb energy imparted during mechanical stress. Consequently, nanoclay fabricated systems gives higher mechanical stress as well as plastic deformation (Yasmin et al., 2006).

Relatively, small amounts of nanoclay in the range of 3-5 wt% can effectively improves the mechanical, thermal and barrier properties of the polymer matrices. Besides, these properties are valuable for providing safety to the aircraft structures during the use or fire impact (Song et al., 2003). Studies are indicating that nanoclays can alter the heat deformation temperature of the polymeric matrix positively (Ray et al., 2003). Moreover, the thixotropic characteristics of nanoclay fillers can soothe the air absorbency of tyres and fuel tanks and hence can serve as a substitute for traditional materials like talc and titanium dioxide (Lin and Chung, 2008; Kawashima et al., 2013).

Surface modification of the nanoclays can further promote the compatibility between the polymer matrixes (Figure 5 (II)). A report by Balakrishnan and Raghvan (2003) indicated the possibility of surface modified nanoclay for fabricating the properties of the epoxy based polymer matrix. Modification of nanoclay leads to proper mixing with polymer matrix further. Author point outs the possibility of produced material in aerospace application. Similarly, in another study Raghavan prepared PNC with DGEBA epoxy system and C19:C18 nanoclay in the presence of MPDA for aerospace application. As compared with the unmodified, modified polymer matrix has been shown prodigious improvement in the tensile properties, in contrast to unmodified polymer (Yebassa et al., 2003).

Remarkable improvement in the tensile modulus of a polymeric material can be manifested with the incorporation of layered silicates into the matrix. On dealing with the nanocomposites, the amount of enhancement in the tensile modulus directly depends upon the aspect ratio of the distributed clay particles (Messersmith and Giannelis, 1995). For instance, Amir et al. applied two roll milling technique to infuse nanoclays into glass fiber reinforced polymers which have been utilized in aerospace structural applications. This research indicates that nanoclays can improve flexural strength and modulus of the polymeric systems with increasing the content of nanoclays (Amir et al., 2015). Zilg et al., (1999) and his co-workers unveiled the interdependence of the structure of the layered silicate and polymer, stiffness and

Figure 5. (I) Modification of polymer matrix with nanoclays (II) effect of surface modified nanoclays on polymer matrix and (III) flame retardency mechanism of nanoclay incorporated polymer matrix

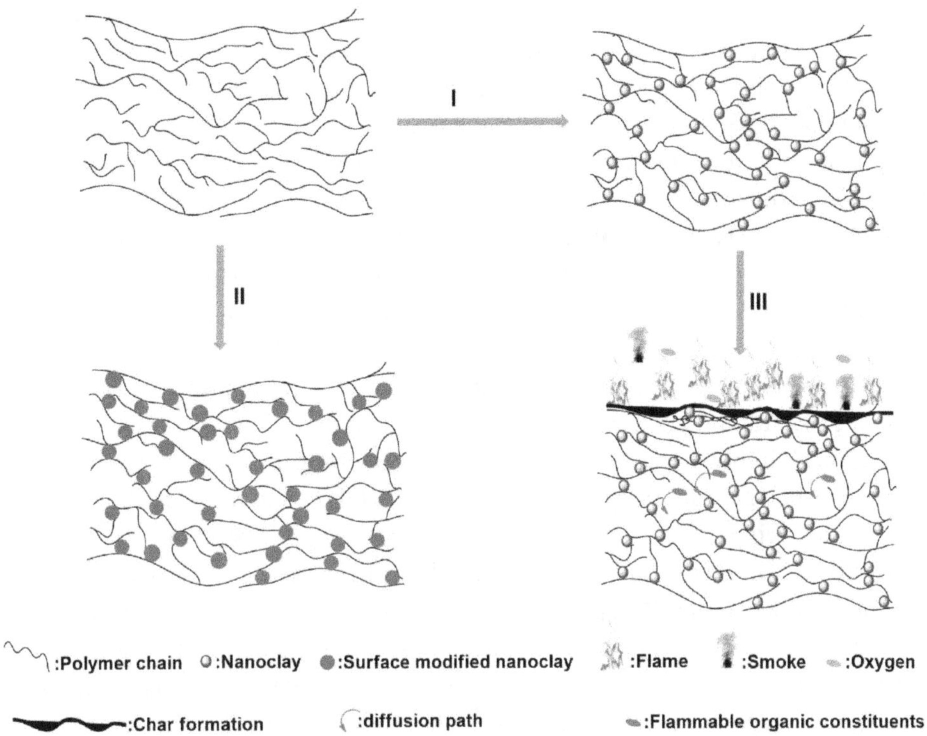

toughness of thermoset nanocomposites in terms of the type of the nanoclay and its components. They reported that the primary factor that affects the enhancement of matrix stiffness relies on the supramolecular construction which is developed as a consequence of the disseminated laminated nanoparticles.

The role of MMT based nanocomposite in the fire retardant applications has been well established since 1976. Later on 1997, Gilman et al., (2000) conducted a detailed study regarding the flammability of the nylon 6/layered silicate nanocomposite. Studies on the flammability of MMT-nanocomposites have shown that the well dispersed nanoclay in the matrix could offer good reinforcement effect that eventually will fall on the flame retardancy in the developed samples. The mechanism of fire retardancy in the clay system is due to the genesis of the carbonaceous-silicate char formed on the surface during the combustion, which can act as an insulating barrier, consequently reduces the mass loss rate during the degradation process (Gilman et al., 2000; Bourbigot et al., 2000). The recommended mechanism involves the formation of char that can serve as a barrier for the mass and energy transfer. The amount of clay in the matrix is a prime factor that modulates the quantity of char formed and the heat release rate gets level off. In this context, one type of nanocomposite got intercalated and the other two systems got exfoliated. The intercalated fraction significantly minimises the heat release rate vis-à-vis other exfoliated fraction (Zhu et al., 2001). This inference adds on the fact that, an intercalated species is more efficient than the exfoliated fraction in the fire performance (Gilman et al., 2000). The possible fire retardency mechanism is indicated in figure 5 (I and III) (Cui et al., 2015). Consequently, many of the researchers adopted nanoclay for improving fire retardant properties of their polymer matrixes which are commonly used in aerospace applications. A report by shan et al., (2013) demonstrated the effect of nanoclay on abaca fiber reinforced polypropylene composite for improving the flammability performance. This research indicates that the tensile and flexural properties can be tuned with nanoclays. Further, the materials developed can be applicable in aerospace components. Additionally, epoxy/clay based nanocomposites are suitable for the generation of lightweight composites with higher cryogenic gas storage capacity. Moreover, such products are highly flame retardant therefore those can be useful in the manufacturing of aerospace and aircraft structures (Rathod et al., 2017; Azeez et al., 2017). Similarly, many researchers have been focussed on the improved thermal stability of the PNC prepared with different types of nanoclay in different polymer matrices. Polyimide/clay nanocomposites are showing superior abrasion resistance, heat resistance and barrier properties as compared with its unmodified part. Therefore, polyimide/clay nanocomposite could be a suitable candidate for floor lining or cargo areas of commercial airlines (Rathod et al., 2017; Azeez et al., 2017).

Nanoclays are proven to influence the barrier properties by means of retarding the acceleration of gas molecules in the polymer matrix via the formed labyrinth. The systematic influence of such type of labyrinth formation in a PNC is obvious in the developed polyimide/MMT composite with improved the barrier properties, where an enhancement in the gas barrier properties with the synchronic minimization of thermal expansion coefficient can be seen (Yano et al., 1993). Various studies indicate the possibilities of the clay-polymer nanocomposites for reducing the permeability of oxygen, H_2O, Helium, CO_2 etc. Exfoliation of the nanomaterials in the polymer matrix affects the barrier properties of the final products. Highly exfoliated materials suppress the diffusive path of the gas molecules. As compared with the conventional nanocomposite nanoclay enabled polymer matrix shows enhanced gas barrier property due to the tortuous path followed by the gas molecule inside the polymer matrix (figure 6). Epoxy/clay nanocomposites are also showing significant gas barrier behaviour. So, they can be used as a replacement for TiO_2 and talc powder in fuel tanks (Stan et al., 2012).

Figure 6. Gas molecule travelling mechanism (a) tortuous path present in polymer/ nanoclay composite and (b) conventional polymer nanocomposite

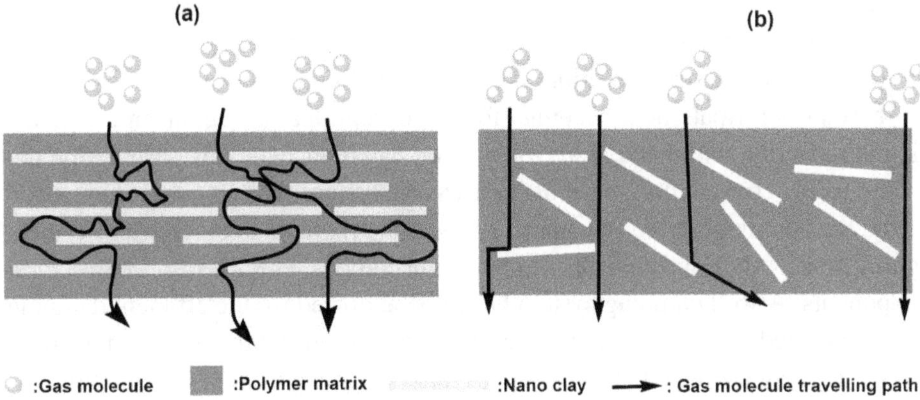

Nanosilica as Constructive Material

Nanosilica reinforcement is aimed to impart improved fracture toughness, impact strength, modulus, abrasion resistance, dielectric properties, heat distortion temperature, chemical resistance, glass transition temperature, toughness and weathering stability to the polymer matrix (Wu et al., 2002; Zhang et al., 2006). Epoxy resin, the most sustainable matrix in the structural applications when reinforced with silica-siloxane structure via *in-situ* sol gel process found that a dramatical increase in the modulus values of the corresponding systems resulting in a uniform dispersion of the inorganic phase in the organic system is still laborious to acquire by means of mechanical amalgamation. Hence, the sol-gel technique has been widely utilized for the effective distribution of silica nanoparticles in the successive polymer matrix systems. With this technique, inorganic or organic-inorganic systems can be productively prepared from the preliminary liquid materials. Making use of the chemical methods can synthesize nanoparticles with lower agglomeration. However, the transfer of these silica nanoparticles from the aqueous medium to the prepolymer of epoxy resin without disturbing the particle distribution is challenging (Matějka et al., 2000). The nanosilica system developed by a German company (Nanopox F, nanoresins AG) is well evaluated to acquire proper dispersion with 40-50 wt% in epoxies. Nanopox F is the spherical, surface modified colloidal silica particles with narrow particle size distribution in the resinous matrix prepared from the aqueous solution of sodium silicate via chemical synthesis. The hydroxyl groups present in the surface of silica particles can react with the bisphenol-A component in the epoxy and thereby a hydrophobic organic surface coating can be achieved (Mahrholz et al., 2003). A well known alloy of aluminum (AL 6061-T6) containing magnesium and silicon elements is often utilized in the aircraft industries due to its superior hardness and weldability. However, the system is susceptible to the friction, the need for lubrication is critical. Sayuti et al., (2014) scrutinizes the advantages of introducing SiO_2 nanolubricant in the cutting zone of the AL6061-T6 system for the stability in the severe conditions filing up of the minimal friction can surmount the thermal deformation. It accounts for the exceptional accomplishment of the friction reduction with nano SiO_2, which itself can reduce the oil consumption too in contrast to conventional pure oil lubrication system that was previously used for the machining process. Mechanical as well as thermal stability are of critical significance for the development of light weight composites for aerospace structural components. A well damping system has to be endowed for the efficient damping of the unwanted vibrations while preserving the thermal stability beyond the entire operational environment. Jang et al., (2011) has been reported the incorporation of 3 wt % of nanosilica with 3 wt % CNF, can augment the damping loss factor by 15.6% at room temperature with a consistency in thermal stability, approximately 15% decrease in the thermal expansion coefficient. The addition of silica nanoparticles can bring better thermal stability to the system with the remarkable advantages of low cost, low density, low thermal expansion coefficient and non-toxic nature. Therefore, it is worth mentioning that nanosilica with a wide continuum of utilities, rules the machining operation as well as development of structural component parts in the aerospace industry.

Other Inorganic Nanomaterials as Constructive Materials

Along with the nanoclays and nanosilicates, some other inorganic nanomaterials are also exploited for the production of PNCs which could be applicable in aerospace applications. In a study by Gilbert et al., (2003) employed nanoalumina for the modification of aerospace grade epoxy based film adhesives. Further, this study indicates that the peel shear strength and fracture toughness of the polymer matrix has been increased with the incorporation of the nanoalumina. Similarly, Qi et al. (2013) prepared fabric self lubricating liner for the tribological application inducing nano-TiO_2 in the diluent of phenolic resin. TiO_2 nanoparticles are also applicable in modification of the mechanical properties of FRP composites. Seshanandan et al., (2016) pointed that TiO_2 could be useful for improving tensile, flexural and shear strength of the hybrid jute glass FRPs. The photovoltaic capacity and electrochemical properties of the high performance polymers can also be altered with the help of the nano-TiO_2 or its modified forms (Prabakaran et al., 2015; Prabakaran et al., 2014b). Besides, another inorganic nano-carbon black (NCB) has been studied by Leng et al., (2009) for the infrared light active shape memory polymer applications.

Similarly, NCB can be also useful for the modification of the interlaminar properties of the phenolic matrix at higher temperatures (Huiping, 2003). Other inorganic nanomaterials like nano-Si_3N_4 and nano-SiC can also be applicable in the aerospace industry since they can provide higher wear resistance to the polymer matrix like PEEK (Dasari et al., 2009).

CURRENT TREND AND CHALLENGES OF PNCS AS CONSTRUCTIVE COMPONENTS IN AEROSPACE INDUSTRY

The PNCs proffer exclusive opportunity for the enhancement and tailorability of physico-mechanical, thermal and structural properties which would be valuable in construction of various constructive materials applied in the aerospace industry. As compared with the earlier periods, recent development in nanotechnology field has been results the advancements in the production of various nanomaterials with higher yield, varied shape and size. Unlike macrofillers, nanofillers can ensure the enhancement of multiple functionalities to the polymer matrix. Among the different nanofillers organic nanomaterials like CNTs, CNFs and graphenes and their various surface modified forms are fructified in current scenario owing to their broad spectra of applications. Among them, surface modified nanomaterials are attractive in recent times since those can augment the compatibility between the polymer matrixes. Moreover, CNTs and its varied form based PNCs are far ahead in developments as compared to other nanomaterials. For instance, the potential of CNTs is exploited in the production of cryogenic fuel tanks sensing applications and advanced structures of aerospace since those can provide higher mechanical strength, chemical resistance, fire retardancy and lower permeability. Moreover, new production techniques chirally sorted CNTs and long CNTs accelerated the possibilities of employing them in aerospace components (Zhang et al., 2013; Ganzhorn et al., 2011).

Several lab studies, simulations and modelling indicated the practicability of the PNCs in aerospace products however their applications are limited. In order to resolve this situation the manufacturers should coincide and advocate the usage of PNCs in their products. Furthermore, difficulties in production of nanomateials and its cost, lower yield, time consumption during its production restricts them in varied application. Moreover, achievement of homogeneous distribution of nanomaterials in polymeric matrixes is also difficult for the undertaking. The heterogeneous distribution of nanomaterials in polymer matrixes adversely affects the properties generated product. Similarly, the release of nanomaterials into the environment during its production or manufacturing of PNCs are also highly hazardous. Such

Table 5. Challenges and possible remedies of application of PNCs in aerospace industry

Challenges	Possible remedies
Limitation in availability of the precise nanomaterials	The adoption of the advanced technologies for the synthesis of the nanomaterials by the manufacturers can be employed for the increased production of nanomaterials.
Specificity of the nanomaterials for the products	The researchers and manufacture should toil together to found the suitable nanomaterials for the existing and formulated components.
Incompatibility of the nanomaterials with polymer matrix	Surface modification of the nanomaterials using various chemicals and advanced techniques such as probe sonication can improve the proper dispersion and thereby compatibility of the nanomaterials in the polymer matrix.
Recycling of developed PNCs	Hazardous behaviour and recycling possibilities of the developed materials should be considered by the manufactures and researchers on its origin stage.

liberated nanoparticles generate negative impact on the human and biota (Farré et al., 2009; Lin et al., 2010). Besides, the recycling techniques of end-of-life PNCs products are still in beginning stage therefore it will also conversely affects the usage of PNC technologies in aerospace applications.

FUTURE PROSPECTIVE AND CONCLUSION

The potential of nanomaterial in the PNCs fabrication and development of novel materials for aerospace applications has been reviewed in this chapter. The current technologies indicates that nanomaterials could be act as a promising candidate for improving the existing struggles in the production of structural and non structural components of aerospace materials. Inorganic and carbon based nanomaterials are suitable for the enhancement in the properties of different polymer composites. Among the different nanomaterials, CNT based components are widely employed due to its peculiar properties. Moreover, the utilization of the PNCs in the aerospace industry is expected to be enhanced in near future due to its specific material characteristics along with light weight nature. The impact of nanotechnology forced to generate different composites with multiple functionalities. The compatibility of the nanomaterials with polymer matrix can be enhanced with the help of techniques like ultrasonication and axial flow impeller. Moreover, surface modifications of the carbon based nanomaterials are also helpful to enhance the compatibility between the polymer matrixes and it can further improve the properties of PNCs. Furthermore, researches over hybrid nanomaterials indicate that they can improve the significant properties than their corresponding nano-counter parts. Therefore, hybrid nanomaterials could be also useful for the fabrication of PNCs.

REFERENCES

Al-Saleh, M. H., & Sundararaj, U. (2011). Review of the mechanical properties of carbon nanofiber/polymer composites. *Composites. Part A, Applied Science and Manufacturing*, *42*(12), 2126–2142. doi:10.1016/j.compositesa.2011.08.005

Allen, M. J., Tung, V. C., & Kaner, R. B. (2009). Honeycomb carbon: A review of graphene. *Chemical Reviews*, *110*(1), 132–145. doi:10.1021/cr900070d PMID:19610631

Altan, M., & Uysal, A. (2016). An Experimental Study on Mechanical Behavior of Nanographene/Epoxy Nanocomposites. *Advances in Polymer Technology*.

Amir, W. W., Jumahat, A., & Mahmud, J. (2015). Effect of nanoclay content on flexural properties of glass fiber reinforced polymer (GFRP) composite. *J. Teknol*, *76*, 31–35.

Araby, S., Meng, Q., Zhang, L., Kang, H., Majewski, P., Tang, Y., & Ma, J. (2014). Electrically and thermally conductive elastomer/graphene nanocomposites by solution mixing. *Polymer*, *55*(1), 201–210. doi:10.1016/j.polymer.2013.11.032

Atif, R., Shyha, I., & Inam, F. (2016). Mechanical, thermal, and electrical properties of graphene-epoxy nanocomposites—A review. *Polymers*, *8*(8), 281. doi:10.3390/polym8080281

Atif, R., Shyha, I., & Inam, F. (2017). Modeling and experimentation of multi-layered nanostructured graphene-epoxy nanocomposites for enhanced thermal and mechanical properties. *Journal of Composite Materials*, *51*(2), 209–220. doi:10.1177/0021998316640060

Azeez, A. A., Rhee, K. Y., Park, S. J., & Hui, D. (2013). Epoxy clay nanocomposites–processing, properties and applications: A review. *Composites. Part B, Engineering*, *45*(1), 308–320. doi:10.1016/j.compositesb.2012.04.012

Balakrishnan, S., & Raghavan, D. (2003). Chemically functionalized clay epoxy nanocomposites for aerospace applications. In *Technical Proceedings of the 2003 Nanotechnology Conference and Trade Show* (*Vol. 3*, pp. 250-253). Academic Press.

Baur, J., & Silverman, E. (2007). Challenges and opportunities in multifunctional nanocomposite structures for aerospace applications. *MRS Bulletin*, *32*(4), 328–334. doi:10.1557/mrs2007.231

Beyer, G. (2002). Carbon nanotubes as flame retardants for polymers. *Fire and Materials*, *26*(6), 291–293. doi:10.1002/fam.805

Bharadwaj, R. K. (2001). Modeling the barrier properties of polymer-layered silicate nanocomposites. *Macromolecules*, *34*(26), 9189–9192. doi:10.1021/ma010780b

Bhuvaneswari, C. M., Kale, S. S., Gouda, G., Jayapal, P., & Tamilmani, K. (2017). Elastomers and Adhesives for Aerospace Applications. In *Aerospace Materials and Material Technologies* (pp. 563–586). Singapore: Springer. doi:10.1007/978-981-10-2134-3_26

Bierwagen, G. P., & Tallman, D. E. (2001). Choice and measurement of crucial aircraft coatings system properties. *Progress in Organic Coatings*, *41*(4), 201–216. doi:10.1016/S0300-9440(01)00131-X

Böger, L., Wichmann, M. H., Meyer, L. O., & Schulte, K. (2008). Load and health monitoring in glass fibre reinforced composites with an electrically conductive nanocomposite epoxy matrix. *Composites Science and Technology*, *68*(7-8), 1886–1894. doi:10.1016/j.compscitech.2008.01.001

Bourbigot, S., Bras, M. L., Dabrowski, F., Gilman, J. W., & Kashiwagi, T. (2000). PA-6 clay nanocomposite hybrid as char forming agent in intumescent formulations. *Fire and Materials*, *24*(4), 201–208. doi:10.1002/1099-1018(200007/08)24:4<201::AID-FAM739>3.0.CO;2-D

Breuer, O., & Sundararaj, U. (2004). Big returns from small fibers: A review of polymer/carbon nanotube composites. *Polymer Composites*, *25*(6), 630–645. doi:10.1002/pc.20058

Chen, T. K., Tien, Y. I., & Wei, K. H. (2000). Synthesis and characterization of novel segmented polyurethane/clay nanocomposites. *Polymer*, *41*(4), 1345–1353. doi:10.1016/S0032-3861(99)00280-3

Coleman, J. N., Khan, U., & Gun'ko, Y. K. (2006). Mechanical reinforcement of polymers using carbon nanotubes. *Advanced Materials*, *18*(6), 689–706. doi:10.1002/adma.200501851

Costa, A. P. D., Botelho, E. C., Costa, M. L., Narita, N. E., & Tarpani, J. R. (2012). A review of welding technologies for thermoplastic composites in aerospace applications. *Journal of Aerospace Technology and Management*, *4*(3), 255–265. doi:10.5028/jatm.2012.040303912

Cox, H. L. (1952). The elasticity and strength of paper and other fibrous materials. *British Journal of Applied Physics*, *3*(3), 72–79. doi:10.1088/0508-3443/3/3/302

Cui, Y., Kumar, S., Kona, B. R., & van Houcke, D. (2015). Gas barrier properties of polymer/clay nanocomposites. *RSC Advances*, *5*(78), 63669–63690. doi:10.1039/C5RA10333A

Dasari, A., Yu, Z. Z., & Mai, Y. W. (2009). Fundamental aspects and recent progress on wear/scratch damage in polymer nanocomposites. *Materials Science and Engineering R Reports*, *63*(2), 31–80. doi:10.1016/j.mser.2008.10.001

DeBerry, D. W. (1985). Modification of the electrochemical and corrosion behavior of stainless steels with an electroactive coating. *Journal of the Electrochemical Society*, *132*(5), 1022–1026. doi:10.1149/1.2114008

Deguchi, R., Nishio, T., & Okada, A. (1992). *U.S. Patent No. 5,102,948*. Washington, DC: U.S. Patent and Trademark Office.

Farré, M., Gajda-Schrantz, K., Kantiani, L., & Barceló, D. (2009). Ecotoxicity and analysis of nanomaterials in the aquatic environment. *Analytical and Bioanalytical Chemistry*, *393*(1), 81–95. doi:10.100700216-008-2458-1 PMID:18987850

Feng, L., Xie, N., & Zhong, J. (2014). Carbon nanofibers and their composites: A review of synthesizing, properties and applications. *Materials (Basel)*, *7*(5), 3919–3945. doi:10.3390/ma7053919 PMID:28788657

Fiedler, B., Gojny, F. H., Wichmann, M. H., Bauhofer, W., & Schulte, K. (2004). Can carbon nanotubes be used to sense damage in composites? *Annales de chimie*, *29*(6), 81–94. doi:10.3166/acsm.29.6.81-94

Finegan, I. C., & Tibbetts, G. G. (2001). Electrical conductivity of vapor-grown carbon fiber/thermoplastic composites. *Journal of Materials Research*, *16*(6), 1668–1674. doi:10.1557/JMR.2001.0231

Fletcher, A., Gupta, M. C., Dudley, K. L., & Vedeler, E. (2010). Elastomer foam nanocomposites for electromagnetic dissipation and shielding applications. *Composites Science and Technology*, *70*(6), 953–958. doi:10.1016/j.compscitech.2010.02.011

Fu, X., & Chung, D. D. L. (1996). Submicron carbon filament cement-matrix composites for electromagnetic interference shielding. *Cement and Concrete Research*, *26*(10), 1467–1472. doi:10.1016/0008-8846(96)00146-9

Ganzhorn, M., Vijayaraghavan, A., Dehm, S., Hennrich, F., Green, A. A., Fichtner, M., ... Kappes, M. M. (2011). Hydrogen sensing with diameter-and chirality-sorted carbon nanotubes. *ACS Nano*, *5*(3), 1670–1676. doi:10.1021/nn101992g PMID:21341751

George, K., Panda, B. P., Mohanty, S., & Nayak, S. K. (2018). Recent developments in elastomeric heat shielding materials for solid rocket motor casing application for future perspective. *Polymers for Advanced Technologies*, *29*(1), 8–21. doi:10.1002/pat.4101

Gilbert, E. N., Hayes, B. S., & Seferis, J. C. (2003). Nano-alumina modified epoxy based film adhesives. *Polymer Engineering and Science*, *43*(5), 1096–1104. doi:10.1002/pen.10093

Gilman, J. W., Jackson, C. L., Morgan, A. B., Harris, R., Manias, E., Giannelis, E. P., ... Phillips, S. H. (2000). Flammability properties of polymer– layered-silicate nanocomposites. Polypropylene and polystyrene nanocomposites. *Chemistry of Materials*, *12*(7), 1866–1873. doi:10.1021/cm0001760

Gkikas, G., Sioulas, D., Lekatou, A., Barkoula, N. M., & Paipetis, A. S. (2012). Enhanced bonded aircraft repair using nano-modified adhesives. *Materials & Design*, *41*, 394–402. doi:10.1016/j.matdes.2012.04.052

Gobi, N., Vijayakumar, D., Keles, O., & Erogbogbo, F. (2017). Infusion of Graphene Quantum Dots to Create Stronger, Tougher, and Brighter Polymer Composites. *ACS Omega*, *2*(8), 4356–4362. doi:10.1021/acsomega.6b00517

Gohardani, O., Elola, M. C., & Elizetxea, C. (2014). Potential and prospective implementation of carbon nanotubes on next generation aircraft and space vehicles: A review of current and expected applications in aerospace sciences. *Progress in Aerospace Sciences*, *70*, 42–68. doi:10.1016/j.paerosci.2014.05.002

Hajibeygi, M., Shabanian, M., Omidi-Ghallemohamadi, M., & Khonakdar, H. A. (2017). Optical, thermal and combustion properties of self-colored polyamide nanocomposites reinforced with azo dye surface modified ZnO nanoparticles. *Applied Surface Science*, *416*, 628–638. doi:10.1016/j.apsusc.2017.04.139

Han, B., Cheng, A., Ji, G., Wu, S., & Shen, J. (2004). Effect of organophilic montmorillonite on polyurethane/montmorillonite nanocomposites. *Journal of Applied Polymer Science*, *91*(4), 2536–2542. doi:10.1002/app.13398

Han, J. H., Zhang, H., Chen, M. J., Wang, D., Liu, Q., Wu, Q. L., & Zhang, Z. (2015). The combination of carbon nanotube buckypaper and insulating adhesive for lightning strike protection of the carbon fiber/epoxy laminates. *Carbon, 94*, 101–113. doi:10.1016/j.carbon.2015.06.026

Huang, G., Wang, S., Song, P. A., Wu, C., Chen, S., & Wang, X. (2014). Combination effect of carbon nanotubes with graphene on intumescent flame-retardant polypropylene nanocomposites. *Composites. Part A, Applied Science and Manufacturing*, *59*, 18–25. doi:10.1016/j.compositesa.2013.12.010

Huiping, C. J. H. L. T. (2003). Interlaminar properties of carbon/phenolic composite modified with nano-carbon black. *Fiber Reinforced Plastics/Composite, 4*.

Hussein, A., & Kim, B. (2018). Graphene/polymer nanocomposites: The active role of the matrix in stiffening mechanics. *Composite Structures*, *202*, 170–181. doi:10.1016/j.compstruct.2018.01.023

Jakubinek, M. B., Ashrafi, B., Zhang, Y., Martinez-Rubi, Y., Kingston, C. T., Johnston, A., & Simard, B. (2015). Single-walled carbon nanotube–epoxy composites for structural and conductive aerospace adhesives. *Composites. Part B, Engineering*, *69*, 87–93. doi:10.1016/j.compositesb.2014.09.022

Jang, J. S., Varischetti, J., Lee, G. W., & Suhr, J. (2011). Experimental and analytical investigation of mechanical damping and CTE of both SiO2 particle and carbon nanofiber reinforced hybrid epoxy composites. *Composites. Part A, Applied Science and Manufacturing*, *42*(1), 98–103. doi:10.1016/j.compositesa.2010.10.008

Jarosz, P., Schauerman, C., Alvarenga, J., Moses, B., Mastrangelo, T., Raffaelle, R., ... Landi, B. (2011). Carbon nanotube wires and cables: Near-term applications and future perspectives. *Nanoscale*, *3*(11), 4542–4553. doi:10.1039/c1nr10814j PMID:21984338

Jen, M. H. R., Tseng, Y. C., & Wu, C. H. (2005). Manufacturing and mechanical response of nanocomposite laminates. *Composites Science and Technology*, *65*(5), 775–779. doi:10.1016/j.compscitech.2004.10.010

Jeon, H., Park, J., & Shon, M. (2013). Corrosion protection by epoxy coating containing multi-walled carbon nanotubes. *Journal of Industrial and Engineering Chemistry*, *19*(3), 849–853. doi:10.1016/j.jiec.2012.10.030

Jin, Y., & Boyd, R. H. (1998). Subglass chain dynamics and relaxation in polyethylene: A molecular dynamics simulation study. *The Journal of Chemical Physics*, *108*(23), 9912–9923. doi:10.1063/1.476430

John, B., Nair, C. R., & Ninan, K. N. (2010). Effect of nanoclay on the mechanical, dynamic mechanical and thermal properties of cyanate ester syntactic foams. *Materials Science and Engineering A*, *527*(21-22), 5435–5443. doi:10.1016/j.msea.2010.05.016

John, B., Reghunadhan Nair, C. P., Mathew, D., & Ninan, K. N. (2008). Foam sandwich composites with cyanate ester based syntactic foam as core and carbon-cyanate ester as skin: Processing and properties. *Journal of Applied Polymer Science*, *110*(3), 1366–1374. doi:10.1002/app.28658

Joshi, M., & Chatterjee, U. (2016). Polymer nanocomposite: An advanced material for aerospace applications. Advanced Composite Materials for Aerospace Engineering, 241-264.

Jouni, M., Faure-Vincent, J., Fedorko, P., Djurado, D., Boiteux, G., & Massardier, V. (2014). Charge carrier transport and low electrical percolation threshold in multiwalled carbon nanotube polymer nanocomposites. *Carbon*, *76*, 10–18. doi:10.1016/j.carbon.2014.04.031

Kashiwagi, T., Du, F., Winey, K. I., Groth, K. M., Shields, J. R., Bellayer, S. P., ... Douglas, J. F. (2005). Flammability properties of polymer nanocomposites with single-walled carbon nanotubes: Effects of nanotube dispersion and concentration. *Polymer*, *46*(2), 471–481. doi:10.1016/j.polymer.2004.10.087

Kausar, A., Rafique, I., & Muhammad, B. (2016). Review of applications of polymer/carbon nanotubes and epoxy/CNT composites. *Polymer-Plastics Technology and Engineering*, *55*(11), 1167–1191. doi:10.1080/03602559.2016.1163588

Kawashima, S., Chaouche, M., Corr, D. J., & Shah, S. P. (2013). Rate of thixotropic rebuilding of cement pastes modified with highly purified attapulgite clays. *Cement and Concrete Research*, *53*, 112–118. doi:10.1016/j.cemconres.2013.05.019

Kelly, A., & Tyson, A. W. (1965). Tensile properties of fibre-reinforced metals: Copper/tungsten and copper/molybdenum. *Journal of the Mechanics and Physics of Solids*, *13*(6), 329–350. doi:10.1016/0022-5096(65)90035-9

Khan, S. U., Li, C. Y., Siddiqui, N. A., & Kim, J. K. (2011). Vibration damping characteristics of carbon fiber-reinforced composites containing multi-walled carbon nanotubes. *Composites Science and Technology*, *71*(12), 1486–1494. doi:10.1016/j.compscitech.2011.03.022

Kim, B. K., Seo, J. W., & Jeong, H. M. (2003). Morphology and properties of waterborne polyurethane/clay nanocomposites. *European Polymer Journal*, *39*(1), 85–91. doi:10.1016/S0014-3057(02)00173-8

Kim, H., Abdala, A. A., & Macosko, C. W. (2010). Graphene/polymer nanocomposites. *Macromolecules*, *43*(16), 6515–6530. doi:10.1021/ma100572e

King, J. A., Klimek, D. R., Miskioglu, I., & Odegard, G. M. (2015). Mechanical properties of graphene nanoplatelet/epoxy composites. *Journal of Composite Materials*, *49*(6), 659–668. doi:10.1177/0021998314522674

Kuan, C. F., Chen, W. J., Li, Y. L., Chen, C. H., Kuan, H. C., & Chiang, C. L. (2010). Flame retardance and thermal stability of carbon nanotube epoxy composite prepared from sol–gel method. *Journal of Physics and Chemistry of Solids*, *71*(4), 539–543. doi:10.1016/j.jpcs.2009.12.031

Kumar, A., Manju, P., Krishnan S, G., & Nayak, S. K. (2017b). Physical, Mechanical, Surface, Thermal and Morphological Studies of PLA-Sepiolite Bionanocomposites. *Journal of Mineral, Metal and Material Engineering, 3*, 103-111.

Kumar, S., Dang, T. D., Arnold, F. E., Bhattacharyya, A. R., Min, B. G., Zhang, X., ... Smalley, R. E. (2002). Synthesis, structure, and properties of PBO/SWNT Composites&. *Macromolecules*, *35*(24), 9039–9043. doi:10.1021/ma0205055

Kumar, S., Samal, S. K., Mohanty, S., & Nayak, S. K. (2017a). Synthesis and Characterization of Nanoclay-Reinforced Trifunctional "Bioresin-Modified" Epoxy Blends Enhanced with Mechanical and Thermal Properties. *ChemistrySelect*, *2*(35), 11445–11455. doi:10.1002lct.201702041

Larson, C., Smith, J. R., & Armstrong, G. J. (2013). Current research on surface finishing and coatings for aerospace bodies and structures–a review. *Transactions of the IMF*, *91*(3), 120-132. 10.1179/0020296713Z.000000000102

Lee, J., & Ramakrishna, S. (2017). Carbon Nanotube Wires and Cables: Near-Term Applications and Future Perspectives. *Nanotechnology for Energy Sustainability*, 485-506.

Leng, J., Wu, X., & Liu, Y. (2009). Infrared light-active shape memory polymer filled with nanocarbon particles. *Journal of Applied Polymer Science*, *114*(4), 2455–2460. doi:10.1002/app.30724

Li, K. Y., Kuan, C. F., Kuan, H. C., Chen, C. H., Shen, M. Y., Yang, J. M., & Chiang, C. L. (2014). Preparation and properties of novel epoxy/graphene oxide nanosheets (GON) composites functionalized with flame retardant containing phosphorus and silicon. *Materials Chemistry and Physics*, *146*(3), 354–362. doi:10.1016/j.matchemphys.2014.03.037

Lin, C., & Chung, D. D. L. (2008). Nanoclay paste as a thermal interface material for smooth surfaces. *Journal of Electronic Materials*, *37*(11), 1698–1709. doi:10.100711664-008-0538-9

Lin, D., Tian, X., Wu, F., & Xing, B. (2010). Fate and transport of engineered nanomaterials in the environment. *Journal of Environmental Quality*, *39*(6), 1896–1908. doi:10.2134/jeq2009.0423 PMID:21284287

Liu, J. M., & Yang, S. C. (1991). Novel colloidal polyaniline fibrils made by template guided chemical polymerization. *Journal of the Chemical Society. Chemical Communications*, (21): 1529–1531. doi:10.1039/c39910001529

Liu, W., Hoa, S. V., & Pugh, M. (2005). Fracture toughness and water uptake of high-performance epoxy/nanoclay nanocomposites. *Composites Science and Technology*, *65*(15-16), 2364–2373. doi:10.1016/j.compscitech.2005.06.007

Llobet, E. (2013). Gas sensors using carbon nanomaterials: A review. *Sensors and Actuators. B, Chemical*, *179*, 32–45. doi:10.1016/j.snb.2012.11.014

Lourie, O., & Wagner, H. D. (1998). Evaluation of Young's modulus of carbon nanotubes by micro-Raman spectroscopy. *Journal of Materials Research*, *13*(9), 2418–2422. doi:10.1557/JMR.1998.0336

Luo, J. J., & Daniel, I. M. (2003). Characterization and modeling of mechanical behavior of polymer/clay nanocomposites. *Composites Science and Technology*, *63*(11), 1607–1616. doi:10.1016/S0266-3538(03)00060-5

Maharsia, R., Gupta, N., & Jerro, H. D. (2006). Investigation of flexural strength properties of rubber and nanoclay reinforced hybrid syntactic foams. *Materials Science and Engineering A*, *417*(1-2), 249–258. doi:10.1016/j.msea.2005.10.063

Mahrholz, T., Mosch, J., Ro¨stermundt, D., Riedel, U., & Herbeck, L. (2003). *Materials for Aerospace Applications*. 20th AAAF Colloquium, Paris, France.

Mallick, P. K. (2007). *Fiber-reinforced composites: materials, manufacturing, and design*. CRC Press. doi:10.1201/9781420005981

Matějka, L., Dukh, O., & Kolařík, J. (2000). Reinforcement of crosslinked rubbery epoxies by in-situ formed silica. *Polymer*, *41*(4), 1449–1459. doi:10.1016/S0032-3861(99)00317-1

Messersmith, P. B., & Giannelis, E. P. (1995). Synthesis and barrier properties of poly (ε-caprolactone)-layered silicate nanocomposites. *Journal of Polymer Science. Part A, Polymer Chemistry*, *33*(7), 1047–1057. doi:10.1002/pola.1995.080330707

Min, C., Shen, X., Shi, Z., Chen, L., & Xu, Z. (2010). The electrical properties and conducting mechanisms of carbon nanotube/polymer nanocomposites: A review. *Polymer-Plastics Technology and Engineering*, *49*(12), 1172–1181. doi:10.1080/03602559.2010.496405

Ming, P., Song, Z., Gong, S., Zhang, Y., Duan, J., Zhang, Q., ... Cheng, Q. (2015). Nacre-inspired integrated nanocomposites with fire retardant properties by graphene oxide and montmorillonite. *Journal of Materials Chemistry. A, Materials for Energy and Sustainability*, *3*(42), 21194–21200. doi:10.1039/C5TA05742F

Mirabedini, S. M., & Kiamanesh, A. (2013). The effect of micro and nano-sized particles on mechanical and adhesion properties of a clear polyester powder coating. *Progress in Organic Coatings*, *76*(11), 1625–1632. doi:10.1016/j.porgcoat.2013.07.009

Monetta, T., Acquesta, A., & Bellucci, F. (2015). Graphene/epoxy coating as multifunctional material for aircraft structures. *Aerospace*, *2*(3), 423–434. doi:10.3390/aerospace2030423

Moniruzzaman, M., & Winey, K. I. (2006). Polymer nanocomposites containing carbon nanotubes. *Macromolecules*, *39*(16), 5194–5205. doi:10.1021/ma060733p

Moriche, R., Jiménez-Suárez, A., Sánchez, M., Prolongo, S. G., & Ureña, A. (2018). Sensitivity, influence of the strain rate and reversibility of GNPs based multiscale composite materials for high sensitive strain sensors. *Composites Science and Technology*, *155*, 100–107. doi:10.1016/j.compscitech.2017.12.002

Njuguna, J., & Pielichowski, K. (2003). Polymer nanocomposites for aerospace applications: Properties. *Advanced Engineering Materials*, *5*(11), 769–778. doi:10.1002/adem.200310101

O'connell, M. J. (2006). *Carbon nanotubes: properties and applications*. CRC Press. doi:10.1201/9781420004212

O'Donnell, S., Sprong, K., & Haltli, B. (2004). Potential impact of carbon nanotube reinforced polymer composite on commercial aircraft performance and economics. In *AIAA 4th Aviation Technology, Integration and Operations (ATIO) Forum* (p. 6402). AIAA. 10.2514/6.2004-6402

Ogasawara, T., Ishida, Y., Ishikawa, T., & Yokota, R. (2004). Characterization of multi-walled carbon nanotube/phenylethynyl terminated polyimide composites. *Composites. Part A, Applied Science and Manufacturing*, *35*(1), 67–74. doi:10.1016/j.compositesa.2003.09.003

Popov, V. N. (2004). Carbon nanotubes: Properties and application. *Materials Science and Engineering R Reports*, *43*(3), 61–102. doi:10.1016/j.mser.2003.10.001

Pora, J. (2001). Composite materials in the airbus A380–from history to future. *Proceedings of ICCM13*. Available on: https://pdfs.semanticscholar.org/31ff/a710 22ce6bb0549d1711facf792b6b171414.pdf

Prabakaran, K., Mohanty, S., & Nayak, S. K. (2014 a). Influence of surface modified nanoclay on electrochemical properties of PVDF-HFP composite electrolytes. *International Journal of Plastics Technology*, *18*(3), 349–361. doi:10.100712588-014-9089-5

Prabakaran, K., Mohanty, S., & Nayak, S. K. (2014b). Influence of surface modified nanoclay on electrochemical properties of PVDF-HFP composite electrolytes. *International Journal of Plastics Technology*, *18*(3), 349–361. doi:10.100712588-014-9089-5

Prabakaran, K., Mohanty, S., & Nayak, S. K. (2015). PEO/PVdF–HFP electrolytes for natural dye sensitized solar cell applications: Effect of modified nano-TiO 2 on electrochemical and photovoltaic performance. *Journal of Materials Science Materials in Electronics*, *26*(6), 3887–3897. doi:10.100710854-015-2916-4

Prabhakaran, R. (2011). Lightning Strikes on Metal and Composite Aircraft and their Mitigation. *Journal of Aerospace Sciences and Technologies*, *63*(3), 208–222.

Qi, X., Jia, Z., & Yang, Y. (2013). Influence of the dispersion of nano titanium dioxide on the tribological performance of fabric self-lubricating liner. *Journal of Applied Polymer Science*, *130*(3), 2100–2105. doi:10.1002/app.39343

Rafiee, M. A., Rafiee, J., Wang, Z., Song, H., Yu, Z. Z., & Koratkar, N. (2009). Enhanced mechanical properties of nanocomposites at low graphene content. *ACS Nano*, *3*(12), 3884–3890. doi:10.1021/nn9010472 PMID:19957928

Rathod, V. T., Kumar, J. S., & Jain, A. (2017). Polymer and ceramic nanocomposites for aerospace applications. *Applied Nanoscience*, *7*(8), 519–548. doi:10.100713204-017-0592-9

Ravichandran, K., Praseetha, P. K., Arun, T., & Gobalakrishnan, S. (2018). Synthesis of Nanocomposites. In Synthesis of Inorganic Nanomaterials (pp. 141-168). Academic Press. doi:10.1016/B978-0-08-101975-7.00006-3

Ray, S. S., Yamada, K., Okamoto, M., & Ueda, K. (2003). New polylactide-layered silicate nanocomposites. 2. Concurrent improvements of material properties, biodegradability and melt rheology. *Polymer*, *44*(3), 857–866. doi:10.1016/S0032-3861(02)00818-2

Rodriguez, A. J., Guzman, M. E., Lim, C. S., & Minaie, B. (2011). Mechanical properties of carbon nanofiber/fiber-reinforced hierarchical polymer composites manufactured with multiscale-reinforcement fabrics. *Carbon*, *49*(3), 937–948. doi:10.1016/j.carbon.2010.10.057

Sadhu, S., & Bhowmick, A. K. (2005). Effect of nanoclay on the dynamic mechanical properties of styrene butadiene and acrylonitrile butadiene rubber vulcanizates. *Rubber Chemistry and Technology*, *78*(2), 321–335. doi:10.5254/1.3547886

Sandler, J., Werner, P., Shaffer, M. S., Demchuk, V., Altstädt, V., & Windle, A. H. (2002). Carbon-nanofibre-reinforced poly (ether ether ketone) composites. *Composites. Part A, Applied Science and Manufacturing*, *33*(8), 1033–1039. doi:10.1016/S1359-835X(02)00084-2

Sang, B., Li, Z. W., Li, X. H., Yu, L. G., & Zhang, Z. J. (2016). Graphene-based flame retardants: A review. *Journal of Materials Science*, *51*(18), 8271–8295. doi:10.100710853-016-0124-0

Sayuti, M., Erh, O. M., Sarhan, A. A., & Hamdi, M. (2014). Investigation on the morphology of the machined surface in end milling of aerospace AL6061-T6 for novel uses of SiO2 nanolubrication system. *Journal of Cleaner Production*, *66*, 655–663. doi:10.1016/j.jclepro.2013.11.058

Schartel, B., Pötschke, P., Knoll, U., & Abdel-Goad, M. (2005). Fire behaviour of polyamide 6/multiwall carbon nanotube nanocomposites. *European Polymer Journal*, *41*(5), 1061–1070. doi:10.1016/j.eurpolymj.2004.11.023

Sebastian, J., Schehl, N., Bouchard, M., Boehle, M., Li, L., Lagounov, A., & Lafdi, K. (2014). Health monitoring of structural composites with embedded carbon nanotube coated glass fiber sensors. *Carbon*, *66*, 191–200. doi:10.1016/j.carbon.2013.08.058

Sengupta, R., Chakraborty, S., Bandyopadhyay, S., Dasgupta, S., Mukhopadhyay, R., Auddy, K., & Deuri, A. S. (2007). A short review on rubber/clay nanocomposites with emphasis on mechanical properties. *Polymer Engineering and Science*, *47*(11), 1956–1974. doi:10.1002/pen.20921

Seshanandan, G., Ravindran, D., & Sornakumar, T. (2016). Mechanical properties of nano titanium oxide particles-hybrid jute-glass FRP composites. *Materials Today: Proceedings*, *3*(6), 1383–1388. doi:10.1016/j.matpr.2016.04.019

Shah, A. U. R., Lee, D. W., Kim, S., Kim, B. S., & Song, J. I. (2013). Tensile and fire retardant properties of nanoclay reinforced Abaca/Polypropylene composite. *Aerospace Science & Engineering (ICASE), 2013 International Conference on IEEE*, 1-5.

Song, M., Hourston, D. J., Yao, K. J., Tay, J. K. H., & Ansarifar, M. A. (2003). High performance nanocomposites of polyurethane elastomer and organically modified layered silicate. *Journal of Applied Polymer Science*, *90*(12), 3239–3243. doi:10.1002/app.12979

Srinath, G., & Gnanamoorthy, R. (2005). Effect of nanoclay reinforcement on tensile and tribo behaviour of Nylon 6. *Journal of Materials Science*, *40*(11), 2897–2901. doi:10.100710853-005-2439-0

Stan, A., Dinca, I., Ban, C., Ilina, S., Donescu, D., Paven, H., . . . Voicu, I. (2012). Epoxy-layered silicate and epoxy MWCNTs nanocomposites. Applied Mechanics and Materials Trans Tech Publications, 146, 160-169.

Suresh, S. S., Mohanty, S., & Nayak, S. K. (2017a). Composition analysis and characterization of waste polyvinyl chloride (PVC) recovered from data cables. *Waste Management (New York, N.Y.)*, *60*, 100–111. doi:10.1016/j.wasman.2016.08.033 PMID:27594574

Suresh, S. S., Mohanty, S., & Nayak, S. K. (2017b). Preparation and characterization of recycled blends using poly (vinyl chloride) and poly (methyl methacrylate) recovered from waste electrical and electronic equipments. *Journal of Cleaner Production*, *149*, 863–873. doi:10.1016/j.jclepro.2017.02.057

Swain, S. S., Unnikrishnan, L., Mohanty, S., & Nayak, S. K. (2017). Carbon nanotubes as potential candidate for separation of H 2 CO 2 gas pairs. *International Journal of Hydrogen Energy*, *42*(49), 29283–29299. doi:10.1016/j.ijhydene.2017.09.152

Tan, H., Huang, Y., Liu, C., & Geubelle, P. H. (2005). The Mori–Tanaka method for composite materials with nonlinear interface debonding. *International Journal of Plasticity*, *21*(10), 1890–1918. doi:10.1016/j.ijplas.2004.10.001

Tibbetts, G. G., Lake, M. L., Strong, K. L., & Rice, B. P. (2007). A review of the fabrication and properties of vapor-grown carbon nanofiber/polymer composites. *Composites Science and Technology*, *67*(7-8), 1709–1718. doi:10.1016/j.compscitech.2006.06.015

Usuki, A., Hasegawa, N., Kato, M., & Kobayashi, S. (2005). Polymer-clay nanocomposites. In *Inorganic polymeric nanocomposites and membranes* (pp. 135–195). Berlin: Springer. doi:10.1007/b104481

Vao-soongnern, V., Doruker, P., & Mattice, W. L. (2000). Simulation of an amorphous polyethylene nanofiber on a high coordination lattice. *Macromolecular Theory and Simulations*, *9*(1), 1–13. doi:10.1002/(SICI)1521-3919(20000101)9:1<1::AID-MATS1>3.0.CO;2-R

Verdejo, R., Saiz-Arroyo, C., Carretero-Gonzalez, J., Barroso-Bujans, F., Rodriguez-Perez, M. A., & Lopez-Manchado, M. A. (2008). Physical properties of silicone foams filled with carbon nanotubes and functionalized graphene sheets. *European Polymer Journal*, *44*(9), 2790–2797. doi:10.1016/j.eurpolymj.2008.06.033

Vodicka, R. (1996). *Thermoplastics for airframe applications: a review of the properties and repair methods for thermoplastic composites*. Academic Press.

Wang, K., Kelly, D., & Dutton, S. (2002). Multi-objective optimisation of composite aerospace structures. *Composite Structures*, *57*(1-4), 141–148. doi:10.1016/S0263-8223(02)00078-8

Wright, W. W. (1991). Polymers in aerospace applications. *Materials & Design*, *12*(4), 222–227. doi:10.1016/0261-3069(91)90169-5

Wu, C. L., Zhang, M. Q., Rong, M. Z., & Friedrich, K. (2002). Tensile performance improvement of low nanoparticles filled-polypropylene composites. *Composites Science and Technology*, *62*(10-11), 1327–1340. doi:10.1016/S0266-3538(02)00079-9

Xie, X. L., Mai, Y. W., & Zhou, X. P. (2005). Dispersion and alignment of carbon nanotubes in polymer matrix: A review. *Materials Science and Engineering R Reports*, *49*(4), 89–112. doi:10.1016/j.mser.2005.04.002

Yang, S., Lozano, K., Lomeli, A., Foltz, H. D., & Jones, R. (2005). Electromagnetic interference shielding effectiveness of carbon nanofiber/LCP composites. *Composites. Part A, Applied Science and Manufacturing*, *36*(5), 691–697. doi:10.1016/j.compositesa.2004.07.009

Yang, Y., Gupta, M. C., Dudley, K. L., & Lawrence, R. W. (2005). A comparative study of EMI shielding properties of carbon nanofiber and multi-walled carbon nanotube filled polymer composites. *Journal of Nanoscience and Nanotechnology*, *5*(6), 927–931. doi:10.1166/jnn.2005.115 PMID:16060155

Yano, K., Usuki, A., & Okada, A. (1997). Synthesis and properties of polyimide-clay hybrid films. *Journal of Polymer Science. Part A, Polymer Chemistry*, *35*(11), 2289–2294. doi:10.1002/(SICI)1099-0518(199708)35:11<2289::AID-POLA20>3.0.CO;2-9

Yano, K., Usuki, A., Okada, A., Kurauchi, T., & Kamigaito, O. (1993). Synthesis and properties of polyimide–clay hybrid. *Journal of Polymer Science. Part A, Polymer Chemistry*, *31*(10), 2493–2498. doi:10.1002/pola.1993.080311009

Yasmin, A., Luo, J. J., Abot, J. L., & Daniel, I. M. (2006). Mechanical and thermal behavior of clay/epoxy nanocomposites. *Composites Science and Technology*, *66*(14), 2415–2422. doi:10.1016/j.compscitech.2006.03.011

Ye, L., Lu, Y., Su, Z., & Meng, G. (2005). Functionalized composite structures for new generation airframes: A review. *Composites Science and Technology*, *65*(9), 1436–1446. doi:10.1016/j.compscitech.2004.12.015

Yebassa, D., Balakrishnan, S., & Raghaνan, D. (2003). Next generation aerospace composites through nanotechnology. *Technical Proceedings of the 2003 Nanotechnology Conference*, 246-249.

Yi, M., & Shen, Z. (2015). A review on mechanical exfoliation for the scalable production of graphene. *Journal of Materials Chemistry. A, Materials for Energy and Sustainability*, *3*(22), 11700–11715. doi:10.1039/C5TA00252D

Young, R. J., Kinloch, I. A., Gong, L., & Novoselov, K. S. (2012). The mechanics of graphene nanocomposites: A review. *Composites Science and Technology*, *72*(12), 1459–1476. doi:10.1016/j.compscitech.2012.05.005

Yu, H., Liu, J., Wen, X., Jiang, Z., Wang, Y., Wang, L., ... Tang, T. (2011). Charing polymer wrapped carbon nanotubes for simultaneously improving the flame retardancy and mechanical properties of epoxy resin. *Polymer*, *52*(21), 4891–4898. doi:10.1016/j.polymer.2011.08.013

Zeng, J., Kumar, S., Iyer, S., Schiraldi, D. A., & Gonzalez, R. I. (2005). Reinforcement of poly (ethylene terephthalate) fibers with polyhedral oligomeric silsesquioxanes (POSS). *High Performance Polymers*, *17*(3), 403–424. doi:10.1177/0954008305055562

Zhang, H., Zhang, Z., Friedrich, K., & Eger, C. (2006). Property improvements of in situ epoxy nanocomposites with reduced interparticle distance at high nanosilica content. *Acta Materialia*, *54*(7), 1833–1842. doi:10.1016/j.actamat.2005.12.009

Zhang, R., Zhang, Y., Zhang, Q., Xie, H., Qian, W., & Wei, F. (2013). Growth of half-meter long carbon nanotubes based on Schulz–Flory distribution. *ACS Nano*, *7*(7), 6156–6161. doi:10.1021/nn401995z PMID:23806050

Zhao, D. L., Qiao, R. H., Wang, C. Z., & Shen, Z. M. (2006). Microstructure and mechanical property of carbon nanotube and continuous carbon fiber reinforced epoxy resin matrix composites. *Advanced Materials Research*, *11*, 517–520. doi:10.4028/www.scientific.net/AMR.11-12.517

Zhu, J., Morgan, A. B., Lamelas, F. J., & Wilkie, C. A. (2001). Fire properties of polystyrene– clay nanocomposites. *Chemistry of Materials*, *13*(10), 3774–3780. doi:10.1021/cm000984r

Zhu, J., Wei, S., Ryu, J., Budhathoki, M., Liang, G., & Guo, Z. (2010). In situ stabilized carbon nanofiber (CNF) reinforced epoxy nanocomposites. *Journal of Materials Chemistry*, *20*(23), 4937–4948. doi:10.1039/c0jm00063a

Zilg, C., Mülhaupt, R., & Finter, J. (1999). Morphology and toughness/stiffness balance of nanocomposites based upon anhydride-cured epoxy resins and layered silicates. *Macromolecular Chemistry and Physics*, *200*(3), 661–670. doi:10.1002/(SICI)1521-3935(19990301)200:3<661::AID-MACP661>3.0.CO;2-4

Chapter 6
Lightweight Nanocomposites Polymers for Shielding Application

Mohamed Darwish
https://orcid.org/0000-0002-4598-8847
Egyptian Petroleum Research Institute, Egypt

ABSTRACT

Electromagnetic waves can have serious effects on human health by long-term exposure. Developing lightweight materials with good electromagnetic radiation shielding (EMS) that could prevent interference is a high desire for protection. Nanocomposites polymers have a wide range of potential applications and offers suggested solutions in environmental and aerospace applications. This chapter will cover the current challenge in the reduction of electromagnetic wave by developing lightweight absorber material with a wide absorption frequency. A wide range of different nanocomposites polymers contain conductive fillers such as metal or magnetic nanoparticles and carbon-based materials will be discussed. In addition, EMS mechanisms of reflection, absorption, and multiple reflections will be discussed. The unique of the chemical and physical properties of nanocomposites polymers are promising for shielding with low-cost environmentally friendly material.

INTRODUCTION

Electromagnetic pollution appears to be inconvenient with advances in technology and use of electronics and telecommunication devices. Electronic equipment and systems emit waves in the range of microwaves and for this reason the need to shield such radiation arise. Shielding efficacy has paying attention in military applications and in civil communication.

DOI: 10.4018/978-1-5225-7921-2.ch006

Electromagnetic interference (EMI) can disrupt electronic devices, equipment, and systems that are used in critical applications. Examples include medical, military, and aerospace electronics as shown in (Figure 1). Typical applications are civil and military manned and unmanned aircraft (fixed and rotary wing), spacecraft, missiles, rockets, blimps, and racing vehicles. Electromagnetic shielding is the practice of reducing the electromagnetic field in a space by blocking the field with barriers made of conductive or magnetic materials. Shielding is typically applied to enclosures to isolate electrical devices from the 'outside world', and to cables to isolate wires from the environment through which the cable runs. EMI shielding effectiveness is heavily dependent on conductivity- higher conductivity results in better shielding performance. The current challenge in the reduction of electromagnetic wave is to develop lightweight absorber material with a wide absorption frequency (Lagarkov, 2009; Sakai, 2010).

The field of astronautics and aerospace presents higher demand for the weight and performance of shielding materials. Shielding materials should meet a series of requirements to suit current demands: and resistant to corrosion and chemicals, lightweight, flexibility and cheap. The fabrication of materials which are capable of shielding incident electromagnetic radiation and meet the criteria mentioned above is nowadays an active field of research.

Figure 1. Electromagnetic shielding in aerospace applications
Source: This picture was taken from https://www.conformalcoating.co.uk/index.php/market-sectors/aerospace

The presence of nanoparticles in a polymer matrix alters the characteristics of the polymer as the mechanical and thermal properties. Polymer composites offer several advantages over traditional absorber used for electromagnetic interference (EMI) shielding and very promising to be the next-generation shielding materials. Due to the extremely high electrical conductivity, efficient and light weight. (Kasagi, 1999; Li, 1993).

Although polymers are electromagnetic transparent, different strategies are used to shift them into active electromagnetic shields. The incorporation of nanomaterials, such as nanocarbons and magnetic nanoparticles, into matrix of polymer is becoming one of the most effective processes to reach this end. Polymers provide many properties over ceramics and metals in the area of shielding.

To obstruct magnetic fields and to control levels of electromagnetic radiation, magnetically permeable particles such as spinel ferrites or ion–nickel alloys should be combined with polymers. In addition to weight reduction, the incorporation of nanoparticles of these magnetic materials might enable to exceed their essential cutoff frequency. Therefore, the tendency of using high aspect ratio particles has grown rapidly. Metallic particles of nickel (Ni), copper (Cu), silver (Ag) and aluminum (Al) flakes incorporated into polymers as polyimidesiloxane, polyurethane resins, or polyethersulfone have been extensively investigated (Li, 1991; Li, 1994; Morgan, 1991).

Next, loading of epoxies and thermoplastics with higher aspect ratio conductive stainless steel fibers or metallized glass, polyester, and carbon fibers shift the attention away from particles (Tan, 1999; Nan, 1986; Shinagawa, 1999; Tzeng, 2001).

Nowadays, conductive polymer-magnetic nanocomposites have drawn significant interest of researchers by combining the magnetic and conducting properties. Because of their extraordinary properties such as their economical importance, superior stability, lighter weight, better workability, resistance to corrosion and satisfactory electrical conductivity (Al-Saleh, 2011; Gelves, 2011; Hu, 2012; Sarvi, 2014).

In particular, magnetic iron oxide nanoparticles can be well dispersed and protected in a polymer mass which, in turn, acquire the characteristic magnetic properties of the oxides. The particle has superparamagnetism, when the diameter of magnetic metal oxide particle is less than 20 nm. The magnetic nanoparticle show strong magnetism in the presence of magnetic field but its magnetism will disappear in the absence of magnetic field. Therefore, this kind of magnetic carrier cannot be magnetized permanently in the magnetic field. The content of the magnetic nanoparticles is one of the important parameters. It can reflect the magnetic response capability of the particles. In addition, flexible polymeric composites loaded with Cu or Ag nanowires have been studied. Structure, geometry, and electrical properties of conductive nanomaterials play an important role in the mechanical and electrical properties of polymer composite. Furthermore, the structure and morphology of nanomaterials in a polymer matrix are found to be essential in product properties. The morphology control of nanomaterials became of attention for researchers in changing the final properties of polymers.

This review is focused on the fabrication methods of wide range of different lightweight nanocomposites polymers. Mechanisms of shielding by reflection, absorption and multiple reflections are also discussed. In addition, the effect of fillers type and filler loadings on the performance of shielding materials was presented.

ELECTROMAGNETIC WAVES

The electromagnetic spectrum contains all known frequencies of electromagnetic (EM) radiation (Figure 2), from radio waves, through the visible light spectrum, up to gamma rays (Kimmel, 1995; Markstein, 1995; O'Shea, 1998).

Radio frequency radiation is low on the EM spectrum. Below it, we find extremely low frequency (ELF) radiation, such as what is emitted by the power lines and electrical circuits we use to supply electricity to our homes, then very low frequency (VLF) radiation such as computers/screens.

Above radio frequency (RF) radiation, we find microwave radiation (MW radiation), which is what your microwave oven and mobile phone uses. Above MW is infrared (IR), such as that emitted by motion sensors or remote controls and then we have visible light. Collectively ELF, VLF, RF and IR are known as non-ionizing radiation. It is a long-held belief that the less energetic EM radiation in the lower end of the EM spectrum is less damaging than radio frequency radiation or MW, and x-rays are more damaging than ELF, VLF, RF and MW. This is why criteria for safety standards are different in each range of the EM spectrum.

Figure 2. Electromagnetic wave spectrum

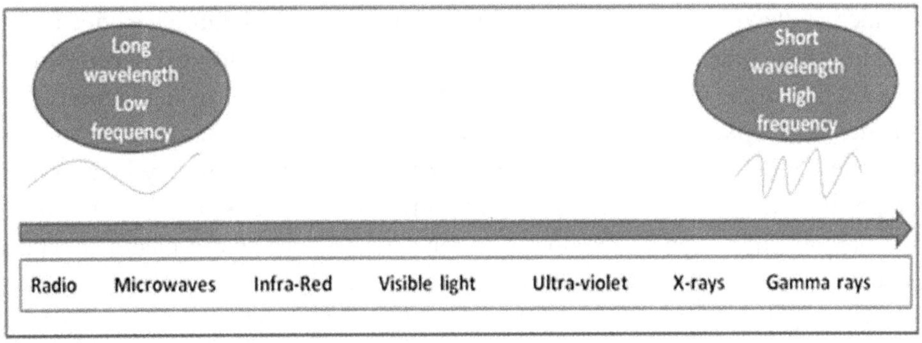

However we know that even low-energy electromagnetic frequency (EMF) can cause bodily damage. A high-powered ELF field can deliver enough current to kill a person (such as a lightning strike or electric chair), while a person does not even feel a low-powered radio signal (such as a transmission from a baby monitor) that is composed of waves that can be a million times more energetic.

PROCESSING METHODS OF LIGHTWEIGHT NANOCOMPOSITES POLYMERS

Typical lightweight nanocomposites polymers are a combination of polymer (matrix) and filler (reinforcement). Fabrication of lightweight nanocomposites polymers can be generally classified in four categories:

The common ones are 1) Direct mixing of polymer and fillers 2) In situ precipitation 3) Polymerization in the presence of fillers and 4) Electrospinning process.

The first simple method is to incorporate the mixing of the filler e.g. metal nanoparticles and polymer materials, which are previously fabricated separately by direct mixing. The adsorption behavior, physical or chemical, is between the filler nanoparticles and the polymer spheres (Bizdoaca, 2002).

The second method is through in situ precipitation of fillers in the presence of polymer, where the fillers are surrounded by a polymer shell (Dumazet-Bonnamour, 2000).

The third method for synthesizing lightweight nanocomposites polymers is monomer polymerization by dispersing the fillers directly in the liquid phase of a polymerizable formulation. Then, polymerizing of the monomer in the presence of the metal nanoparticles to form lightweight nanocomposites polymers. Several processes have been developed including emulsion polymerization, dispersion polymerization, suspension polymerization, microemulsion polymerization and miniemulsion polymerization (Lu, 2009; Hoffmann, 2001).

The fourth and common route for fabrication of lightweight nanocomposites polymers is electrospinning. Electrospinning has been used to fabricate wide types of hybrid composite nanofibers by incorporating nanoparticles into different polymer matrixes. By embedding functional materials into nanofiber, it is also possible to provide electrospun nanofiber with magnetic, optical, electrical, and antibacterial properties (Sadchikov, 1997; Jackson, 1998; Kumar, 1997). Commonly, electrospun composite nanofiber is fabricated by two methods. The first is done by direct incorporation of nanoparticles into the polymer solution before electrospinning process. As shown in Figure 3, represent of electrospun functionalized magnetic polyamide 6 composite nanofiber (Darwish, 2017). The second method is performed via a post-treatment the surface of the fiber with nanoparticles (Bhatgadde, 1997).

Figure 3. Electrospun functionalized magnetic polyamide 6 composite nanofiber
Source: This picture was taken from Darwish (2017)

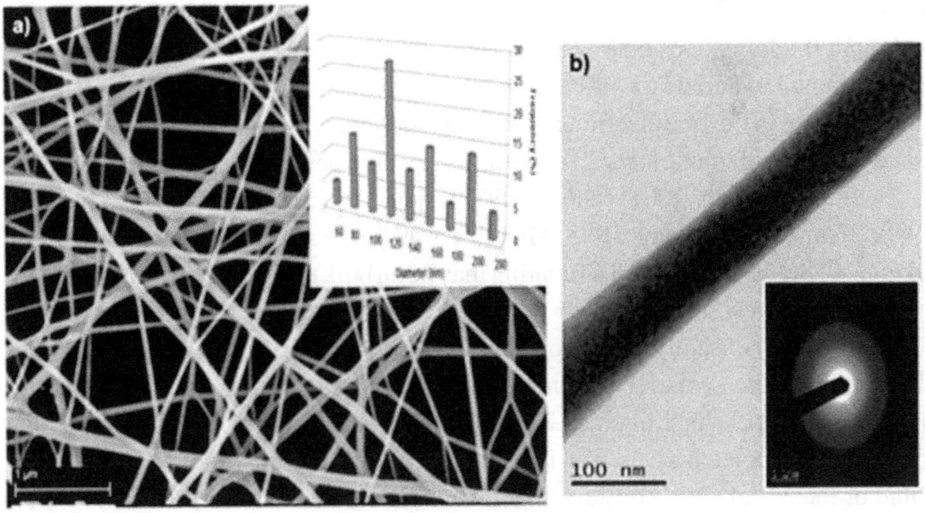

Electrospinning is a broadly applied technology used to fabricate nanofibers from a wide range of materials of synthetic and natural polymers (Bhardwaj, 2010). A high voltage electrostatic field is used to induce the ejection of a liquid jet, which stretches into continuous fibers in the submicron range (Kang, 2005). The electrospun material exhibits unique features such as uniform ultrafine fibers, high surface-to-volume ratio, tunable porous structures, and controllable composition. Making it capable to achieve the desired result from its properties and functionality. These qualities allow the materials to be successfully applied in various fields such as biosensor (Liu, 2009), tissue engineering scaffold (Agarwal, 2008), drug delivery (Huang, 2006), wound dressing (Jia, 2007), and protective clothing (Ramakrishna, 2006).

TYPES OF MECHANISMS

Mechanisms of shielding can be generally classified in three categories: 1) Reflection 2) Absorption 3) Multiple reflections.

Reflection is the primary mechanism of EMI shielding. For reflection of the radiation by the shield, the shield must have mobile charge carriers of holes or electrons, which interact with the electromagnetic fields in the radiation. As a result, the shield tends to be electrically conducting, although a high conductivity is not required. For example, a volume resistivity of the order of 1 ohm cm is typically sufficient. However, electrical conductivity is not the scientific criterion for shielding, as conduction requires connectivity in the conduction path (percolation in case of a composite material containing a conductive filler), whereas shielding does not. Although shielding does not need connectivity, it is enhanced by connectivity. The most common materials for EMI shielding are the metals. They function mainly by reflection due to the free electrons in them. Metal sheets are bulky, so metal coatings made by electroplating, electroless plating or vacuum deposition are commonly used for shielding (Chung, 2001; Sidhu, 1997; Hajdu, 1997; Klemmer, 1996; Gwinner, 1996; Bhatia, 1995; Zhang, 1995; Mandich, 1994; Jackson, 1994; Nagasawa, 1999; Dixon, 1998; Mason, 1994). The coating may be on bulk materials, fibers or particles. Coatings tend to suffer from their poor wear or scratch resistance.

Absorption is usually the secondary mechanism of EMI shielding. For major absorption of the radiation by the shield, the shield must have electric and/or magnetic dipoles which interact with the electromagnetic fields in the radiation. The electric dipoles may be provided by BaTiO or other materials having a high value of the 3 dielectric constant. The magnetic dipoles may be provided by FeO or other materials having a high value of the magnetic permeability, which may be improved by reducing magnetic domain walls number by using of a multilayer of magnetic films (Grimes, 1994).

The third a mechanism of shielding is multiple reflections, which refer to the reflections at a variety of surfaces or interfaces in the shield. This mechanism requires the presence of a large surface area or interface area in the shield. An example of a shield with a large surface area is a porous or foam material. An example of a shield with a large interface area is a composite material containing filler, which has a large surface area. The loss due to multiple reflections can be neglected when the distance between the reflecting surfaces or interfaces is large compared to the skin depth.

The sum of all the losses, whether due to reflection, absorption or multiple reflections, is commonly expressed in shielding effectiveness (dB). The absorption loss is proportional to the thickness of the shield. When an electromagnetic wave with incident energy (E_I) impacts on a material (Figure 4) two waves are created on the surface: A reflected energy (E_R), due to the impedance mismatch between the two mediums, and a transmitted wave into the material (E_{I-R}) (Gonza´lez, 2018).

Inside the material, a fraction of the wave (E_{I-R}) may be dissipated as heat (E_A) by Joule effect until it reaches the second surface of the material. At this point, two new waves appear: One that is transmitted through the surface (E_T) and a new wave that is reflected into the material. This process is repeated successively until it meets the criteria stated by equation (1):

$$E_I = \sum E_R + \sum E_A + \sum E_T \qquad (1)$$

Figure 4. Attenuation mechanisms of the incident electromagnetic radiation

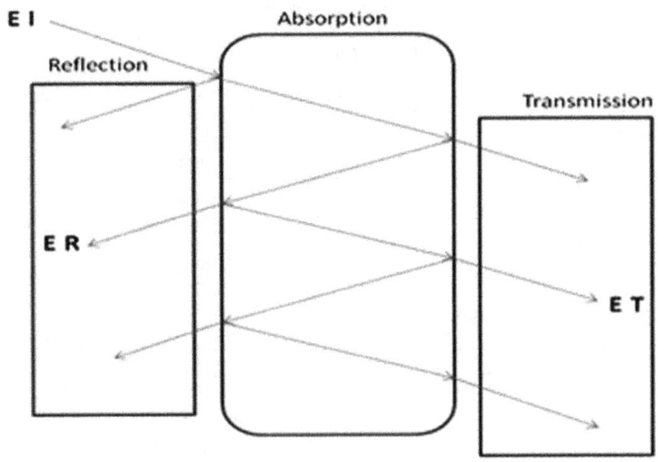

Both in the reflection and transmission process, waves generated at each step, may cause constructive and destructive interferences depending on the sample thickness and frequency. The reflection process on each plane of the material is what is called multiple reflections. Therefore, the electromagnetic shielding efficiency of a material can be quantified as the sum of three contributions: reflection, absorption and multiple reflections equation (2):

$$SE = SE_R + SE_A + SE_T \tag{2}$$

MATERIALS FOR ELECTROMAGNETIC SHIELDING

The current challenges are to develop lightweight materials with wide absorption frequency, high thermal stability, and resistance to oxidation, for the next generation of EM absorbers. In Table 1, represent the shielding performance of various lightweight nanocomposites polymers

Magnetic materials are good potential candidates as microwave shielding due to their combination of dielectric and magnetic loss as well as controllable morphologies (Wang, 2013; Al-Ghamdi, 2012; Kim, 2012). Specially, among the available various magnetic materials, magnetite (Fe_3O_4) nanoparticles are one of the most promising conduction allotropes, on the basis of their superior electrical and thermal properties comparing to other magnetic materials (Wei, 2011; Ma, 2012). Fe_3O_4 nanoparticles have recently attracted great attention because of their valuable properties, including extraordinary magnetic properties, low toxicity, good electrical conductivity, and

Table 1. Shielding performance of various lightweight nanocomposites polymers

Ref.	Shielding (dB)	Thickness d (mm)	Filler	Matrix
(Chiscan, 2012)	6	4.5	Fe_3O_4	Poly(vinyl alcohol)
(Gupta, 2013; Hoang, 2011)	20-29	0.1-0.2	Multi-walled Carbon nanotubes	Polyurethane
(Pande, 2009)	27	0.06	Multi-walled Carbon nanotubes	Poly (methyl methacrylate)
(Nayak, 2013)	45	1	Carbon nanofibers	Polysulfone
(Al-Ghamdi, 2013)	80-90	2	Fe_3O_4	Nitrile butadiene rubber
(Saleh, 2013)	40	2.8	Carbon black	Polypropylene
(Kashi, 2016)	15	1.5	Graphene	Polylactide
(Li, 2008)	25	1.5	Multi-walled Carbon nanotubes	Polyacrylate
(Das, 2009)	40	4.5	Single-walled carbon nanotubes	Poly (methyl methacrylate)
(Yu, 2012)	45.1	2.5	Graphene	Polyaniline
(Ling, 2013)	44	2.3	Graphene	Polyetherimide

nearly full spin polarization at room temperature; these properties have great potential for applications in EMI shielding/absorbing materials (Miyauchi, 2011). At present, Fe_3O_4 nanoparticles have been intensively examined for their superior properties, which make them an excellent choice to fabricate magnetic polymer composite for EMI shielding at very low loading. Various techniques have been developed to fabricate Fe_3O_4/polymer nanocomposites (Wang, 2010). But, among the various methods, electrospinning is recognized as an economical method for manufacturing magnetic composite with nanofibrous structure, when compared with other techniques. Some literatures (Zhang, 2009; Wang, 2008; Wang,2010) have reported the electrospinning process of Fe_3O_4/ polymer composite nanofibrous structure, there is almost no report studied on simultaneous morphology controlling and EMI shielding of a Fe_3O_4/polymer nanofibrous mat. Polyvinylpyrrolidone (PVP), an important synthetic polymer, is very biocompatible and nontoxic. The PVP has high tensile strength, excellent thermal stability, low chemical toxicity, and good spinnability. Therefore, PVP has been considered as a promising polymer for EMI shielding application (Nasouri, 2016; Yang, 2004; Nasouri, 2015).

Magnetite (Fe_3O_4) nanoparticles/polyvinylpyrrolidone (PVP) composite nanofibers (FCNFs) have been fabricated to evaluate the potential of FCNFs as electromagnetic interference (EMI) shielding material with frequency 8.2–12.4 GHz.

The scanning electron microscope and viscosity analyses confirmed the presence of good dispersion Fe_3O_4 nanoparticles encapsulated within the electrospun nanofibers and showed FCNF morphologies with diameters of 150–500 nm. The magnetic properties and electrical conductivity of FCNFs were found to be dependent on Fe_3O_4 nanoparticles concentration and showed an increase with increasing Fe_3O_4 nanoparticles loading. The EMI shielding efficiency of FCNFs increased up to approximately 22 dB. The EMI shielding results for FCNFs showed that absorption was the major shielding mechanism and reflection was the secondary shielding mechanism. The present study has shown the possibility of utilizing magnetic FCNFs as EMI shielding/absorption materials (Nasouri, 2017).

Incorporation of metal oxides into electrospun fibers can further enhance their potential applications. Metal oxide/carbon nanocomposite fibers can be prepared either by electrospinning a polyacrylonitrile (PAN)/N, N-dimethyl formamide (DMF) solution containing a dissolved metal oxide precursor (such as acetylacetonate salt, $Fe(acac)_3$) for magnetite (Fe_3O_4) followed by carbonization, or by the dispersion of previously synthesized metal oxide nanoparticles in PAN/DMF solution just prior to electrospinning and carbonization (Wang, 2008; Panels, 2008). Most of the electrospun magnetic fibrous mats reported in the literature have been produced based on the first method using DMF as the solvent, because the preparation of a colloidally-stable suspension of magnetite nanoparticles in a strong polar solvent is still problematic. Some Fe_3O_4/polymer nanocomposite fibers have been developed based on the second method, however, especially with polymers that are soluble in aqueous or weakly polar organic solvents (such as chloroform, CH_3Cl). Zhang et al. (Zhang, 2009) prepared magnetic Fe_3O_4/PAN nanocomposites fibers from DMF/ CH_3Cl solutions with suspended Fe_3O_4 nanoparticles, and examined the influence of the operating parameters and the rheological behavior of the solution on the fibers produced. Since DMF is perhaps the most widely used solvent in electrospinning, methods for the preparation of welldispersed magnetic nanoparticle suspensions in DMF would greatly expand and facilitate the production of other magnetic polymer membranes. Wang et al., for instance, produced electrospun fibers using colloidally-stable suspensions of magnetite nanoparticles in polyethylene oxide (PEO) and polyvinyl alcohol (PVA) solutions (Yang, 2004).

Electromagnetic interference shielding composites based on acrylic resin matrix (AR) have been prepared by incorporation of up to 30 wt% activated charcoal (AC) loading. X-ray diffraction (XRD) patterns and Raman studies confirm the incorporation of AC particles inside AR matrix and suggest possible interactions between phases. The scanning electron microscope (SEM) images show that incorporation of AC particles leads to systematic change in the morphology of composites especially the formation of porous structure. The dielectric measurements show that 30 wt% AC loading composite display higher relative permittivity value

(~79) compared to pure AR (~5). Further, the porous structure, electrical conductivity, and permittivity value contribute towards EMI shielding effectiveness value of −36 dB for these composites, thereby demonstrating their suitability for making efficient EMI shielding coatings (Khan, 2014).

Porous polymer nanocomposite with a density less than 0.26 g cm^{-3} has been prepared by ionic self-assembly of gold nanoparticles (NPs) on the charged polymer skeleton made of poly(pyridobisimidazole)-grafted-poly(dimethyl diallyl ammonium chloride) (PIPD-g-PDDA) composite nanofibers. Toward electromagnetic interference (EMI) shielding, a shielding effectiveness of over -64.9 dB in the frequency range of 250 MHz–1.5 GHz was demonstrated in the as-obtained nanocomposites with a thickness of only 20 mm. The equivalent gold thickness of these films was 10–50 times thinner than the skin depth of 6–12 mm for the bulk gold in the same frequency range. The electrical conductivity of these composites was around 15890 S cm^{-1} with the volume fraction of gold nanoparticles at 40.5%. The added nanoparticles enhanced the tensile strength by 12.4% as compared to the pure polymer films. No obvious degradation in electrical conductivity was observed even after repeatedly bending 1000 times with a bending radius of 1 cm. The shielding mechanism was disclosed by comparatively analyzing the reflection from the material surface (SER), the absorption of electromagnetic energy, and the multiple internal reflection of electromagnetic radiation (Li, 2017).

Polyaniline nanofiber graphite composites have been fabricated by a facile chemical polymerization process. Nanofiber show high electromagnetic interference (EMI) shielding of above 80 dB with frequency of 8.2–18 GHz. EMI shielding fabrics of 0.1 mm thickness based on Polyaniline nanofibers and their composite have been developed by an in situ polymerization route. These fabrics combine the properties of polyaniline nanofibers as well as their composite and fabrics (cotton and nylon). The developed functional fabrics with 0.1 mm thickness exhibit EMI shielding effectiveness in the range of 11–15 dB in the 8.2–18 GHz frequency range. Optical and scanning electron microscopy studies indicate the uniform coating of the polyaniline nanofibers over the individual fibers and interweave regions. Thin and flexible shielding materials suitable for a broad range of millimeter-wave shielding applications have been developed using this simple and potentially profitable method (Joseph, 2017).

The electrical properties and electromagnetic shielding effectiveness (EMSE) of nanocomposites consisting of heat-treated carbon nanofibers (CNF) in a linear low density polyethylene (LLDPE) matrix were assessed. Heat treatment (HT) of carbon nanofibers at 2500°C significantly improved their graphitic crystallinity and intrinsic transport properties, thereby increasing the electromagnetic shielding efficiency of the nanocomposites. Nanocomposites containing 11 vol% (20 wt%) HT displayed a DC electrical conductivity of about 1.0 6 0.1 3 101 S/m (n ¼ 4), about 10 orders of magnitude better than that of as-received PR-19 CNF nanocomposites. Over a frequency range of 30 MHz to 1.5 GHz, nanocomposites (2.5 mm thick) containing HT displayed EMSE average values of about 14 6 2 dB (n ¼ 4). Absorption was determined to be the main EMSE mechanism for the heat-treated CNF nanocomposites. The nanocomposites possessed a modulus of 632 6 36 MPa (n ¼ 6) (nominally twice that of pure LLDPE) and a strain-to-failure of 180 6 98% (n ¼ 6), indicating that a significant ductility is retained in the nanocomposites. Such nanocomposites display potential as absorptive electromagnetic interference shielding materials for thin films and micromolding (Villacorta, 2013).

Carbon nanotube (CNT) possesses outstanding mechanical and physical properties. It has very low density, ultrahigh strength (tensile strength of 200 GPa, Elastic modulus of 1.3 TPa), and high thermal conductivity and electrical (2*107 S/m). By inserting CNT film into fiber reinforced polymer composites, it has shown great potential in electromagnetic shielding application. A 2 mm thick single-walled carbon nanotube (SWCNT)/epoxy composites with 15% SWCNTs shows a 20–30 dB EMI shielding efficiency (Huang, 2007). Higher shielding obtaining by introducing highly conductive metal particles as 4 μm CNT film with approximately 35wt% iron showed shielding effectiveness as high as 61–67 dB (Wu, 2011). In another report, the five-layer CNT film shows extraordinary shielding performance in X-band with shielding effectiveness ranging from 67 dB to 78 dB (Wei, 2015).

The EMI shielding performance of a composite material depends on many factors, such as the filler's aspect ratio and intrinsic conductivity. The small diameter, high aspect ratio, high conductivity, and mechanical strength of carbon nanofibers (CNFs) have made them appear to be an excellent option for high performance EMI shielding materials even with low filler loading (Lee, 2001; Lee, 2002; Liying, 2013). A new carbon nanofiber (CNF)-reinforced syntactic foam (CNFRSF) was fabricated. The electrical conductivity and electromagnetic interference (EMI) shielding effectiveness (SE) of CNFRSF with various volume fraction of CNFs (0–2.0 vol%) were investigated. The measurement of EMI SE was performed with a frequency range from 30 MHz to 1.2 GHz. The experimental results showed that the SE of CNFRSF increases with increase of CNFs content. CNF shows more valuable in providing EMI shielding compared to chopped carbon fiber and long carbon fiber due to the larger aspect ratio. The CNFRSF having 2.0 vol% CNFs has a SE of 25 dB, which is good enough for most practical applications. The EMI SE of CNFRSF has been improved compared to CNFs composites due to the presence of hollow carbon microspheres.

Multi-axial fabrics containing steel yarns and carbon filaments, and their polyester (PES) resin-reinforced composites have been prepared for electromagnetic shielding applications. The electromagnetic shielding effectiveness (EMSE) of these structures was determined by using coaxial transmission line measurement technique. There were eight different multi-axial fabrics constructed. It was observed that the amount and the orientation of carbon and stainless steel yarns influenced the EMSE performances of multi-axial fabrics and their reinforced PES composites. The structures containing both carbon filaments and stainless steel yarns exhibited enhanced EMSE than the ones including only one type of conductive yarns or filaments. In addition, the EMSE performance of multi-axial fabrics was found better than their reinforced composites. The best EMSE results were obtained for the fabric, including two layers of yarns (steel and carbon) on top of each other in the centre with the angle of 45 and –45o (Erdem, 2016).

The shielding effectiveness of composite matrix materials at filler loadings near or above the percolation threshold was investigated. Such a model has practical applications in electromagnetic interference and radio frequency interference, and is validated here for Fortafil 243 carbon fiber within nylon 6,6. At very high filler loadings, experimental results show a weak dependence on the frequency of the wave to be shielded, which may be attributed to enhanced reflection from multiple, coherent scatterers (carbon fiber network). This model performs very well over an electrical resistivity range from 1015 ohm-cm (at low filler loading levels below the percolation threshold) down to 10-1 ohm-cm (at high filler loading levels well above the percolation threshold), and can be used to determine filler loadings needed to provide a certain level of shielding of electromagnetic waves (Keith, 2005). Figure 5 shows the mean transmitted and mean reflected power versus frequency for the composite containing 7 wt% (4.6 vol%) carbon fiber (Keith, 2005).

The electromagnetic (EM) shielding effectiveness (SE) was presented by reducing graphene oxide sheets which interleaved between polyetherimide films fabricated by electrophoretic deposition (Sanghoon, 2014). When the polymer fiber diameters are changed from micrometers to nanometers or submicrons, they have numerous wonderful features such as huge surface area-to-volume ratio (this ratio for a nanofiber can be as large as 103 times of that of a microfiber), flexibility in surface functionalities, and superior mechanical performance compared with any other recognized materials form. These exceptional things make the polymer nanofibers to be optimal applicants for several significant uses (Zheng, 2003).

Electromagnetic (EM) wave absorption materials have attracted much attention recently for expanded EM interference problems. In general, the performance of EM absorption materials depends on both their dielectric loss and magnetic loss properties (Yang, 2013; Petrov, 2001).

Promising structural carbon-based EM absorption materials that are both load bearing and can absorb EM radiation are receiving increasing attention owing to their superior chemical inertness and thermal stability relative to other materials (Micheli, 2010).

Figure 5. The mean transmitted and mean reflected power versus frequency for the composite containing 7 wt% carbon fiber
(This picture was taken from ref. Keith, 2005)

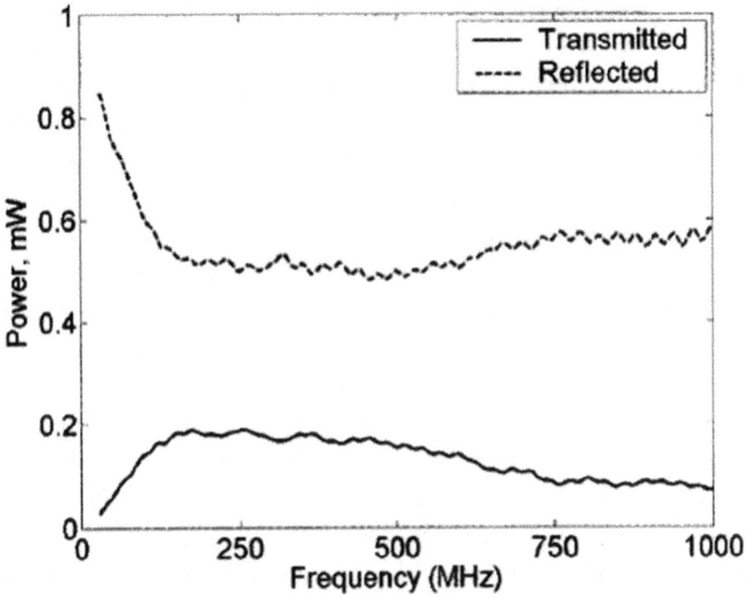

Magnetic losses in existing dielectric absorbers can be mediated by incorporating magnetic nanoparticles (Chen, 2010).

Electromagnetic interference (EMI) is an offshoot of explosive growth of electronics and telecommunication in the modern society (Saini, 2012; Ott, 2009; Saini, 2013). The EMI among electronic instruments/appliances may lead to degradation of device performance and may even adversely affect human health (Olmedo, 1997; Saini, 2011). Due to possible hazards of EMI only, the use of EM wave receipting/emitting electronic gadgets is prohibited inside sensitive zones, for example, during flight or inside hospital's ICUs (Yang, 2005). Therefore, systematic strategies and suitable counter measures are essential to prevent/suppress EMI so as to ensure uninterrupted performance of appliances (Colaneri, 1992; Saini, 2009).

The primary mechanism of shielding is based on reflection and the material used for shielding by reflection requires mobile charge carriers; that is, shield should have conducting property. Consequently, metals (in the form of filler, coatings, or laminates) are the most common shielding material which uses primarily reflection mechanism for shielding along with minor absorption component. However, metals suffer from problems like poor wear/scratch resistance, corrosion susceptibility, high density, difficult processing, and high cost. The secondary shielding mechanism is absorption for which shield material should have electrical or magnetic dipoles (Abbas, 2005) along-with finite electrical conductivity. For such purpose materials with high dielectric constant like ZnO, SiO_2, TiO_2, $BaTiO_3$, or high magnetic permeability, for example, carbonyl iron, Ni, Co, or Fe metals, -Fe_2O_3, Fe_3O_4, and so forth are used (Chung, 2000; Saini, 2012). However, these materials or their composites possess problems like low permittivity or permeability at gigahertz frequencies, weight penalties, narrow-band action, and processing difficulties (Singh, 1999) Other than reflection and absorption, another mechanism of shielding is multiple reflections, which refer to the reflections at various surfaces or interfaces in the shield. This mechanism requires the presence of a large interfacial area and porous structure. An example of a shield with a large surface area is conducting composite material containing filler, foamed composites, and honeycomb structures. Interestingly, the properties of such composites are highly dependent on nature and concentration constituent phases. Carbon based materials (carbon black (CB), graphite, activated carbon (AC), fullerenes, carbon nanofibers (CNFs), cabon nanotubes (CNTs), and recently graphene) are attractive choice as a filler, due to good electrical/thermal conductivity, low density, excellent corrosion resistance, and reinforcing capability (Li, 2006 ;Saini, 2013; Tripathi, 2013; Saini, 2009; Wu, 2002; Varrla, 2011; Yang, 2005). Particularly, AC with high electrical conductivity and large surface area is an economical alternative. Similarly, acrylics are extremely popular matrix materials due to low cost, solution/thermal processability, high filler reception capacity and ability of acrylic resins (AR) to form hard, abrasion resistant and mechanically strong/tough impermeable coatings (Vallittu, 1999; Park, 2001; Khan, 2014).

Cellulose is the most abundant material in nature and features cheap, high crystallinity, and excellent mechanical performance (Deepa, 2011; Moran, 2008; Fahma, 2011; Rosa,2010).

A radically new innovation was established for development of electromagnetic interference shielding. The innovation emphasis synthesis of carboxymethyl cellulose (CMC), carboxymethyl cellulose composite containing different metal nanoparticles (CMCMNPs), and carboxymethyl cellulose nanofiber mat (CMC-NF) and carboxymethyl cellulose containing metal nanofiber mat (CMC-MNPs nanofiber mat) by electrospinning technique. Metal nanoparticles used include copper nanoparticles, iron nanoparticles, zinc nanoparticles, cadmium nanoparticles, and cobalt nanoparticles. Synthesized CMC–MNPs were characterized by using scanning electron microscopy coupled with high-energy dispersive X-ray and UV–visible spectroscopy that was used for confirmation of nanoparticles formation. The scanning electron microscopy images clearly showed regular flat shape with semiporous surface. All metal nanoparticles were well distributed inside the backbone of the cellulose without aggregation. The average particle diameter was 29–39 nm for zinc nanoparticles, 29–33 nm for cadmium nanoparticles, 25–33 nm for cobalt nanoparticles, 23–27 nm for copper nanoparticles, and 22–26 nm for iron nanoparticles. Electrospun carboxymethyl cellulose and CMC–MNPs nanofiber mats were synthesized by electrospinning technique and characterized using scanning electron microscopy, energy dispersive X-ray, and transmission electron microscopy. Scanning electron microscopy images of electrospun carboxymethyl cellulose and CMC–MNPs nanofibers reveal smooth and uniformly distributed nanofibers without bead formation with average fiber diameters in the range of 300–450 nm. Moreover, the diameters of electrospun carboxymethyl cellulose nanofiber mat were not affected by the presence of metal nanoparticles. Metal nanoparticles' content inside the electrospun CMC–MNPs nanofibers was investigated by using atomic absorptions pectroscopy. Electromagnetic interference shielding of electrospun carboxymethyl cellulose and CMC–MNPs nanofiber mats was evaluated. Data showed that the EMI-SE was increased in presence of metal nanoparticles and depending on both the metal nanoparticle contents and the electrical conductivity of metal nanoparticles (Gouda, 2016). Figure 6 shows EMI shielding of electrospun cellulose nanofiber containing different nanoparticle contents (Gouda, 2016).

CONCLUSION

The unique of the chemical and physical properties of lightweight nanocomposites polymers are promising for shielding with low-cost environmentally-friendly material. Shielding materials should meet a series of requirements to suit current

Figure 6. EMI shielding of electrospun cellulose nanofiber containing different nanoparticle contents
Source: This picture was taken from Gouda (2016)

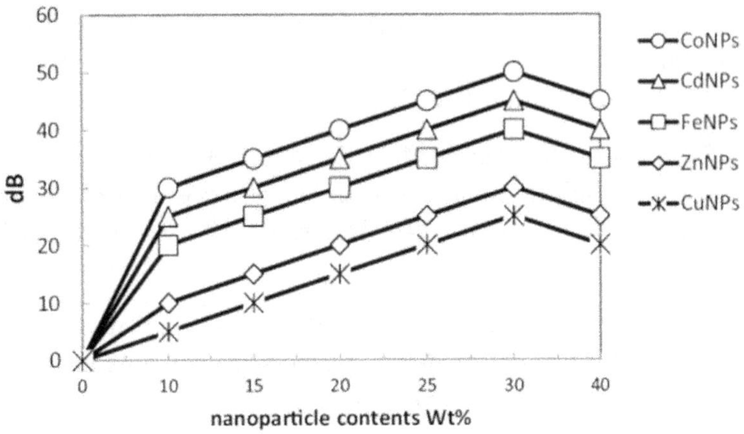

demands: resistant to corrosion and chemicals, lightweight, flexibility and cheap. These nanocomposites combine the advantageous properties of both polymers and fillers in one system. The presence of nanoparticles in a polymer matrix alters the characteristics of the polymer as the mechanical and thermal properties. It offer several advantages over traditional materials used for electromagnetic interference (EMI) shielding and very promising to be the next-generation shielding material. The type of filler and its ratio can determine the final performance of these lightweight nanocomposites polymers.

REFERENCES

Abbas, S. M., Dixit, A. K., Chatterjee, R., & Goel, T. C. (2005). Complex permittivity and microwave absorption properties of $BaTiO_3$- polyaniline composite. *Materials Science and Engineering B*, *123*(2), 167–171. doi:10.1016/j.mseb.2005.07.018

Agarwal, S., Wendorff, J. H., & Greiner, A. (2008). Use of electrospinning technique for biomedical applications. *Polymer*, *49*(26), 5603–5621. doi:10.1016/j.polymer.2008.09.014

Al-Ghamdi, A. A., Al-Hartomy, O. A., Al-Salamy, F., Al-Ghamdi, A. A., El-Mossalamy, E. H., Abdel Daiem, A. M., & El-Tantawy, F. (2012). Novel electromagnetic interference shielding effectiveness in the microwave band of magnetic nitrile butadiene rubber/magnetite nanocomposites. *Journal of Applied Polymer Science*, *125*(4), 2604–2613. doi:10.1002/app.36371

Al-Saleh, M. H., Gelves, G., & Sundararaj, U. (2011). Copper nanowire/polystyrene nanocomposites: Lower percolation threshold and higher EMI shielding. *Compos Part A*, *42*(1), 92–97. doi:10.1016/j.compositesa.2010.10.003

Bhardwaj, N., & Kundu, S. C. (2010). Electrospinning: A fascinating fiber fabrication technique. *Biotechnology Advances*, *28*(3), 325–347. doi:10.1016/j.biotechadv.2010.01.004 PMID:20100560

Bizdoaca, E., Spasova, M., Farle, M., Hilgendorff, M., & Caruso, F. (2002). Magnetically directed self-assembly of submicron spheres with a Fe_3O_4 nanoparticle shell. *Journal of Magnetism and Magnetic Materials*, *240*(1-3), 44–46. doi:10.1016/S0304-8853(01)00724-7

Chen, X., Wei, S., Gunesoglu, C., Zhu, J., Southworth, C. S., Sun, L. Y., ... Guo, Z. (2010). Electrospun Magnetic Fibrillar Polystyrene Nanocomposites Reinforced with Nickel Nanoparticles. *Macromolecular Chemistry and Physics*, *211*(16), 1775–1783. doi:10.1002/macp.201000153

Chiscan, O., Dumitru, I., Postolache, P., Tura, V., & Stancu, A. (2012). Electrospun PVC/Fe_3O_4 composite nanofibers for microwave absorption applications. *Materials Letters, 68*, 251–254. doi:10.1016/j.matlet.2011.10.084

Chung, D. D. L. (2000). Materials for electromagnetic interference shielding. *Journal of Materials Engineering and Performance, 9*(3), 350–354. doi:10.1361/105994900770346042

Chung, D. L. (2001). Review: Electromagnetic interference shielding effectiveness of carbon materials. *Carbon, 39*(2), 279–285. doi:10.1016/S0008-6223(00)00184-6

Colaneri, N. F., & Shacklette, L. W. (1992). EMI shielding measurements of conductive polymer blends. *IEEE Transactions on Instrumentation and Measurement, 41*(2), 291–297. doi:10.1109/19.137363

Darwish, M. S., Bakry, A., Kolek, O., Martinová, L., & Stibor, I. (2019). Electrospun functionalized magnetic polyamide 6 composite nanofiber: Fabrication and stabilization. *Polymer Composites, 40*, 296–303. doi:10.1002/pc.24647

Das, N. C., Liu, Y., Yang, K., Peng, W., Maiti, S., & Wang, H. (2009). Single-walled carbon nanotube/poly(methyl methacrylate) composites for electromagnetic interference shielding. *Polymer Engineering and Science, 49*(8), 1627–1634. doi:10.1002/pen.21384

Deepa, B., Abraham, E., & Cherian, B. M. (2011). Cellulose-based bio- and nanocomposites. *Bioresource Technology, 102*, 1988–1997. doi:10.1016/j.biortech.2010.09.030 PMID:20926289

Dumazet-Bonnamour, I., & Perchec, P. (2000). Colloidal dispersion of magnetite nanoparticles via in situ preparation with sodium polyoxyalkylene di-phosphonates. *Carbon, 45*(8), 1614–1621.

Erdem, R. (2016). Evaluation of electromagnetic shielding effectiveness of multi-axial fabrics and their reinforced PES composites. *Bulletin of Materials Science, 39*(4), 963–970. doi:10.100712034-016-1234-7

Fahma, F., Iwamoto, S., Hori, N., Iwata, T., & Takemura, A. (2011). Effect of pre-acid-hydrolysis treatment on morphology and properties of cellulose nanowhiskers from coconut husk. *Cellulose (London, England), 18*(2), 443–450. doi:10.100710570-010-9480-0

Gelves, G. A., Al-Saleh, M. H., & Sundararaj, U. (2011). Highly electrically conductive and high performance EMI shielding nanowire/polymer nanocomposites by miscible mixing and precipitation. *Journal of Materials Chemistry*, *21*(3), 829–836. doi:10.1039/C0JM02546A

Gonza'lez, M., Pozuelo, J., & Baselga, J. (2018). Electromagnetic Shielding Materials in GHz Range. *Chemical Record (New York, N.Y.)*, *18*, 1–11. PMID:29380939

Gouda, M., Hebeish, A. A., & Aljaafari, A. I. (2016). New route for development of electromagnetic shielding based on cellulosic nanofibers. *Journal of Industrial Textiles*, *46*(8), 1598–1615. doi:10.1177/1528083715627166

Grimes, C. A. (1994). EMI shielding characteristics of permalloymultilayer thin films. In *IEEE Aerospace Applications Conf. Proc,* IEEE.

Gupta, T. K., Singh, B. P., Dhakate, S. R., Singh, V. N., & Mathur, R. B. (2013). Improved nanoindentation and microwave shielding properties of modified MWCNT reinforced polyurethane composites. *Journal of Materials Chemistry. A, Materials for Energy and Sustainability*, *1*(32), 9138. doi:10.1039/c3ta11611e

Hajdu, J. (1997). New challenges for electroless plating technologies. *Trans Inst Metal Finish*, *75*(1), B7–B10.

Hoang, A. S. (2011). Electrical conductivity and electromagnetic interference shielding characteristics of multiwalled carbon nanotube filled polyurethane composite films. *Adv. Nat. Sci. Nanosci. Nanotechnol.*, *2*(2), 025007. doi:10.1088/2043-6262/2/2/025007

Hoffmann, D., Landfester, K., & Antonietti, M. (2001). Encapsulation of magnetite in polymer particles via the miniemulsion polymerization process. *Magnetohydrodynamics*, *37*(3), 217–221.

Hu, M., Gao, J., Dong, Y., Li, K., Shan, G., Yang, S., & Li, R. K.-Y. (2012). Flexible transparent PES/silver nanowires/PET sandwichstructured film for high efficiency electromagnetic interference shielding. *Langmuir*, *28*(18), 7101–7106. doi:10.1021/la300720y PMID:22533864

Huang, Z. M., He, C. L., Yang, A. Z., Zhang, Y., Han, X.-J., Yin, J., & Wu, Q. (2006). Encapsulating drugs in biodegradable ultrafine fibers through co-axial electrospinning. *Journal of Biomedical Materials Research. Part A*, *77*(1), 169–179. doi:10.1002/jbm.a.30564 PMID:16392131

Jia, J., Duan, Y. Y., & Wang, S. H. (2007). Preparation and characterization of antibacterial silver containing nanofibers for wound dressing applications. *J US-China Med Sci*, *22*, 452–454.

Joseph, N., Varghese, J., & Sebastian, M. T. (2017). In situ polymerized polyaniline nanofiber-based functional cotton and nylon fabrics as millimeter-wave absorbers. *Polymer Journal*, *49*(4), 391–399. doi:10.1038/pj.2016.121

Kang, T. S., Lee, S. W., Joo, J., & Lee, J. Y. (2005). Electrically conducting polypyrrole fibers spun by electrospinning. *Synthetic Metals*, *153*(1-3), 61–64. doi:10.1016/j.synthmet.2005.07.135

Kasagi, T., Tsutaoka, T., & Hatakeyama, K. (1999). Particle size effect on the complex permeability for perm-alloy composite materials. *IEEE Transactions on Magnetics*, *35*(5), 3424–3426. doi:10.1109/20.800545

Kashi, S., Gupta, R. K., Baum, T., Kao, N., & Bhattacharya, S. N. (2016). Morphology, electromagnetic properties and electromagnetic interference shielding performance of poly lactide/graphene nanoplatelet nanocomposites. *Materials & Design*, *95*, 119–126. doi:10.1016/j.matdes.2016.01.086

Keith, J. M., Janda, N. B., King, J. A., Perger, W. F., & Oxby, T. J. (2005). Shielding Effectiveness Density Theory for Carbon Fiber/Nylon 6,6 Composites. *Polymer Composites*, *26*(5), 671–678. doi:10.1002/pc.20139

Khan, S. D., Arora, M., Puri, C., Wahab, M. A., & Saini, P. (2014). Synthesis and characterization of acrylic resin/activated carbon composites. *Indian Journal of Pure and Applied Physics*, *52*(4), 251–254.

Khan, S. D., Arora, M., Wahab, M. A., & Saini, P. (2014). *Permittivity and Electromagnetic Interference Shielding Investigations of Activated Charcoal Loaded Acrylic Coating Compositions*. Journal of Polymers. doi:10.1155/2014/193058

Kim, H. R., Kim, B. S., & Kim, I. S. (2012). Fabrication and EMI shielding effectiveness of Ag decorated highly porous poly (vinyl alcohol) /Fe_2O_3 nanofibrous composites. *Materials Chemistry and Physics*, *135*(2-3), 1024–1029. doi:10.1016/j.matchemphys.2012.06.008

Kimmel, W. D., & Gerke, D. D. (1995). Controlling EMI with cable shields. *Medical Device & Diagnostic Industry*, *17*(7), 112–115.

Lagarkov, A. N., & Rozanov, K. N. (2009). High-frequency behavior of magnetic composites. *Journal of Magnetism and Magnetic Materials*, *321*(14), 2082–2092. doi:10.1016/j.jmmm.2008.08.099

Lee, B. O., Woo, W. J., & Kim, M. S. (2001). EMI shielding effectiveness of carbon nanofiber filled poly (vinyl alcohol) coating materials. *Macromolecular Materials and Engineering*, *286*(2), 114–118. doi:10.1002/1439-2054(20010201)286:2<114::AID-MAME114>3.0.CO;2-8

Lee, B. O., Woo, W. J., Park, H. S., Hahm, H. S., Wu, J. P., & Kim, M. S. (2002). Influence of aspect ratio and skin effect on EMI shielding of coating materials fabricated with carbon nanofiber/PVDF. *Journal of Materials Science*, *37*(9), 1839–1843. doi:10.1023/A:1014970528482

Li, J., Liu, H., Guo, J., Hu, Z., Wang, Z., Wang, B., ... Guo, Z. (2017). Flexible, conductive, porous, fibrillar polymer– gold nanocomposites with enhanced electromagnetic interference shielding and mechanical properties. *Journal of Materials Chemistry. C, Materials for Optical and Electronic Devices*, *5*(5), 1095–1105. doi:10.1039/C6TC04780G

Li, L., & Chung, D. D. L. (1991). Electrically conducting powder filled polyimidesiloxane. *Composites*, *22*(3), 211–218. doi:10.1016/0010-4361(91)90321-7

Li, L., & Chung, D. D. L. (1993). Effect of viscosity on the electrical properties of conducting thermoplastic composites made by compression molding of a powder mixture. *Polymer Composites*, *14*(6), 467–472. doi:10.1002/pc.750140604

Li, L., & Chung, D. D. L. (1994). Electrical and mechanical properties of electrically conductive polyethersulfone composites. *Composites*, *25*(3), 215–224. doi:10.1016/0010-4361(94)90019-1

Li, N., Huang, Y., Du, F., He, X., Lin, X., Gao, H., ... Eklund, P. C. (2006). Electromagnetic Interference (EMI) shielding of single-walled carbon nanotube epoxy composites. *Nano Letters*, *6*(6), 1141–1145. doi:10.1021/nl0602589 PMID:16771569

Li, Y., Chen, C., Zhang, S., Ni, Y., & Huang, J. (2008). Electrical conductivity and electromagnetic interference shielding characteristics of multiwalled carbon nanotube filled polyacrylate composite films. *Applied Surface Science*, *254*(18), 5766–5771. doi:10.1016/j.apsusc.2008.03.077

Ling, J., Zhai, W., Feng, W., Shen, B., Zhang, J., & Zheng, W. G. (2013). Facile Preparation of Lightweight Microcellular Polyetherimide/Graphene Composite Foams for Electromagnetic Interference Shielding. *ACS Applied Materials & Interfaces*, *5*(7), 2677–2684. doi:10.1021/am303289m PMID:23465462

Liu, Y., Teng, H., Hou, H., & You, T. (2009). Nonenzymatic glucose sensor based on renewable electrospun Ni nanoparticle-loaded carbon nanofiber paste electrode. *Biosensors & Bioelectronics*, *24*(11), 3329–3334. doi:10.1016/j.bios.2009.04.032 PMID:19450966

Liying, Z., Lin, B. W., Kye, Y. S., & Ma, J. (2013). Effect of carbon nanofiber reinforcement on electromagnetic interference shielding effectiveness of syntactic foam. *Journal of Materials Science*, *48*(21), 7757–7763. doi:10.100710853-013-7597-x

Lu, S., Qu, R., & Forcada, J. (2009). Preparation of Magnetic Polymeric Composite Nanoparticles by Seeded Emulsion Polymerization. *Materials Letters*, *63*(9-10), 770–772. doi:10.1016/j.matlet.2008.12.045

Ma, Q., Yu, W., Dong, X., Wang, J., Liu, G., & Xu, J. (2012). Electrospinning preparation and properties of Fe_3O_4/Eu(BA)$_3$ phen /PVP magnetic-photoluminescent bifunctional composite nanofibers. *Journal of Nanoparticle Research*, *14*(10), 1203–1209. doi:10.100711051-012-1203-z

Mandich, N. V. (1994). EMI shielding by electroless planting of ABSplastics. *Plating and Surface Finishing*, *81*(10), 60–63.

Markstein, H. W. (1995). Effective shielding defeats EMI. Electronic. *Packaging & Production*, *35*(2), 4.

Micheli, D., Apollo, C., Pastore, R., & Marchetti, M. (2010). X-Band microwave characterization of carbon-based nanocomposite material, absorption capability comparison and RAS design simulation. *Composites Science and Technology*, *70*(2), 400–409. doi:10.1016/j.compscitech.2009.11.015

Miyauchi, M., Simmons, T. J., Miao, J., Gagner, J. E., Shriver, Z. H., Aich, U., ... Linhardt, R. J. (2011). Electrospun polyvinylpyrrolidone fibers with high concentrations of ferromagnetic and superparamagnetic nanoparticles. *ACS Applied Materials & Interfaces*, *3*(6), 1958–1964. doi:10.1021/am200187x PMID:21561090

Moran, J. I., Alvarez, V. A., Cyras, V. P., & Vázquez, A. (2008). Extraction of cellulose and preparation of nanocellulose from sisal fibers. *Cellulose (London, England)*, *15*(1), 149–159. doi:10.100710570-007-9145-9

Morgan, N. (1991). *Corrosion inhibiting EMI/RFI shielding coating and method of its use.* EP Patent application EP19,900,314,408.

Nagasawa, C., Kumagai, Y., Urabe, K., & Shinagawa, S. (1999). Electromagnetic shielding particleboard with nickel-plated wood particles. *Journal of Porous Materials*, *6*(3), 247–254. doi:10.1023/A:1009692232398

Nan, L. I., & Van der Meer, R. (1986). *Synergistic effect of metal flake and metal or metal coated fiber on EMI shielding effectiveness of thermoplastics*. US Patent 4,566,990.

Nasouri, K., & Shoushtari, A. M. (2017). Fabrication of magnetite nanoparticles/ Polyvinylpyrrolidone composite nanofibers and their application as Electromagnetic interference shielding material. *Journal of Thermoplastic Composite Materials*, *31*(4), 431–446. doi:10.1177/0892705717704488

Nasouri, K., Shoushtari, A. M., & Mojtahedi, M. R. M. (2015). Thermodynamic studies on polyvinylpyrrolidone solution systems used for fabrication of electrospun nanostructures: Effects of the solvent. *Advances in Polymer Technology*, *34*(3), 21495–21502. doi:10.1002/adv.21495

Nasouri, K., Shoushtari, A. M., & Mojtahedi, M. R. M. (2016). Theoretical and experimental studies on EMI shielding mechanisms of multi-walled carbon nanotubes reinforced high performance composite nanofibers. *Journal of Polymer Research*, *23*(4), 71–78. doi:10.100710965-016-0943-3

Nayak, L., Khastgir, D., & Chaki, T. K. (2013). A mechanistic study on electromagnetic shielding effectiveness of polysulfone/carbon nanofibers nanocomposites. *Journal of Materials Science*, *48*(4), 1492–1502. doi:10.100710853-012-6904-2

O'Shea, P. (1998). How to meet the shielding needs of a 500-MHz PC. *Evaluation Engineer*, *37*(6), 40–46.

Olmedo, L., Hourquebie, P., & Jousse, F. (1997). *Handbook of Organic Conductive Molecules and Polymers* (Vol. 2). Chichester, UK: John Wiley & Sons.

Ott, H. W. (2009). *Electromagnetic Compatibility Engineering*. Hoboken, NJ: John Wiley & Sons. doi:10.1002/9780470508510

Pande, S., Singh, B. P., Mathur, R. B., Dhami, T. L., Saini, P., & Dhawan, S. K. (2009). Improved Electromagnetic Interference Shielding Properties of MWCNT–PMMA Composites Using Layered Structures. *Nanoscale Research Letters*, *4*(4), 327–334. doi:10.100711671-008-9246-x PMID:20596500

Panels, J. E., Lee, J. W., Park, K. Y., Kang, S. Y., Marquez, M., Wiesner, U., & Joo, Y. L. (2008). Synthesis and characterization of magnetically active carbon nanofiber/iron oxide composites with hierarchical pore structure. *Nanotechnology*, *19*(45), 455612–455618. doi:10.1088/0957-4484/19/45/455612 PMID:21832789

Park, H. S., Yang, I. M., Wu, J. P., Kim, M.-S., Hahm, H.-S., Kim, S.-K., & Rhee, H.-W. (2001). Synthesis of siliconeacrylic resins and their applications to superweatherable coatings. *Journal of Applied Polymer Science*, *81*(7), 1614–1623. doi:10.1002/app.1592

Ramakrishna, S., Fujihara, K., Teo, W. E., Yong, T., Ma, Z., & Ramaseshan, R. (2006). Electrospun nanofibers: Solving global issues. *Materials Today*, *9*(3), 40–50. doi:10.1016/S1369-7021(06)71389-X

Rosa, M. F., Medeiros, E. S., Malmonge, J. A., Gregorski, K. S., Wood, D. F., Mattoso, L. H. C., ... Imam, S. H. (2010). Cellulose nanowhiskers from coconut husk fibers: Effect of preparation conditions on their thermal and morphological behavior. *Carbohydrate Polymers*, *81*(1), 83–92. doi:10.1016/j.carbpol.2010.01.059

Sadchikov, V. V., & Prudnikova, Z. G. (1997). Amorphous materials in electromagnetic shields. *STAL. Sciences et Techniques de l'Animal de Laboratoire*, *4*, 66–69.

Saini, P. (2013). Electrical properties and electromagnetic interference shielding response of electrically conducting thermosetting nanocomposites. In Thermoset Nanocomposites. Wiley-VCH Verlag GmbH & Co. KGaA. doi:10.1002/9783527659647.ch10

Saini, P., & Arora, M. (2012). *Microwave absorption and EMI shielding behavior of nanocomposites based on intrinsically conducting polymers, graphene and carbon nanotubes. In New Polymers for Special Applications*. InTech.

Saini, P., Arora, M., Gupta, G., Gupta, B. K., Singh, V. N., & Choudhary, V. (2013). High permittivity polyaniline-barium titanate nanocomposites with excellent electromagnetic interference shielding response. *Nanoscale*, *5*(10), 4330–4336. doi:10.1039/c3nr00634d PMID:23563991

Saini, P., & Choudhary, V. (2013). Enhanced electromagnetic interference shielding effectiveness of polyaniline functionalized carbon nanotubes filled polystyrene. Journal of Nanoparticle Research, 15. doi:10.100711051-012-1415-2

Saini, P., Choudhary, V., Singh, B. P., Mathur, R. B., & Dhawan, S. K. (2009). Polyaniline MWCNT nanocomposites for microwave absorption and EMI shielding. *Materials Chemistry and Physics*, *113*(2-3), 919–926. doi:10.1016/j.matchemphys.2008.08.065

Saini, P., Choudhary, V., Singh, B. P., Mathur, R. B., & Dhawan, S. K. (2011). Enhanced microwave absorption behavior of polyaniline-CNT/polystyrene blend in 12.4-18.0 GHz range. *Synthetic Metals*, *161*(15-16), 1522–1526. doi:10.1016/j.synthmet.2011.04.033

Saini, P., Choudhary, V., Sood, K. N., & Dhawan, S. K. (2009). Electromagnetic interference shielding behavior of polyaniline/ graphite composites prepared by in situ emulsion pathway. *Journal of Applied Polymer Science*, *113*(5), 3146–3155. doi:10.1002/app.30183

Saini, P., Choudhary, V., Vijayan, N., & Kotnala, R. K. (2012). Improved electromagnetic interference shielding response of poly (aniline)-coated fabrics containing dielectric and magnetic nanoparticles. *The Journal of Physical Chemistry C*, *116*(24), 13403–13412. doi:10.1021/jp302131w

Sakai, K., Asano, N., Wada, Y., & Yoshikado, S. (2010). Composite electromagnetic wave absorber made of soft magnetic material and polystyrene resin and control of permeability and permittivity. *Journal of the European Ceramic Society*, *30*(2), 347–353. doi:10.1016/j.jeurceramsoc.2009.05.044

Saleh, M. H., & Sundararaj, U. (2013). X-band EMI shielding mechanisms and shielding effectiveness of high structure carbon black/polypropylene composites. *Journal of Physics. D, Applied Physics*, *46*, 1–7.

Sanghoon, K., Joon-Suk, O., & Myeong-Gi, K. (2014). Electromagnetic interference (EMI) transparent shielding of reduced graphene oxide (RGO) interleaved structure fabricated by electrophoretic deposition. *Appl Mater Interfaces*, *6*(20), 17647–17653. doi:10.1021/am503893v PMID:25238628

Sarvi, A., & Sundararaj, U. (2014). Electrical permittivity and electrical conductivity of multiwall carbon nanotube-polyaniline (MWCNT-PANi) core-shell nanofibers and MWCNT-PANi/polystyrene composites. *Macromolecular Materials and Engineering*, *299*(8), 1013–1020. doi:10.1002/mame.201300406

Shinagawa, S., Kumagai, Y., & Urabe, K. (1999). Conductive papers containing metalized polyester fibers for electromagnetic interference shielding. *Journal of Porous Materials*, *6*(3), 185–190. doi:10.1023/A:1009619711017

Singh, P., Babbar, V. K., Razdan, A., Srivastava, S. L., & Puri, R. K. (1999). Complex permeability and permittivity, and microwave absorption studies of Ca(CoTi) Fe12–2O19 hexaferrite composites in X-band microwave frequencies. *Materials Science and Engineering B*, *67*(3), 132–138. doi:10.1016/S0921-5107(99)00328-1

Tan, S. T., Zhang, M. Q., Rong, M. Z., & Zeng, H. (1999). Effect of interfacial modification on metal fiber filled polypropylene composites and property balance. *Polymer Composites*, *20*(3), 406–412. doi:10.1002/pc.10366

Tripathi, S. N., Saini, P., Gupta, D., & Choudhary, V. (2013). Electrical and mechanical properties of PMMA/reduced graphene oxide nanocomposites prepared via in situ polymerization. *Journal of Materials Science*, *48*(18), 6223–6232. doi:10.100710853-013-7420-8

Tzeng, S. S., & Chang, F. Y. (2001). EMI shielding effectiveness of metal-coated carbon fiberreinforced ABS composites. *Materials Science and Engineering A*, *302*(2), 258–267. doi:10.1016/S0921-5093(00)01824-4

Vallittu, P. K. (1999). Flexural properties of acrylic resin polymers reinforced with unidirectional and woven glass fibers. *The Journal of Prosthetic Dentistry*, *81*(3), 318–326. doi:10.1016/S0022-3913(99)70276-3 PMID:10050121

Varrla, E., Venkataraman, S., & Sundara, R. (2011). Functionalized graphene-PVDF foam composites for EMI shielding. *Macromolecular Materials and Engineering*, *296*(10), 894–898. doi:10.1002/mame.201100035

Villacorta, B. S., Ogale, A. A., & Hubing, T. H. (2013). Effect of Heat Treatment of Carbon Nanofibers on the Electromagnetic Shielding Effectiveness of Linear Low Density Polyethylene Nanocomposites. *Polymer Engineering and Science*, *53*(2), 417–423. doi:10.1002/pen.23276

Wang, B., Sun, Y., & Wang, H. (2010). Preparation and properties of electrospun PAN/Fe_3O_4 magnetic nanofibers. *Journal of Applied Polymer Science*, *115*(3), 1781–1786. doi:10.1002/app.31288

Wang, G., Chang, Y., Wang, L., Liu, L., & Liu, C. (2013). Facilely preparation and microwave absorption properties of Fe_3O_4 nanoparticles. *Materials Research Bulletin*, *48*(3), 1007–1012. doi:10.1016/j.materresbull.2012.11.089

Wang, L., Yu, Y., Chen, P. C., Zhang, D. W., & Chen, C. H. (2008). Electrospinning synthesis of C/Fe_3O_4 composite nanofibers and their application for high performance lithium-ion batteries. *Journal of Power Sources*, *183*(2), 717–723. doi:10.1016/j.jpowsour.2008.05.079

Wang, L., Yu, Y., Chen, P. C., Zhang, D. W., & Chen, C. H. J. (2008). Electrospinning synthesis of C/Fe3O4 composite nanofibers and their application for high performance lithium-ion batteries. *Power Sources*, *183*(2), 717–723. doi:10.1016/j.jpowsour.2008.05.079

Wang, M., Singh, H., Hatton, T. A., & Rutledge, G. C. (2004). Field-responsive superparamagnetic composite nanofibers by electrospinning. *Polymer*, *45*(16), 5505–5514. doi:10.1016/j.polymer.2004.06.013

Wang, S., Wang, C., Zhang, B., Sun, Z., Li, Z., Jiang, X., & Bai, X. (2010). Preparation of Fe_3O_4/PVA nanofibers via combining in-situ composite with electrospinning. *Materials Letters*, *64*(1), 9–11. doi:10.1016/j.matlet.2009.09.043

Wei, Y., Zhang, X., Song, Y., Han, B., Hu, X., Wang, X., ... Deng, X. (2011). Magnetic biodegradable Fe_3O_4/CS/PVA nanofibrous membranes for bone regeneration. *Biomedical Materials (Bristol, England)*, *6*(5), 1–16. doi:10.1088/1748-6041/6/5/055008 PMID:21893702

Wei, Z., Huagang, X., Shaokai, W., Min, L., & Yizhuo, G. (2015). Electromagnetic characteristics of carbon nanotube film materials. *Chinese Journal of Aeronautics*, *28*(4), 1245–1254. doi:10.1016/j.cja.2015.05.002

Wu, J., & Chung, D. D. L. (2002). Increasing the electromagnetic interference shielding effectiveness of carbon fiber polymermatrix composite by using activated carbon fibers. *Carbon*, *40*(3), 445–447. doi:10.1016/S0008-6223(01)00133-6

Wu, Z. P., Li, M. M., Hu, Y. Y., Li, Y. S., Wang, Z. X., Yin, Y. H., ... Zhou, X. (2011). Electromagnetic interference shielding of carbon nanotube macrofilms. *Scripta Materialia*, *64*(9), 809–812. doi:10.1016/j.scriptamat.2011.01.002

Yang, Q., Li, Z., Hong, Y., Zhao, Y., Qiu, S., Wang, C., & Wei, Y. (2004). Influence of solvents on the formation of ultrathin uniform poly (vinyl pyrrolidone) nanofibers with electrospinning. *Journal of Polymer Science. Part B: Polymer Letters*, *42*(20), 3721–3726. doi:10.1002/polb.20222

Yang, Y., Guo, Z., Zhang, H., Huang, D., Gu, J., Huang, Z., ... Rutledge, G. C. (2013). Electrospun Magnetic Carbon Composite Fibers: Synthesis and Electromagnetic Wave Absorption Characteristics. *Journal of Applied Polymer Science*, *127*(6), 4288–4295. doi:10.1002/app.38027

Yang, Y., Gupta, M. C., Dudley, K. L., & Lawrence, R. W. (2005). Novel carbon nanotube—Polystyrene foam composites for electromagnetic interference shielding. *Nano Letters*, *5*(11), 2131–2134. doi:10.1021/nl051375r PMID:16277439

Yang, Y., Gupta, M. C., Dudley, K. L., & Lawrence, R. W. (2005). Conductive carbon nanofiber-polymer foam structures. *Advanced Materials*, *17*(16), 1999–2003. doi:10.1002/adma.200500615

Yu, H., Wang, T., Wen, B., Lu, M., Xu, Z., Zhu, C., ... Cao, M. (2012). Graphene/ polyaniline nanorod arrays: Synthesis and excellent electromagnetic absorption properties. *Journal of Materials Chemistry*, *22*(40), 21679. doi:10.1039/c2jm34273a

Zhang, D., Karki, A. B., Rutman, D., Young, D. P., Wang, A., Cocke, D., ... Guo, Z. (2009). Electrospun polyacrylonitrile nanocomposite fibers reinforced with Fe_3O_4 nanoparticles: Fabrication and property analysis. *Polymer*, *50*(17), 4189–4198. doi:10.1016/j.polymer.2009.06.062

Zhang, L., Li, W., Liu, J., & Ren, B. (1995). Process and performance of electroless Cu/Ni–P double film. Cailiao Gongcheng/ *J. Materials & Design*, *7*, 38–41.

Zheng-Ming, H., Zhang, Y. Z., & Kotakic, M. (2003). A review on polymer nanofibers by electrospinning and their applications in nanocomposites. *Composites Science and Technology*, *63*(15), 2223–2253. doi:10.1016/S0266-3538(03)00178-7

Chapter 7
Nanomaterials and Nanocomposites Thermal and Mechanical Properties Modelling

Siddhartha Kosti
Rajkiya Engineering College, India

ABSTRACT

This chapter deals with the modelling of nanomaterial and nanocomposite mechanical and thermal properties. Enrichment in the technology requires materials having higher thermal properties or higher structural properties. Nanomaterials and nanocomposites can serve this purpose accurately for aerospace or thermal applications and structural applications respectively. The thermal system requires materials having high thermal conductivity while structural system requires materials having high strength. Selection of the material for particular application is very critical and requires knowledge and experience. Al, Cu, TiO2, Al2O3, etc. are considered for thermal applications while epoxy-glass, FRP, etc. are considered for structural applications. Modelling of these nanomaterials and nanocomposites is done with the help of different mathematical models available in the literature. Results show that addition of the nanoparticle/composite in the base material can enhance the thermal and structural properties. Results also show that amount of weight percentage added also affects the properties.

DOI: 10.4018/978-1-5225-7921-2.ch007

INTRODUCTION

Nanomaterials are defined as a combination of nanoparticles reinforced into the base material. Nanoparticles are nothing but the powder form of a solid metal in small size. Nanoparticles are of very small sizes (1 nm = 10^{-9}m), they also have very high thermal properties, thermal conductivity heat capacity etc. They also possesses higher surface area to volume ratio, this property increases the heat transfer rate as amount of surface area available gets increased.

A lot of research is going in the field of nanoparticles behaviour analysis under different conditions or applications. They have been widely utilized in different applications, like in energy, construction, biomedical, chemical, electronics, agriculture, paints and cosmetics etc.

Energy

Nanomaterials are very helpful in energy field, as they possess enhanced thermal properties compared to the conventional materials. Nanomaterials are utilizing in solar and other energy conservation devices. Addition of nanoparticles enhances the thermal conductivity of the material which increases the rate of molecular momentum inside the material which results in increment in the heat transfer rate from the material. This increment in the heat transfer rate increases the efficiency of the system. As nanoparticle's surface area to volume ratio is very high they provide higher surface area for transfer of heat compared to conventional material for available volume. This property further enhances the heat transfer rate and efficiency of the system. Literature study done in the field of solar cooling states that use of nanoparticles/nanomaterials increases the rate of solar cooling by considerable amount (Al-Shamani et al, 2014). Nanoparticles like Cu and Al are reinforced into the base material like water (H_2O) or ethylene which enhances the effective thermal conductivity and heat transfer rate. This reinforcement can be done for different nanoparticle weight percentage, generally weight percentage vary from 0 to 20, as adding more nanoparticles can cause the problem of agglomeration or clustering and this reduces the performance. Use of nano technology in combustion engines, solar cells and capacitors has results in device improvements. A company named as ConsERV has developed a device, made of nano polymer layer which increases the cooling and heating efficiency of the system. Applied NanoWorks is company established in New York is working in the field of white LED, this can give this area a new dimension. There are other numerous applications of the nanomaterials which has results in the start of new ventures/companies.

Carbon nanotubes (CNTs) are the widely studied nanoparticles in the literature; they are very much suitable, for application in the aerospace as they have high thermal conductivity, high tensile strength and high modulus of elasticity. CNTs have 500 times better strength to weight ratio, electrical conductivity compared to the conventional material like steel, Al and titanium (Zhang, 2003; CMP Cientifica, 2002). CNTs are very flexible as they have Carbon-Carbon covalent bond. Nanotechnology can be utilized in other aerospace industry applications like in sensors, composites tires, radar technology etc.

Construction

Strength of the construction materials can be improved by nanotechnology. Nanocomposites can enhance the strength, load bearing capability and can also reduce the overall weight. In the construction industries nano fibres, carbon nanotubes and nano particles have shown prominent improvement in the buildings strength, durability and toughness. Different nanoparticles like SiO_2, TiO_2 and $CaCo_3$ have been utilized in paints/coatings and sealants in the form of additives or fillers (Zhang, 2003; Bigley & Greenwood, 2003). (Masciangioli, & Zhang, 2003; Kuzumaki, 1998) studied the coating of TiO_2 and Fe & Pd based nanoparticle coatings respectively to reduce the environmental contaminants by absorbing them.

LITERATURE REVIEW

Xiao & Liu, 2018, conducted review study of the 2D nanomaterials as lubricant additive. Nanomaterials because of their molecular structure and lubricating properties have been widely utilized in the tribological applications. Singh, Lara, & Tlali, 2017 conducted enthalpy, specific heat and melting entropy of the nanomaterials for varying shape and size. They found that specific heat increases with decrement in the particle size while melting entropy and enthalpy decreases with particle size. Sharma et al, 2018, conducted review study in advance applications of nanomaterials. Santosh et al, 2016, conducted review study on the nanomaterials applications in the water treatment. Moustafa, 2017, conducted review study on isothermal reduction process and kinetics of nanomaterials in reducing the atmospheric temperature. Kabir et al, 2018, conducted analysis on the impact of nanomaterials on the environment. Bernard, Ulm, & Lemarchand, 2003, conducted the multiscale micromechanics-hydration model for the early-age elastic properties of cement-based materials. Kishi & Maekawa, 1997, conducted the modelling of the hydration heating of blended cement for concrete performance. Garboczi & Bentz, 1996, conducted the modelling of transport and microstructure properties of concrete.

Rubel & Landis, 1969; Khanafer, Vafai, & Lightstone, 2003; Turan, Poole, & Chakraborty, 2012; Lai & Yang, 2011; Bilgen & Muftuoglu, 2008; Tiwari & Das,

2007; Talebi, Mahmoudi, Shahi, 2010; Kahafer, Al-Amiri & Pop, 2007, Kosti, Das & Saha, 2013; Kosti, 2014; Santra, Sen, & Chakraborty, 2008; Abu-Nada & Oztop, 2009; conducted numerical analysis of convection flow problems inside rectangular and square cavity for types of nanoparticles. They analysed the effect of nanoparticles on the flow structure generating inside the cavity, they also discusses about the temperature and streamline contours inside the cavity. They all concluded that addition of nanoparticle inside the cavity changes the flow structure and affect the final performance of the problem.

Enrichment in the technology is resulting in miniaturisation of the equipment used in the aerospace or thermal systems and, but these small equipment releases high heat and requires efficient cooling. Due to the small sizes only passive cooling methods are appropriate which can either

MATHEMATICAL MODELS

To conduct the modelling of the nanomaterials different mathematical models are considered in the present study. Wide varieties of mathematical models are available in the literature (Lee et al, 2010) to calculate the effective properties of the nanomaterials like thermal conductivity, dynamic viscosity, heat transfer coefficient, thermal expansion coefficient and density, but most of the studied are devoted towards the finding of the effective thermal conductivity as thermal conductivity is the parameter which largely affect the heat transfer rate. In the present study thermal conductivity model, viscosity model and density model are considered and analysed for different nanoparticles.

Mixture Rule

Physical mixture rule (Rohatagi, 1993) is a weighted mean method to calculate the effective thermal conductivity of the nanofluid materials. Above rule can also be used to calculate the other properties like density, viscosity, heat capacity, thermal diffusivity and thermal expansion coefficient. Mathematically mixture rule can be expressed as,

$$k_{nf} = k_{np}\phi + k_{bf}\left(1-\phi\right)$$

where: ϕ is nanoparticle weight percentage and it can be expressed as the ratio of the volume of nanoparticle added to the total volume. Mathematically it can be represented as,

$$\phi = \frac{V_{np}}{V_{np} + V_{bf}}$$

Above method utilized to calculate the effective properties of the nanofluid is also called as the upper bound limit.

Thermal Conductivity Models

Four thermal conductivity models are considered in the present study based on amount of weight percentage or volume fraction addition of nanoparticle (Maxwell, 1873), thickness of the nanoparticle layers (Yu & Choi, 2003), nanoparticle shape factor (Hamilton & Crosser, 1962) and physical mixture rule (Rohatagi, 1993).

Maxwell Model

Maxwell model is the first classical thermal conductivity model based on the effective medium theory (EMT). This classical model comes with some assumptions like, particles are motionless and heat transfer by diffusion takes place in dispersed phase and continuous matrix phase and calculates the effective thermal conductivity of liquid-solid mixture. The formula to calculate the effective thermal conductivity developed by the author can be expressed as,

$$k_{nf} = \left[\frac{\left(k_{np} + 2k_{bf}\right) + 2\phi\left(k_{np} - k_{bf}\right)}{\left(k_{np} + 2k_{bf}\right) - \phi\left(k_{np} - k_{bf}\right)} \right] k_{bf}$$

Where:

k: Thermal conductivity
ϕ: Amount of nanoparticle weight percentage added
nf, np, bf: Nanofluid, nanoparticle and base fluid

Above model illustrates that effective thermal conductivity of the nanofluid depends on the thermal conductivity of the base fluid, thermal conductivity of the nanoparticle added and nanoparticle weight percentage/volume fraction. Maxwell model is only valid for small amount of addition of nanoparticle weight percentage into the base material. Above model is also only valid for spherical shape nanoparticles.

Yu and Choi Model

Yu and Choi model is the first thermal conductivity model including the effect of nanolayer thickness to calculate the effective thermal conductivity. During establishing the model they assume that solid-like layer or nanolayer are formed of thickness h, these layer have larger thermal conductivity compared to bulk liquid, these nanolayers can be united with nanoparticle and develop equivalent particle and low weight percentage of nanoparticle results in no overlapping of equivalent particle. The above equation is valid for nanoparticle having radius below 5mm. They also concluded that if the nanoparticle radius is above 5mm effect of change in the thermal conductivity becomes very small and can be neglected. The equivalent nanoparticle effective radius is 'r+h' and effective weight percentage is ϕ_e. This effective weight percentage can be calculated as,

$$\phi_e = \frac{4}{3}\pi\left(r+h\right)^3 = \frac{4}{3}\pi r^3\left(1+\frac{h}{r}\right)^3 = \frac{4}{3}\pi r^3\left(1+\beta\right)^3$$

Considering all these assumptions, model established by them can be expressed as,

$$k_{nf} = \left[\frac{\left(k_{ep}+2k_{bf}\right)+2\phi\left(k_{ep}-k_{bf}\right)\left(1+\beta\right)^3}{\left(k_{ep}+2k_{bf}\right)-\phi\left(k_{ep}-k_{bf}\right)\left(1+\beta\right)^3}\right]k_{bf}$$

where k_{ep} is the thermal conductivity of equivalent nanoparticle and can be calculated as (Schwartz, Garboczi, & Bentz, 1995)

$$k_{ep} = \left[\frac{\left\{\left(1+\beta\right)^3\left(1+2\gamma\right)+2\left(1-\gamma\right)\right\}\gamma}{\left\{\left(1+\beta\right)^3\left(1+2\gamma\right)-\left(1-\gamma\right)\right\}}\right]k_{np}$$

Where:

γ: Ratio of thermal conductivity of nanolayer to nanoparticle thermal conductivity
β: Ratio of nanolayer thickness to nanoparticle radius
ep: Equivalent particle combined with nanolayer

Rest of the notations have usual meanings as explained in the Maxwell model. For extreme cases when thickness of nanolayer is same as that of nanoparticle (i.e. $\gamma=1$) will give $k_{ep}=k_{np}$. The above model is reduced to Maxwell model for ($\gamma=1$ and $\beta=0$)

Hamilton and Crosser Model

They extended the Maxwell model by considering the effect of irregular shape of the nanoparticle to calculate the effective thermal conductivity. They developed the model for heterogeneous two-component mixtures for continuous and discontinuous phase. Model established by them can be expressed as,

$$k_{nf} = \left[\frac{k_{np} + (n-1)k_{bf} + (n-1)(k_{np} - k_{bf})\phi}{k_{np} + (n-1)k_{bf} - (k_{np} - k_{bf})\phi}\right] k_{bf}$$

where: n: Shape factor

Nanoparticle shape factor is defined as $3/\psi$, where ψ is the sphericity of the nanoparticle, which is defined as the surface area of the nanoparticle to its volume. For sphericity equal to one Hamilton & Crosser model is reduced to Maxwell model for spherical nanoparticle. The above model is valid for the nanoparticles which have 100 times higher thermal conductivity as compared to the base fluid.

Viscosity Models

To analyse the viscosity of the nanoparticle (Einstein, 1906; Einstein, 1956) and (Brinkman, 1952) models are considered. The above considered models have been widely utilized in the literature.

Einstein Model

It is the first model developed to calculate the effective viscosity of the nanomaterials. This model is valid for spherical solid nanoparticles. This model has been developed considering the small particles movement analysis in a stationary liquid according to the molecular kinetic theory of heat. Mathematically the model can be expressed as,

$$\mu_{nf} = (1 + 2.5\phi)\mu_{bf}$$

where: μ: Dynamic viscosity or viscosity

Other notations have usual meaning as explained in the above paragraphs.

Brinkman Model

They developed the dynamic viscosity of the concentrated solutions and suspensions. Their model is also valid for spherical shape nanoparticle. The model developed by them can be expressed mathematically as,

$$\mu_{nf} = \frac{1}{\left(1-\phi\right)^{2.5}}\mu_{bf}$$

THERMAL ANALYSIS

Thermal analysis of the nanomaterials is conducted on the basis of weight percentage of nanoparticles and models for calculating effective thermal conductivity. Five different nanoparticles considered here for the analyses are copper (Cu), boron carbide (B_4C), aluminium oxide (Al_2O_3), titanium oxide (TiO_2) and silicon carbide (SiC).

Nanoparticles Properties

Wide verities of nanomaterials are available in the literature, as present deals with applications of nanomaterials in the thermal/aerospace industry and in the structural/construction industry. Nanoparticles like Cu, Carbon-nanotubes, B_4C, Al_2O_3 and TiO_2 are considered for thermal analysis. Weight percentage of these nanoparticles is varied from 0 to 30%. While composites like SiC, Fibre reinforced polymer and epoxy-glass are considered for structural analysis. Table 1 represents the properties of these nanoparticles.

Table 1. Properties of nanoparticles

Nanoparticle/ Material	Thermal conductivity (W/m-K)	Specific heat (J/kg-K)	Density (kg/m^3)
Cu	400	383	8954
B_4C	42	1288	2550
Al_2O_3	36	773	3880
TiO_2	8.9538	686.2	4250
H_2O	0.6	4179	997.1
SiC	100	1300	3200

Weight Percentage/Volume Fraction (ϕ)

Copper (Cu)

First the analysis of thermal conductivity is done, for this analysis H_2O is considered as a base material. Table 2 shows the thermal conductivity values of Copper nanoparticles suspended into the water. Weight percentage of the Copper nanoparticles is varied up to 30. The values of shape factor are taken from (Timofeeva, Routbort & Singh, 2009).

From Table 2 it can be seen that Mixture rule predicts largest thermal conductivity compared to other thermal conductivity models. But this high prediction is vague as mixture rule does not involve any physical phenomena. Results from the above table are also plotted in a figure to have a clear understanding of the variation between the different models considered.

Figure 1 above shows the comparison of the effective thermal conductivity of Cu-H_2O nanofluids calculated for different model considered. On the horizontal axis weight percentage or volume fraction of the Cu nanoparticle is plotted while on the vertical axis thermal conductivity is plotted. Effective thermal conductivity results of the Mixture rule are not included in the figure, as addition of them was making the other results unreadable. It can be seen from the figure that Yu & Choi model predicts the largest thermal conductivity for all weight percentage addition. While Maxwell model predicts the lowest thermal conductivity of Ch-H_2O nanofluid for all weight percentage addition. While thermal conductivity predicts by Hamilton & Crosser model lies between the Yu & Choi model and Maxwell model. Thermal conductivity predicted by Yu & Choi model is largest because the model considers the nanolayer thickness while calculating the effective thermal conductivity, this

Table 2. Thermal conductivity values of Cu nanoparticles for different thermal conductivity models

Material		Maxwell	Yu & Choi			Hamilton & Crosser			Mixture rule
H_2O	Cu		$\beta = 0$	$\beta = 0.2$	$\beta = 0.4$	n = 3	n = 4.9	n = 5.7	
1.0	0	0.6	0.6	0.6	0.6	0.6	0.6	0.6	0.6
0.95	0.5	0.6943	0.6943	0.7694	0.8847	0.6943	0.7535	0.7784	20.57
0.9	0.1	0.7990	0.7990	0.974	1.2765	0.799	0.9240	0.9764	40.54
0.85	0.15	0.9160	0.9160	1.226	1.8496	0.916	1.1144	1.1975	60.51
0.8	0.2	1.0475	1.0475	1.544	2.7677	1.0475	1.3283	1.4459	80.48
0.75	0.25	1.1964	1.1964	1.9582	4.4768	1.1964	1.5705	1.7271	100.45
0.7	0.3	1.3665	1.3665	2.5196	8.7727	1.3665	1.8469	2.0480	120.42

Figure 1. Thermal conductivity vs. volume fraction for Cu-H_2O nanoparticle

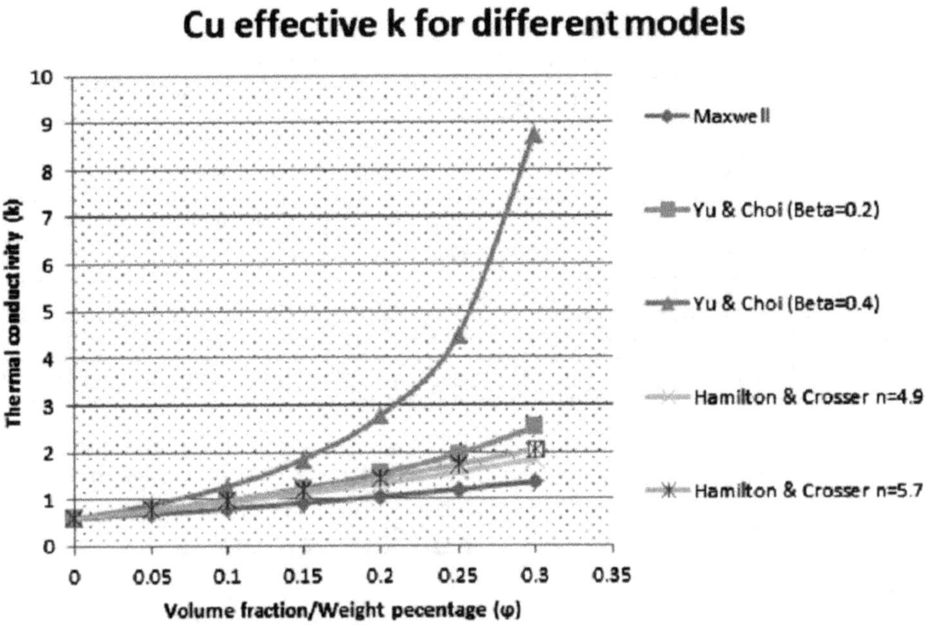

inclusion of nanolayer thickness increases the effective weight percentage of the nanoparticle added in the base material, this increment in the weight percentage increases the overall effective thermal conductivity.

Boron Carbide (B_4C)

Table 3 shows the thermal conductivity values of boron carbide (B_4C) nanoparticles suspended into the water (H_2O). Weight percentage of the B_4C nanoparticles is varied up to 30. Values of shape factor and nanolayer thickness are considered from the literature.

Figure 2 shows the comparison of the effective thermal conductivity of the B_4C-H_2O nanofluid. As it can be perceived from the figure that same trend of effective thermal conductivity of B_4C-H_2O nanofluid is observed as observed for Cu-H_2O nanofluid.

Aluminum Oxide (Al_2O_3)

Table 4 shows the thermal conductivity values of Al2O3 nanoparticles for different thermal conductivity models.

Table 3. Thermal conductivity values of B_4C nanoparticles for different thermal conductivity models

Material		Maxwell	Yu & Choi			Hamilton & Crosser			Mixture rule
H_2O	B4C		$\beta = 0$	$\beta = 0.2$	$\beta = 0.4$	n = 3	n = 4.9	n = 5.7	
1.0	0	0.6	0.6	0.6	0.6	0.6	0.6	0.6	0.6
0.95	0.5	0.6906	0.6906	0.7625	0.8725	0.6906	0.744	0.7656	2.67
0.9	0.1	0.7908	0.7908	0.9572	1.2422	0.7908	0.9028	0.9480	4.74
0.85	0.15	0.9022	0.9022	1.1949	1.7725	0.9022	1.0788	1.1500	6.81
0.8	0.2	1.0268	1.0268	1.4914	2.5969	1.0268	1.2751	1.375	8.88
0.75	0.25	1.1671	1.1671	1.8717	4.0542	1.1671	1.4952	1.6269	10.95
0.7	0.3	1.3263	1.3263	2.3771	7.3268	1.3263	1.7439	1.911	13.02

Figure 2. Thermal conductivity vs. volume fraction for B_4C-H_2O nanoparticle

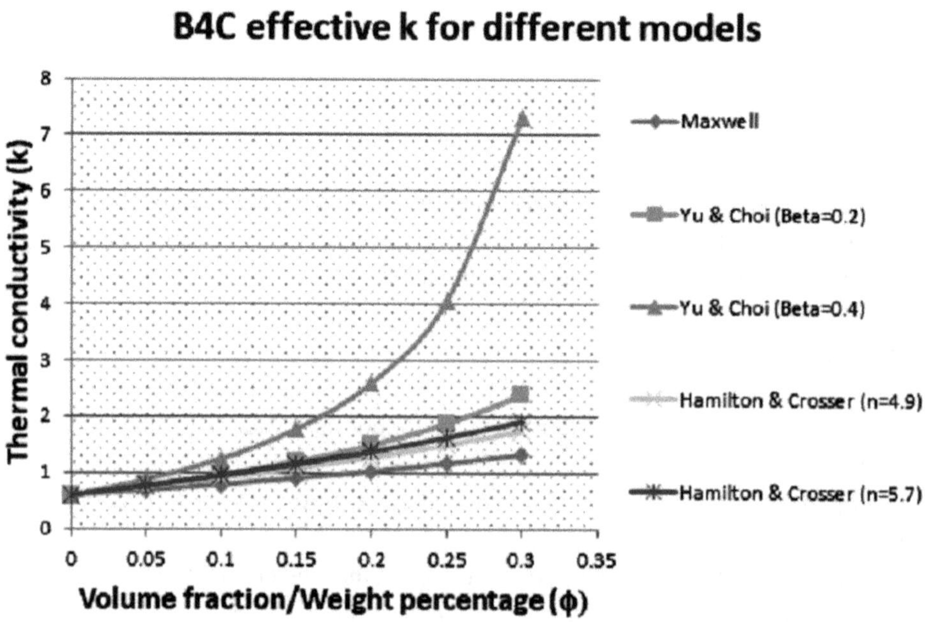

Titanium Oxide (TiO_2)

Table 5 shows the thermal conductivity values of TiO_2 nanoparticles for different thermal conductivity models.

Figure 3 and 4 show the effective thermal conductivity comparison of Al_2O_3-H_2O and TiO_2-H_2O nanofluids for different thermal conductivity models.

Table 4. Thermal conductivity values of Al_2O_3 nanoparticles for different thermal conductivity models

Material		Maxwell	Yu & Choi			Hamilton & Crosser			Mixture rule
H_2O	Al2O3		β = 0	β = 0.2	β = 0.4	n = 3	n = 4.9	n = 5.7	
1.0	0	0.6	0.6	0.6	0.6	0.6	0.6	0.6	0.6
0.95	0.5	0.6899	0.6899	0.7612	0.8703	0.6899	0.7423	0.7634	2.37
0.9	0.1	0.7893	0.7893	0.9542	1.2361	0.7893	0.8991	0.9432	4.14
0.85	0.15	0.8997	0.8997	1.1893	1.7589	0.8997	1.0726	1.1419	5.91
0.8	0.2	1.0231	1.0231	1.4821	2.5676	1.0231	1.2659	1.3629	7.68
0.75	0.25	1.1619	1.1619	1.8565	3.9844	1.1619	1.4823	1.6099	9.45
0.7	0.3	1.3192	1.3192	2.3525	7.109	1.3192	1.7264	1.888	11.22

Table 5. Thermal conductivity values of TiO_2 nanoparticles for different thermal conductivity models

Material		Maxwell	Yu & Choi			Hamilton & Crosser			Mixture rule
H_2O	TiO2		β = 0	β = 0.2	β = 0.4	n = 3	n = 4.9	n = 5.7	
1.0	0	0.6	0.6	0.6	0.6	0.6	0.6	0.6	0.6
0.95	0.5	0.6772	0.6772	0.7377	0.8290	0.6772	0.7129	0.7258	1.0177
0.9	0.1	0.7614	0.7614	0.8983	1.1248	0.7614	0.8348	0.8612	1.4354
0.85	0.15	0.8534	0.8534	1.0879	1.5216	0.8534	0.9669	1.0073	1.8531
0.8	0.2	0.9545	0.9545	1.3151	2.0817	0.9545	1.1104	1.1656	2.2708
0.75	0.25	1.0661	1.0661	1.5925	2.9321	1.0661	1.267	1.3374	2.6884
0.7	0.3	1.1899	1.1899	1.9386	4.3774	1.1899	1.4384	1.5248	3.1061

Silicon Carbide (SiC)

Table 6 shows the thermal conductivity values of SiC nanoparticles for different thermal conductivity models

Thermal Conductivity Models

Here analysis of the thermal conductivity models is conducted for different kind of nanomaterials considered in the present study. To analyse this, amount of weight of nanoparticles added in the water is varied up to 30%. To analyse the Hamilton & Crosser model and Yu & Choi model maximum values of shape factor and nanolayer thickness factor are considered as maximum values of these factors will

Figure 3. Thermal conductivity vs. volume fraction for Al_2O_3-H_2O nanoparticle

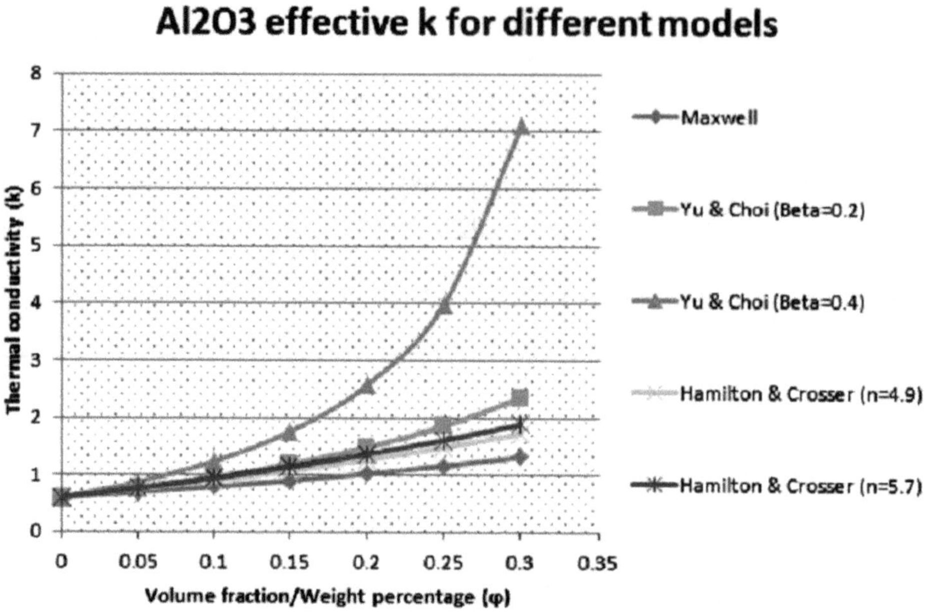

Figure 4. Thermal conductivity vs. volume fraction for TiO_2-H_2O nanoparticle

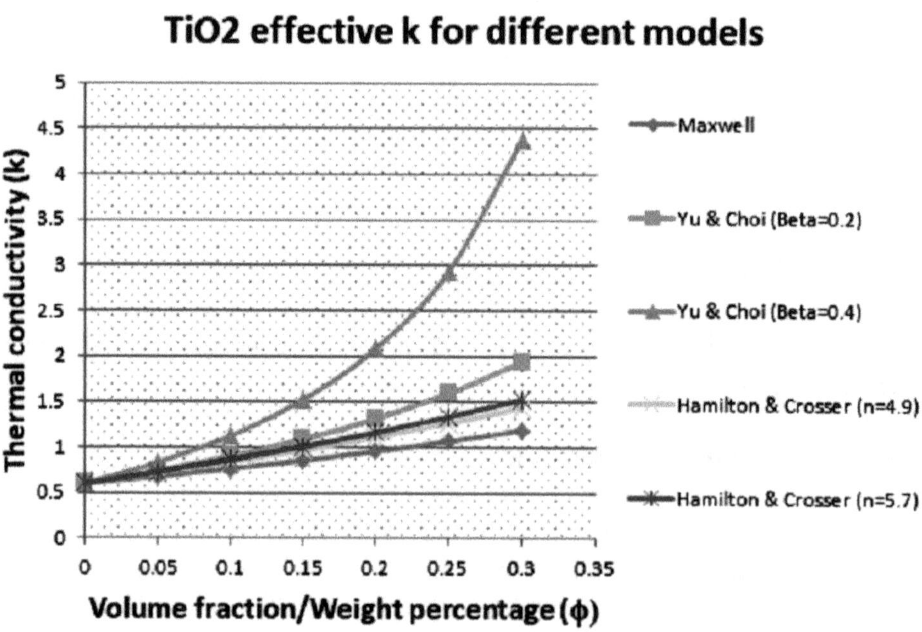

Table 6. Thermal conductivity values of SiC nanoparticles for different thermal conductivity models

Material		Maxwell	Yu & Choi			Hamilton & Crosser			Mixture rule
H_2O	SiC		$\beta = 0$	$\beta = 0.2$	$\beta = 0.4$	n = 3	n = 4.9	n = 5.7	
1.0	0	0.6	0.6	0.6	0.6	0.6	0.6	0.6	0.6
0.95	0.5	0.693	0.693	0.7669	0.8803	0.693	0.7501	0.7737	5.57
0.9	0.1	0.7961	0.7961	0.968	1.2641	0.7961	0.9163	0.9660	10.54
0.85	0.15	0.9110	0.9110	1.2148	1.8215	0.9110	1.1014	1.1800	15.51
0.8	0.2	1.040	1.040	1.5250	2.7049	1.040	1.3088	1.4197	20.48
0.75	0.25	1.1858	1.1858	1.9267	4.3181	1.1858	1.5428	1.6899	25.45
0.7	0.3	1.352	1.352	2.4673	8.2023	1.352	1.8089	1.9970	30.42

Figure 5. Thermal conductivity vs. volume fraction for SiC-H_2O nanoparticle

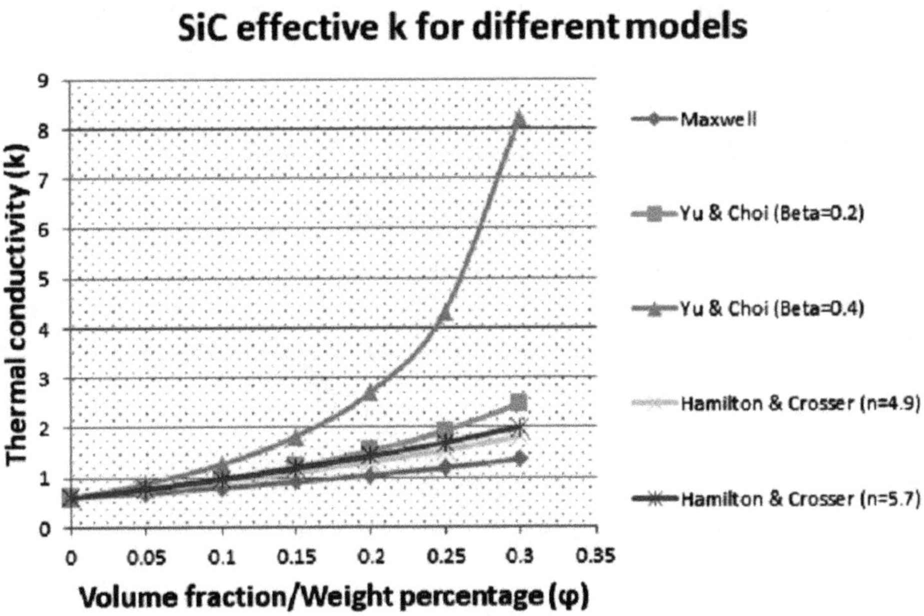

give the clear understanding of the results. Figure 6-9 represents the results of Maxwell model, Hamilton & Crosser model, Yu & Choi model and Mixture rule respectively. It can be seen from the figures that thermal conductivity of all the nanomaterials increases with increment in the weight percentage of nanoparticles. Maximum thermal conductivity is predicted by Mixture rule which is found to be inaccurate with the literature results [Maxwell, Hamilton & Crosser, Yu & Choi].

Maxwell Model

Figure 6 shows thermal conductivity vs. volume fraction for Maxwell model.

Hamilton and Crosser Model

Figure 7 shows thermal conductivity vs. volume fraction for Hamilton and Crosser model.

Yu and Choi Model

Figure 8 shows thermal conductivity vs. volume fraction for Yu and Choi model.

Mixture Rule

Figure 9 shows thermal conductivity vs. volume fraction for Mixture rule.

Figure 6. Thermal conductivity vs. volume fraction for Maxwell model

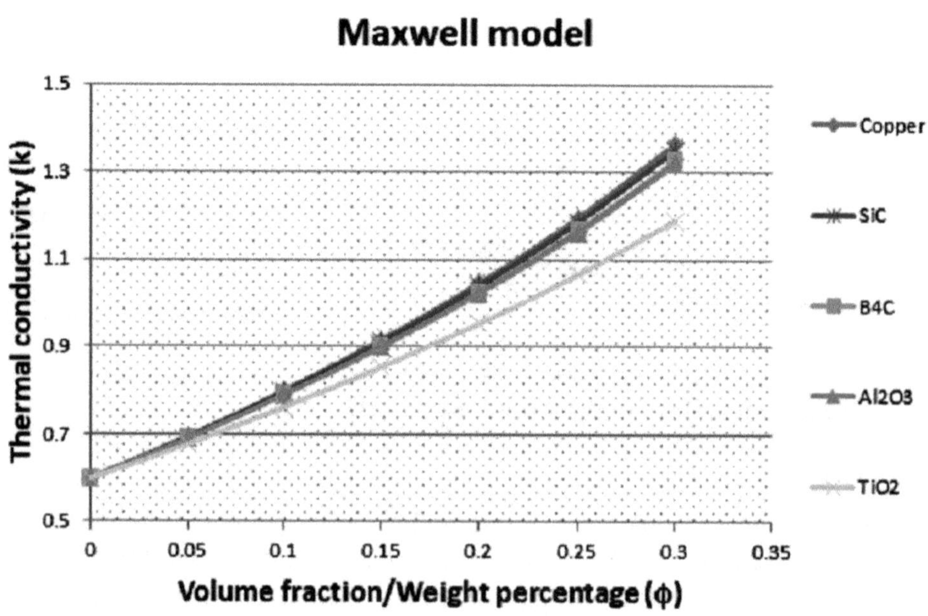

Figure 7. Thermal conductivity vs. volume fraction for Hamilton and Crosser model

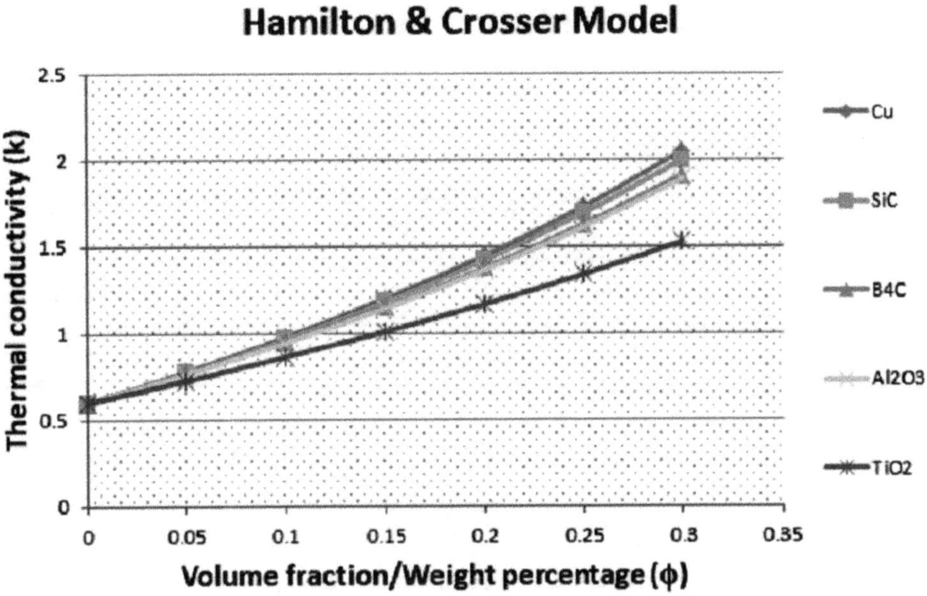

Figure 8. Thermal conductivity vs. volume fraction for Yu and Choi model

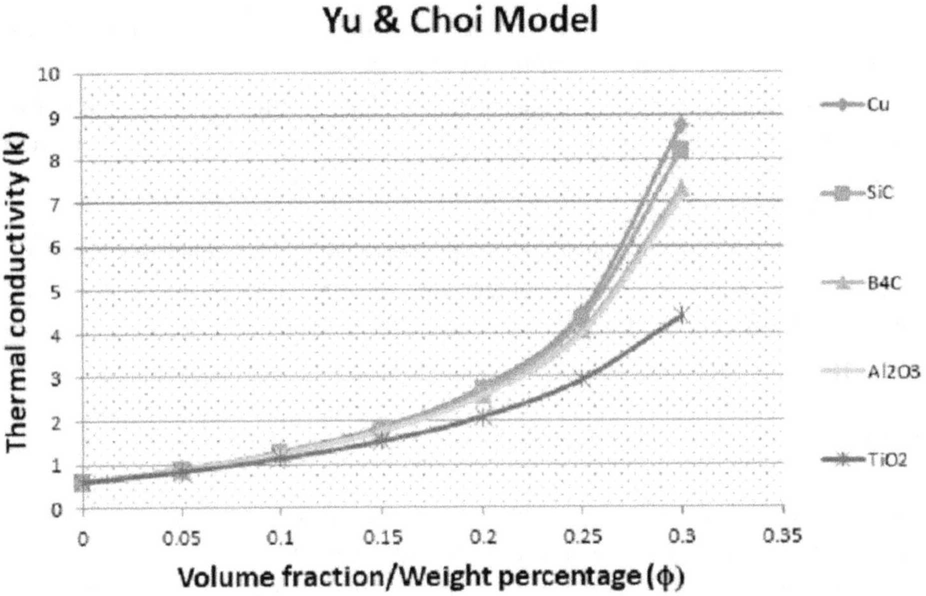

Figure 9. Thermal conductivity vs. volume fraction for Mixture rule

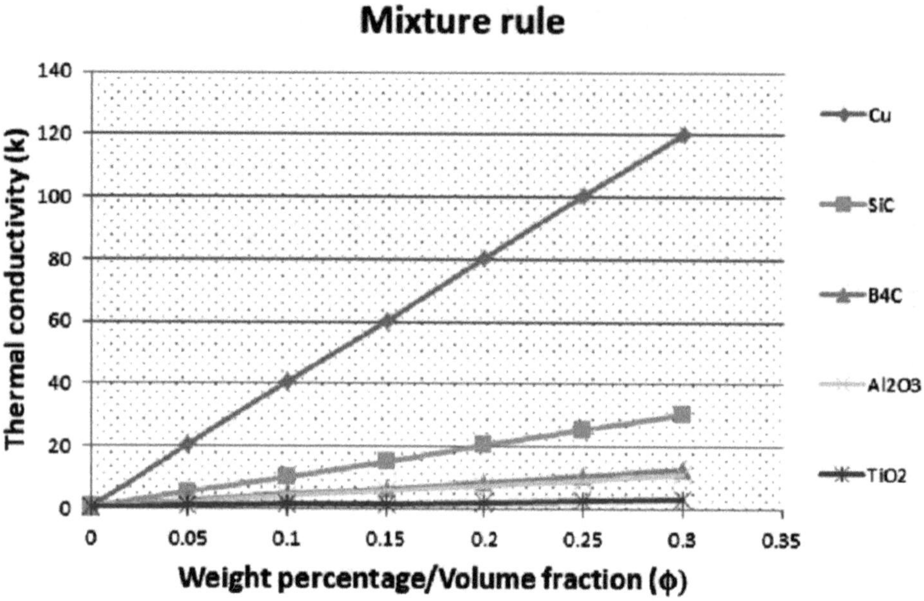

STRUCTURAL ANALYSIS

In the present work structural analysis of nanocomposites is conducted using two methods. In one method density of the nanocomposites is calculated and analysed using the mixture rule. While in the second method, a simple geometry is drawn and effect of load applied on the geometry is studied by analysing the deformation generated on the geometry.

For structural analysis, those materials have been considered which have applications in the field of constructions or automobiles. Materials considered are epoxy E-glass, silicon carbide, boron carbide, resin epoxy, aluminium oxide and conventional structural steel. Table 7 represents the density of the different composites considered in the present work.

Method 1

In this method, nanoparticles listed above are reinforced into the base material to form new nanocomposites. These nanocomposites also called metal matric composites (MMCs) (Kosti & Malvi, 2018; Kosti & Pathak, 2018). MMCs are the combination of two or more materials, one of which is base material (metal alloys) while others can be ceramics (nanoparticles). Weight percentage of these nanoparticles is varied

Table 7. Density of the nanocomposites

Material	Density (kg/m³)
Al6061	2700
Epoxy E-glass	2000
Silicon carbide (SiC)	3100
Boron carbide (B_4C)	2400
Resin epoxy	1160
Aluminium oxide (Al_2O_3)	3960
Structural steel	7850

up to 30%. Base material considered in this method is aluminium alloy (Al6061) while nanoparticles considered are silicon carbide (SiC), boron carbide (B_4C) and aluminium oxide (Al_2O_3).

Mixture rule is considered to analyse the effective density of the MMC. The results have been plotted in the figure 10. From the figures it can be observed that addition of nanocomposites in the base material affect the effective density. Figure reveals that addition of silicon carbide and boron carbide increases the nanocomposites

Figure 10. Density vs. volume fraction for mixture rule

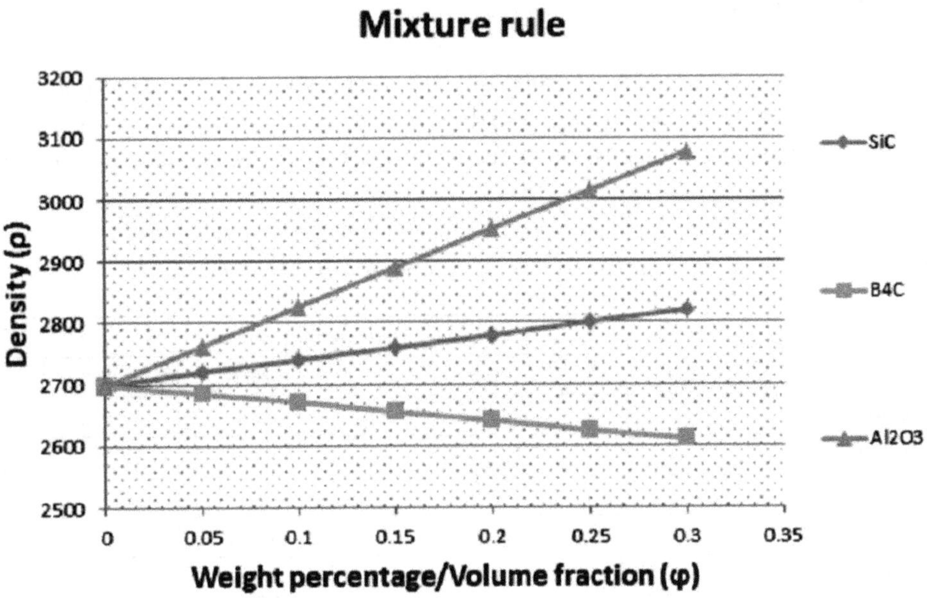

density while addition of aluminium oxide decreases the nanocomposites effective density. Higher the density means tighter the intermolecular packing of the atoms compared to the lower density material.

Method 2

In this method geometry is drawn then a load is applied on the geometry. Material of the geometry is changed and comparative analysis is conducted by analysing the deformation. Nanocomposites considered in the present method are epoxy E-glass, resin epoxy, boron carbide, silicon carbide and structural steel. Leaf spring geometry is drawn to analyse the effect of material on the deflection generated for different values of load.

From Table 8 it can be seen that use of nanocomposites in place of conventional material like helps in reducing the deflection generated on the geometry under the same loading conditions. This decrement in the deflection, under the same application of load shows that the capability of the material to bear large amount of load is increased. It can be noticed from the table that leaf spring made of resin epoxy shows almost 200 times less deflection when compared with leaf spring made of structural steel. This type of materials can be utilized in the construction industry to enhance the strength of buildings.

CONCLUSION

Nanocomposites have enhanced properties suitable for wide verities of thermal and structural applications. Depending upon the type of application, nanocomposites can be utilized either by reinforced it into the conventional material or by replacing it by the

Table 8. Comparison of the deflection generated for different materials

Material Load	Deflection (mm)		
	100 N	1000 N	10000 N
Resin epoxy	3.6558×10^{-6}	3.6558×10^{-5}	3.6558×10^{-4}
Epoxy E-glass	7.8180×10^{-7}	7.8180×10^{-6}	7.8180×10^{-5}
Boron carbide	3.0649×10^{-8}	3.0649×10^{-7}	3.0649×10^{-6}
Silicon carbide	3.4457×10^{-8}	3.4457×10^{-7}	3.4457×10^{-6}
Structural steel	6.9072×10^{-8}	6.9072×10^{-7}	6.9072×10^{-6}

conventional material. Present work shows that reinforcement of the nanocomposites into the base material (water) gives nanofluids like Cu-H_2O, Sic-H_2O, B_4C-H_2O, Al_2O_3- H_2O and TiO_2- H_2O which possess enhanced thermal properties and can be utilized in verities of thermal applications. While nanocomposites like epoxy E-glass, resin-epoxy, SiC, B_4C and Al_2O_3 can be utilized in the construction or automobile industry as a replacement of the conventional materials. These nanocomposites provide high strength and load bearing capacity to the structure/vehicle.

REFERENCES

Abu-Nada, E., & Oztop, H. F. (2009). Effects of Inclination Angle on Natural Convection in Enclosures Filled With Cu–Water Nano-Fluid. *International Journal of Heat and Fluid Flow*, *30*(4), 669–678. doi:10.1016/j.ijheatfluidflow.2009.02.001

Al-Shamani, A. N., Yazdi, M. H., Alghoul, M. A., Abed, A. M., Ruslan, M. H., Mat, S., & Sopian, K. (2014). Nanofluids for Improved Efficiency in Cooling Solar Collectors: A Review. *Renewable & Sustainable Energy Reviews*, *38*, 348–367. doi:10.1016/j.rser.2014.05.041

Bernard, O., Ulm, F. J., & Lemarchand, E. (2003). A Multiscale Micromechanics-Hydration Model for the Early-Age Elastic Properties of Cement-Based Materials. *Cement and Concrete Research*, *33*(9), 1293–1309. doi:10.1016/S0008-8846(03)00039-5

Bigley, C., & Greenwood, P. (2003). Using Silica to Control Bleed and Segregation in Self-Compacting Concrete. *Concrete (London)*, 43–45.

Bilgen, E., & Muftuoglu, A. (2008). Natural Convection in an Open Square Cavity with Slots. *International Communications in Heat and Mass Transfer*, *35*(8), 896–900. doi:10.1016/j.icheatmasstransfer.2008.05.001

Brinkman, H. C. (1952). The Viscosity of Concentrated Suspensions and Solutions. *The Journal of Chemical Physics*, *20*(4), 571. doi:10.1063/1.1700493

Cientifica, C. M. P. (2002). *The Nanotechnology Opportunity Report*. Retrieved from www.cmp-cientifica.com/

Einstein, A. (1906). Eine Neue bestimmung der molekuldimensionen. *Annalen Der Physik, Leipzig*, *19*(2), 289–306. doi:10.1002/andp.19063240204

Einstein, A. (1956). *Investigation on Theory of Brownian Motion*. Dover.

Garboczi, E. J., & Bentz, D. P. (1997). Analytical formulas for Interfacial Transition Zone Properties. *Journal of Advanced Cement Based Materials*, *6*(3-4), 99–108. doi:10.1016/S1065-7355(97)90016-X

Hamilton, R. L., & Crosser, O. K. (1962). Thermal Conductivity of Heterogeneous Two-Component System. *Industrial & Engineering Chemistry Fundamentals*, *1*(3), 187–191. doi:10.1021/i160003a005

Kabir, E., Kumar, V., Kim, K. H., Yip, A. G. K., & Sohn, J. R. (2018). Environmental Impacts of Nanomaterials. *Journal of Environmental Management*, *225*, 261–271. doi:10.1016/j.jenvman.2018.07.087 PMID:30096714

Khanafer, K., Vafai, L., & Lightstone, M. (2003). Buoyancy-Driven Heat Transfer Enhancement in a Two-Dimensional Enclosure Utilizing Nano-Fluids. *International Journal of Heat and Mass Transfer*, *46*(19), 3639–3653. doi:10.1016/S0017-9310(03)00156-X

Khanafer, K. M., Al-Amiri, A. M., & Pop, I. (2007). Numerical Simulation of Unsteady Mixed Convection in a Driven Cavity using an Externally Excited Sliding Lid. *European Journal of Mechanics - B/Fluids*, *26*(5), 669–687. doi:10.1016/j.euromechflu.2006.06.006

Kishi, T., & Maekawa, K. (1997). Multi-Component Model for Heat of Hydration Heating of Blended Cement with Blast Furnace Slag and Fly Ash. *Concrete Library of JSCE*, *30*(12), 125–139.

Kosti, S. (2014). Numerical Study of Heat Flux Boundary in Nanofluid-Filled Cavity. *Nanomaterials and Energy*, *3*(6), 193–205. doi:10.1680/nme.14.00017

Kosti, S., Das, M. K., & Saha, A. K. (2013). Buoyancy-Driven Flow and Heat Transfer in a Nano-Fluid-Filled Enclosure. *Nanomaterials and Energy*, *2*(4), 200–211. doi:10.1680/nme.13.00013

Kosti, S., & Malvi, C. S. (2018). Cumulative Influence of Nanoparticles on MMCs Time-Temperature History Curve. *Nanomaterials and Energy*, *7*(1), 1–10. doi:10.1680/jnaen.17.00011

Kosti, S., & Pathak, P. (2018). Genetic Algorithm Based Finite Difference Simulation of Solidification Process of MMC's. *Materials Today: Proceedings*, *5*(2), 8271–8279. doi:10.1016/j.matpr.2017.11.518

Kuzumaki, T., Miyazawa, K., Ichinose, H., & Ito, K. (1998). Processing of Carbon Nanotube Reinforced Aluminium Composites'. *Journal of Materials Research*, *13*(9), 2445–2449. doi:10.1557/JMR.1998.0340

Lai, F. H., & Yang, Y. T. (2011). Lattice Boltzmann Simulation of Natural Convection Heat Transfer of Al2O3/Water Nano-Fluids in a Square Enclosure. *International Journal of Thermal Sciences*, *50*(10), 1930–1941. doi:10.1016/j.ijthermalsci.2011.04.015

Lee, J.-H., Lee, S.-H., Choi, C. J., Jang, S. P., & Choi, S. U. S. (2010). A Review of Thermal Conductivity Data, Mechanism and Models for Nanofluids. *International Journal of Micro-Nano Scale Transport*, *1*(4), 269–322. doi:10.1260/1759-3093.1.4.269

Masciangioli, T., & Zhang, W. X. (2003). Environment Technologies at the Nano-scale. *Environmental Science & Technology*, *37*(5), 102A–109A. doi:10.1021/es0323998 PMID:12666906

Maxwell, J. C. (1873). *A Treatise on Electricity and Magnetism* (1st ed.; Vol. 1). Clarendoon Press.

Moustafa, A. F. (2017). Isothermal Reduction Process and Kinetic of Nanomaterials in Reducing Atmosphere: A review. *Journal of Analytical and Applied Pyrolysis*, *127*, 126–139. doi:10.1016/j.jaap.2017.08.015

Rohatagi, P. K. (1993). Metal Matrix Composites. *Defence Science Journal*, *43*(4), 323–349. doi:10.14429/dsj.43.4336

Rubel, A., & Landis, F. (1969). Numerical Study of Natural Convection in a Vertical Rectangular Enclosure. *Physics of Fluids*, *12*, 208–213.

Santosh, C., Velmurugan, V., Jacob, G., Jeong, S. K., Grace, A. N., & Bhatnagar, A. (2016). Role of Nanomaterials in Water Treatment Applications: A Review. *Chemical Engineering Journal*, *306*, 1116–1137. doi:10.1016/j.cej.2016.08.053

Santra, A. K., Sen, S., & Chakraborty, N. (2008). Study of Heat Transfer Augmentation in a Differentially Heated Square Cavity using Copper–Water Nano-Fluid. *International Journal of Thermal Sciences*, *47*(9), 1113–1122. doi:10.1016/j.ijthermalsci.2007.10.005

Schwartz, L., Garboczi, E., & Bentz, D. (1995). Interfacial Transport in Porous Media: Applications to DC Electrical Conductivity of Mortars. *Journal of Applied Physics*, *78*(10), 5898–5908. doi:10.1063/1.360591

Sharma, V. P., Sharma, U., Chattopadhyay, M., & Shukla, V. N. (2018). Advance Applications of Nanomaterials: A review. *Materials Today: Proceedings*, *5*(2), 6376–6380. doi:10.1016/j.matpr.2017.12.248

Singh, M., Lara, S., & Tlali, S. (2017). Effects of Size and Shape on the Specific Heat, Melting Entropy and Enthalpy of Nanomaterials. *Journal of Taibah University for Science Journal, 11*(6), 922–929. doi:10.1016/j.jtusci.2016.09.011

Talebi, F., Mahmoudi, A. H., & Shahi, M. (2010). Numerical Study of Mixed Convection Flows in a Square Lid-Driven Cavity Utilizing Nano-Fluid. *International Communications in Heat and Mass Transfer, 37*(1), 79–90. doi:10.1016/j.icheatmasstransfer.2009.08.013

Timofeeva, E. T., Routbort, J. L., & Singh, D. (2009). Particle shape effect on thermophysical properties of alumina nanofluids. *Journal of Applied Physics, 106*(1), 1–10. doi:10.1063/1.3155999

Tiwari, R. K., & Das, M. K. (2007). Heat Transfer Augmentation in a Two-Sided Lid-Driven Differentially Heated Square Cavity Utilizing Nano-Fluids. *International Journal of Heat and Mass Transfer, 50*(9-10), 2002–2018. doi:10.1016/j.ijheatmasstransfer.2006.09.034

Turan, O., Poole, R. J., & Chakraborty, N. (2012). Influences of Boundary Conditions on Laminar Natural Convection in Rectangular Enclosures with Differentially Heated Side Walls. *International Journal of Heat and Fluid Flow, 33*(1), 131–146. doi:10.1016/j.ijheatfluidflow.2011.10.009

Xiao, H., & Liu, S. (2018). 2D Nanomaterials a Lubricant Additive: A Review. *Materials & Design, 135*, 319–332. doi:10.1016/j.matdes.2017.09.029

Yu, W., & Choi, S. U. S. (2003). The Role of Interfacial Layers in the Enhanced Thermal Conductivity of Nanofluids: Arenovated Maxwell Model. *Journal of Nanoparticle Research, 6*(4), 167–171. doi:10.1023/A:1024438603801

Zhang, G. D. (2003). Electrical Properties of Nano-ceramics Reinforced with Ropes of Single-Walled Carbon Nanotubes. *Applied Physics Letters, 83,* 1228-1230. doi:10.1063/1.1600511

Chapter 8
Simulation and Modeling of Nanotechnology Aircraft Using MATLAB

Indradeep Kumar
Vels Institute of Science, Technology & Advanced Studies, India

ABSTRACT

The design methods based on aerospace model have been widely used in aircraft conceptual design for decades and proven very effective when restricted to simple problems with very approximate analyses. These monolithic, large, design and analysis codes are genuinely multidisciplinary, but as analyses become more complex, such codes have grown so large as to be incomprehensible and hence difficult to maintain. This chapter deals with the computational modeling of nanoparticles. Nanomaterials constitute a prominent sub-discipline in the materials and chemical sciences. Conventional materials like glass, ceramic, metals, polymers, or semiconductors can be acquired with nanoscale proportions. Nanomaterials have various microstructural distinctive attributes such as nanodiscs, nanotubes, nanocoatings, quantum dots, nanocomposites, and nanowires. The unique properties of nanoparticle-based materials and devices depend directly on size and structure dependent properties.

INTRODUCTION

This chapter deals with the computational modeling of nanoparticles. Nanomaterials constitute a prominent sub-discipline in the materials and chemical sciences. Conventional materials like glass, ceramic, metals, polymers or semiconductors can be acquired with nanoscale proportions. Nanomaterials have various micro

DOI: 10.4018/978-1-5225-7921-2.ch008

structural distinctive attributes such as nanodiscs, nanotubes, nanocoatings, quantum dots, nanocomposites, and nanowires. The unique properties of nanoparticle based materials and devices depend directly on size and structure dependent properties. Nanoparticle size must be firmly controlled to take all advantage of effects of quantum size in technological applications, and agglomeration must be averted. It can only be done if the flock is controlled, which requires, the rate of new particle formation is quantitatively resolute.

In Real-time measurements, distributions of particle size and particle structure are, thus, validate techniques for the evolution of nanotechnology. Precise and reliable models for simulating transport, coagulation, deposition, and dispersion of nanoparticles and their cluster are needed for the development of design tools for applications in technological, including nanoparticle instrumentation, sensing, dilution, sampling, and focusing nanoparticle behavior in the chemically reactive framework. Computational techniques allow us to validate and explore hypotheses about the experimentally observation that may otherwise not be approachable through conventional experimental techniques. Additionally, simulations with computer allow for the theory to propose areas of interest in which experimental techniques can be applied. The most common computer simulation methods are Monte Carlo (MC) and molecular dynamics (MD), have been extensively applied in a variety of model framework.

Computer science offers more opportunities for nanotechnology. Soft computing techniques such as cellular automata, genetic algorithms and swarm intelligence, can impart required emergent properties like self-repair, growth, and complex networks to the framework. Many books have successfully applied such techniques to real-world problems, including complex control systems in manufacturing and control in aircraft system. Dealing in nanoscale systems involves the understanding and development of nanotechnology computing techniques as well as the application of these techniques in real-world tasks, often to the problems of other Chapter areas. The techniques in nanotechnology systems comprise algorithms in machine learning, artificial intelligence (AI), knowledge representation reasoning, and natural computing. With some improvement in nanotechnology characteristics, these techniques can be applied to control a flock of a trillion nano assemblers. It is expected that normal computing methods such as these will overcome concerns about harmful implications of nanotechnology and prevent the notorious scenario of self-replicating nanorobots multiplying uncontrollably.

Presents micromagnetics: finite element analysis of nano-sized magnetic materials using MATLAB. Fabrication techniques developed over the past decades have allowed engineers and scientists to form magnetic materials in the nanometer regime. Such materials exhibit very interesting magnetic properties that are significantly different from bulk materials. At such scales, the behaviors of the magnetic materials are inadequately described by Maxwell's equations alone. The theory of micromagnetics deals with the behaviors of these nano-sized magnetic materials by including the quantum mechanical effects that are significant at this scale. Micromagnetics theory has been successful in predicting the formation of domain walls in magnetic materials and also the formation of interesting magnetic states such as the vortex and leaf states. It also finds its application in many engineering aspects such as the digital data storage technology. Chapter 3 will introduce the basic theory of micromagnetics, show how the finite element method can be applied to the governing equation of micromagnetics (Landau–Lifshitz–Gilbert equation), and finally, show how all these can be done using MATLAB.

System-level modeling of N/MEMS; it discusses some of the latest advances in designing and modeling complex nano-and micro-electromechanical (N/MEMS) systems at the system level. These advances have been applied to an open source tool called PSugar at Purdue University. For design, the author discusses a graphical user interface (GUI) that allows users to quickly configure complex systems in 3D using a computer mouse or pen at a rate faster than what might be drawn with pencil and paper. This GUI is coupled to a powerful netlist language for design flexibility. For modeling, the chapter applies for recent advances in analytical system dynamics and differential–algebraic equations (DAEs) into a framework that facilitates the systematic modeling of multidisciplinary systems that may comprise static or dynamic constraints. Where appropriate, we verify lumped models against their distributed model counterparts. Several test cases are discussed that demonstrate a wide range of utility of system-level analysis for N/MEMS. The test cases include simple to complex MEMS, and NEMS using molecular dynamics.

This chapter explains MATLAB applications in behavior analysis of systems consisting of CNTs through molecular dynamics simulation. This chapter presents the applications of MATLAB in calculations for nano-systems, which mostly contain CNTs and consist of five sections. Also, some specialized applications of MATLAB for studying the behavior of nano-oscillators are introduced. In this the authors study the behavior of (8, 2) single-walled CNTs and functionalized CNTs (FCNTs) with four functional groups in water using the molecular dynamic simulation method. Glutamine, as a long chain functional group, and carboxyl, as a short chain functional group, have been used as functional groups in FCNTs. Four functional groups in each FCNT were localized at two positions: (i) all four functional groups were in the sidewalls of the nanotube and (ii) two functional groups were at the ends and two functional groups were in the sidewalls of the nanotube. The intermolecular interaction energies between CNTs or FCNTs and water molecules, the plots of radial distribution function, and the diffusion coefficients of CNTs and FCNTs in water were computed for investigating the effects of type and position of functional groups on the behavior of FCNTs in water. The obtained results from three methods are consistent with each other. Results showed that the position of the functional groups in FCNTs has an important role in the interaction of hydrophilic groups of FCNTs with water molecules. Furthermore, the authors also investigated the behavior of FCNTs with 16 carboxyl functional groups in water. The presence of these large numbers of carboxyl functional groups on CNTs prevents water molecules from moving toward hydrophilic carboxyl functional groups. This well demonstrates the advantage of using a lower number of functional groups, each containing many hydrophilic groups like the glutamine functional group. In Section 7.4, the molecular dynamics simulations were presented to investigate the interfacial binding between the single-walled CNTs and conjugated polymers including.

In these days Nanotechnology and nonmaterial's are hot topics in Aerospace. Simply defined, nanotechnology is the study of manipulating matter on an atomic and molecular scale to develop materials, devices, and other structures at a scale of 1 to 100 nanometers.

In this chapter you will see "NanoInterface" evaluated the use of modeling and simulation with MATLAB, in conjunction with experimental data, to understand the interfacial behavior of the metal-oxide-polymer interfaces for nanoelectronic devices. This chapter introduces new technology for the design of complex systems and describes tools that may aid in the implementation of these various schemes. The approaches describe here include, first, the simplification and decomposition of analyses using *numerical optimization* and, next, the transformation of the design problem itself into parallel, collaborative tasks.

WHAT IS OPTIMIZATION?

Optimization and Predictability In virtually every branch of science, there is a need for calculations to estimate the parameters of a system, to find the extremum of an objective function such as minimum energy or maximum entropy states of a system, and to design and control systems. Since mathematical models are often used to guide experimentation and to predict or adapt to behaviors, there is an equally strong need to understand the errors in these models and to be able to make mathematically sound and precise statements regarding the accuracy and precision of the predictions. Since all of these needs pervade science, it is natural to ask in what way nanoscience is special. The degree of detail and dynamic complexity including wide-ranging, nonseparable scales in both space and time make the formulation and solution of problems in this area fundamentally more challenging than most and arguably all problems successfully tackled to date. The payoff to developing a new set of computational models and tools would be better understanding, prediction, design, optimization, and control of complex nanosystems with confidence in the reliability of the results. Successes in this field would almost certainly have an impact on other fields that share the complexity of a large range of complex behavior on nonseparable temporal and spatial scales.

Why Optimization Is Required

Many of the problems mentioned above rely on formulating and solving optimization problems. For example, in order to gain a quantitative understanding of nanosystems at a fundamental level, it is essential to have the capability to calculate the ground state of the system. This can be formulated as a minimum energy problem where the solution gives the configuration of the particles of the system at the lowest energy, given an energy functional. Complicated nanosystems of interest can have millions to billions of particles, resulting in huge optimization problems characterized by an equally huge number of local minima with energy levels close to the ground state. There is no hope of solving this problem with brute force methods. To make any headway on such problems requires careful formulation of the objective function and the constraints. For example, progress in the protein-folding problem has been achieved by understanding that certain sequences of amino acids in the unfolded molecule almost always end up in a standard form, e.g., an alpha helix. By inducing the optimization path towards this conformation, better results have been attained. Constraints, based on additional knowledge can also be formulated to further reduce the search space. Optimization methods that exploit physical insight could have a dramatic impact on efficiency. Monte Carlo methods are discussed elsewhere in this report, but other stochastic or deterministic optimization methods can also be accelerated by the domain knowledge. For example, when trying to simulate a novel material, knowledge about the structural properties and energetic of related materials could be used to preferentially steer the computation towards promising regions of the state space. Such methods are likely to be much more efficient than general optimization strategies, but quite limited in their applicability. Self-assembly is a central feature of nanoscience. Understanding, predicting, and controlling this process is crucial for the design and manufacture of viable nanosystems. Clearly, the subsystems involved in this process are assembling themselves according to some minimum energy principle. Once an understanding of the underlying physics is attained, optimization problems can be formulated to predict the final configurations. Since these systems are also huge and likely to have many local minima, a careful development of the models, the constraints, and the algorithms will also be required here. Given the necessity to create models that incorporate many scales, it has been suggested that a hierarchy of models be created to span this range, i.e., a connected set of models that capture the quantum effects through the micromorphic effects. In this case, the parameters at a given level should be attainable from the next lowest level using parameter-determining techniques. Optimization formulations for such parameter estimation problems are ubiquitous, but constructing the correct error models and the associated optimization problem is not obvious. Simply using least squares may not be appropriate; the more general maximum likelihood estimators will likely be required. As alluded to above, the design and control of nanosystems are vital to the eventual manufacture of products. The problem here is to optimize the design to achieve the desired characteristic or to control the process so that a given state is reached. In these cases, the constraints of the optimization problem contain the hierarchy of models describing these systems. In particular, the constraints contain coupled systems of partial differential equations (PDE). There is considerable interest in such problems in other fields, e.g., aerodynamics, and some efficient algorithms have been created, but the number of levels involved in nanoscience is far higher than in other fields. Progress has been made by understanding that these coupled systems of PDEs do not have to be satisfied at each step of the optimization process, but only need to be satisfied in the limit. This has resulted in the optimized solution in a modest multiple of the time required for one nonlinear solve. Such approaches, however, require that the formulation of the optimization problem incorporate some or all of the state variables in addition to the control variables. Effective optimization methods in these cases must have more control over the PDE software than is typically the case, e.g., gradients and/or adjoints must be computed along with other quantities. Thus, the formulation of the models, the algorithms, and the software with the ultimate goal of design and optimization is crucial. One can easily envision optimization problems arising in many other aspects of nanoscience. For example, methods for shape optimization for determining the optimal the geometry of both tools and products, nonsmooth optimization due to derivative discontinuities between the models at adjoining levels, and discrete optimization for particles restricted to discrete locations will needed. All of the developments in these areas, as with those mentioned above, should be done collaboratively between optimizers and nanoscientists.

Why Simulation Is Required

In Real-time measurements, distributions of particle size and particle structure are, thus, validate techniques for the evolution of nanotechnology. Precise and reliable models for simulating transport, coagulation, deposition, and dispersion of nanoparticles and their cluster are needed for the development of design tools for applications in technological, including nanoparticle instrumentation, fuselage, wings control surfaces and other parts focusing nanoparticle behavior in the chemically and environmentally reactive framework. Computational techniques allow us to validate and explore hypotheses about the experimental observation that may otherwise not be approachable through conventional experimental techniques. Additionally, simulations with computer allow for the theory to propose areas of interest in which experimental techniques can be applied. The most common computer simulation methods are Monte Carlo (MC) and molecular dynamics (MD), have been extensively applied in a variety of model framework.

HISTORY

Computer science offers more opportunities for nanotechnology. Soft computing techniques such as cellular automata, genetic algorithms and swarm intelligence, can impart required emergent properties like self-repair, growth, and complex networks to the framework. Many Books have successfully applied such techniques to real-world problems, including complex control systems in manufacturing and control in aircraft system. Chapter in nanoscale systems involves the understanding and development of nanotechnology computing techniques as well as the application of these techniques in the aerospace world, often to the problems of other Chapter areas. The techniques in nanotechnology systems comprise algorithms in machine learning, artificial intelligence (AI), knowledge representation reasoning, and natural computing. With some improvement in nanotechnology characteristics, these techniques can be applied to control a flock of a trillion nano assemblers. It is expected that normal computing methods such as these will overcome concerns about harmful implications of nanotechnology and prevent the notorious scenario of self-replicating nanorobots multiplying uncontrollably.

HOW THIS SERVES

Advancement in computational methods have increasingly enabled the simulation of heat transfer, acoustics complex flows, heat transfer, acoustics, and fluid-structure interaction phenomena. Furthermore, advances in high-performance computing (HPC) have also allowed more complex simulations to be performed at shorter turn-around times. The design and condition monitoring challenge increasingly require an understanding of complex phenomena and conditions. Reducing the uncertainty in engineering simulations by increasing the accuracy, while making computations more efficient still remains a scientific challenge. In this presentation, we will show results from our recent Chapter with respect to the development and application of high-order CFD and multi-scale/multi-physics methods for aerospace and nanotechnology applications.

Optimization analysis of nano-sized magnetic materials using MATLAB, Fabrication techniques developed over the past decades have allowed engineers and scientists to form magnetic materials in the nanometer regime. Such materials exhibit very interesting magnetic properties that are significantly different from bulk materials. At such scales, the behaviors of the magnetic materials are inadequately described by Maxwell's equations alone. The theory of micromagnetics deals with the behaviors of these nano-sized magnetic materials by including the quantum mechanical effects that are significant at this scale. Micromagnetics theory has been successful in predicting the formation of domain walls in magnetic materials and also the formation of interesting magnetic states such as the vortex and leaf states. It also finds its application in many aerospace engineering aspects such as the digital data storage technology, formation of aircraft parts and navigation systems. This chapter describes the system-level modeling of N/MEMS; it discusses some of the latest advances in designing and modeling complex nano- and micro-electromechanical (N/MEMS) systems at the system level. These advances have been applied to an open source tool called PSugar at Purdue University. For design, the author discusses a graphical user interface (GUI) that allows users to quickly configure complex systems in 3D using a computer mouse or pen at a rate faster than what might be drawn with pencil and paper. This GUI is coupled to a powerful netlist language for design flexibility. For modeling, the chapter applies for recent advances in analytical system dynamics and differential–algebraic equations (DAEs) into a framework that facilitates the systematic modeling of multidisciplinary systems that may comprise static or dynamic constraints.

In aircraft production, it is crucial to understand and evaluate performance, safety and behavior, other aspects of the systems before and after they are available for testing. Simulated models are used to gain knowledge in order to make decisions at all development stages.

Modeling and simulation (M&S) in aircraft system development, for example of fuel, hydraulic and electrical power systems, is today an important part of the design process. Through M&S a problem in a function or system is found early on in the process. An increasing part of the aircraft end system verification depends on results obtained from simulation models rather than expensive testing in flight tests. Hence, the need of integrated models of complex systems, and their validation is increasing day by day. It's not only one model is needed, but many interacting models with known and high accuracies and validity ranges are required. The development of modeling and simulation and computer performance, tools has enabled large-scale simulation.

This Chapter includes two parts related to these topics. The first part describes a modeling technique, hosted simulation, how to simulate a complete system with models from different tools, e.g. control software from one tool and the equipment

Figure 1. The M&S process and its interaction between actors and artifacts for an unreleased and a released
Source: Karlson, 2010

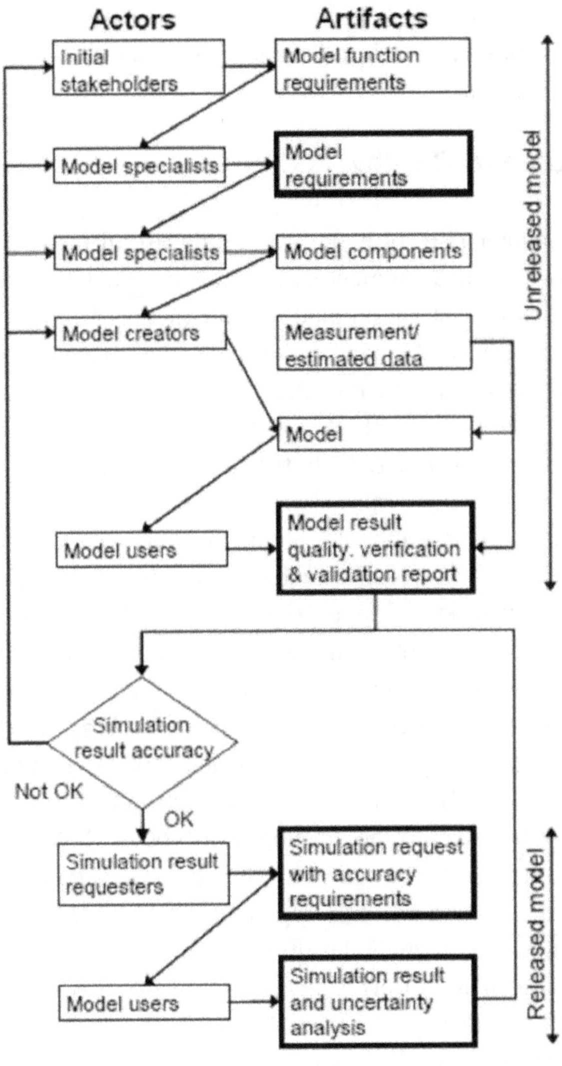

model from another. The second part describes the use of M&S in the development of an aircraft. The use of M&S in early system development phases has increased. This has led to the emerging need for uncertainty management and the complete validation methods that complement the traditional validation based on measurement data, knowledge of a model's maturity, and simulation result accuracy. A solution to this requirement is dominant to advance in system development where early design decisions have to be made supported by simulation.

Sensitivity analyses can be made to point out model parameters/inputs that have a strong influence on the simulation result or compare the assumed uncertainty data influence with similar model accuracy experience. Sensitivity analyses can, therefore, support the process of increasing knowledge of the models' validity.

The ability to handle uncertain information and lack of information is the key to success in early design.

Aim, Limitation and Context

The aim of the Chapter is to provide industrially useful methods for supporting model validation when measurement data is lacking. First, simulation techniques are needed that enable system simulation with models from different domains, Second, as a model constantly evolves through the development process, these validation methods must also fit into the entire development process. Third, the industrial applicability of the methods and metrics using uncertainty data must be investigated. One limitation that reduces the number of methods is that the selected model category, mathematically based aircraft system models, is computationally demanding. Others are that the model equations are not explicitly accessible and that the number of model parameters affected by uncertainty can be relatively large. The context in which the methods for supporting model validation must be developed to be useful is in area 'A' in Figure 1. Figure 1 shows a top-level view of domains that are always present when using the sought methods that support model validation and also shows the consequences if some domain is lacking. The areas should be considered in order to obtain a sound and balanced context for the problem-solving approach. To achieve an efficient development environment with useful methods, consideration must be taken to the development process, M&S methods, and the applied system. Area A is characterized by good balance and high efficiency. If for example, the process is not present during system development, area 'B', methods implementation in the company's development process and its repeatability will be underdeveloped, which will result in disorder. If M&S is suppressed, area 'C', assumptions must be made regarding the M&S properties/behavior. If the system is suppressed, area 'D', some data/tasks will be missing. Areas C and D both lead to shortcomings in method development and validation. As always, human behavior must also be considered when evaluating and selecting methods.

CHAPTER OUTLINE

This Chapter is divided into three main sections:

- Section 1 Introduction
- Section 2 Frame of reference. An introduction to M&S, model-based design, and aircraft systems.
- Section 3 Modeling and simulation of aircraft systems. An overview of how aircraft systems are designed in an M&S context. One simulation technique, hosted simulation, is described in detail.

Introduction

Earlier described

Frame of Reference

THIS SECTION PROVIDES an overview of model classification, simulation methods and their relation to engineering and aircraft system classification. The focus is on mathematical models for aircraft systems.

Models

Everything that is a representation of something can be seen as a model. In other words, a model is a simplification of something, e.g. anything from a screw to an airplane. Even nontechnical areas can be modeled as living organisms, abstract social behavior, and economic progress. A model represents a system. A general definition of a system is given by (Gains, 1979).

"The largest possible system of all is the universe. Whenever we decide to cut out a piece of the universe such that we can clearly say what is inside that piece (belongs to that piece), and what is outside (does not belong to that piece), we define a new system. A system is characterized by the fact that we can say what belongs to it and what does not, and by the fact that we can specify how it interacts with its environment. System definitions can furthermore be hierarchical. We can take the piece from before, cut out a yet smaller part of it, and we have a new system." Examples of models related to a pump might be a thought or idea for its performance and type, a piece of plastic that shows its geometric extent, a blueprint, a textual description, mathematical equations that predict its performance, or a prototype. The model simplifies and helps the user to better understand the real object. Models can fall into at least four categories (Ljung et al, 1991; Smith et al, 2007).

- Physical
- Schematic

- Verbal/Mental
- Mathematical

Physical models attempt to recall the system in terms of appearance. Typical physical models are scale models used in e.g. wind tunnels, test rigs, mock-ups or prototypes etc. Note that the term ‘physical models’ is somewhat confusingly also used in the field of physics, for example for mathematical models.

Verbal models are simpler than the other three model categories and provide textual or oral descriptions. Verbal models are often used early in projects when very little is known. The weakness of verbal models is their ambiguity, which may lead to difficulties in interpretation, which programming and modeling languages, for example, do not suffer from. An example of a mental model might be a driver’s

Figure 2. Models and their physical interactions (the bus network is excluded) in a flight simulator
Source: Andersson et al, 2009

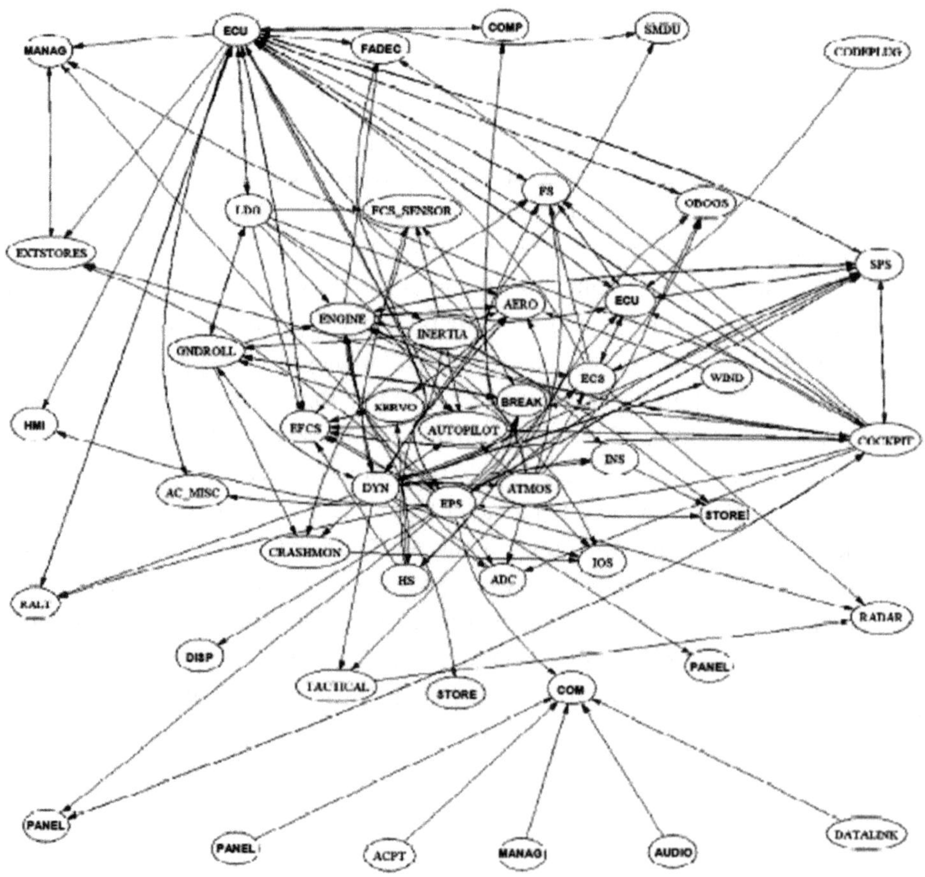

capacity to predict a car's behavior achieved through experience, education, and training. Mental models are often difficult to communicate verbally. Mathematical models describe the system using equations, e.g. in physics, chemistry, mechanics, economics, and biology. Typical outputs from mathematical models in the aerospace industry are pressure, temperature, and flow of fuel in a system. Mathematical models attempt to recall the system in terms of performance or behavior and are therefore also called behavior models. Which kind of model is meant, however, is usually read from the context. Another, more technical, way to divide the models is as specifying/descriptive and analytical models, (Andersson et al, 20009). The purpose of specifying/descriptive models is to provide a manufacturing schedule, for example, a drawing of a detail, a verbal specification, or a modeled software specification. The analytical models' purpose, however, is to contribute system analysis to be able to take necessary design decisions that are input to the specifying models, such as the necessary analytical models, such as CFD models, and thus design iterations are necessary. If much of the design occurs in the analytical models, such as the selection and evaluation of system concepts, the analytical model corresponding to the selected system can then be considered a specifying/descriptive model in terms of system layout. The specifying model may have different resolutions and still be useful from day one of its construction, while the analytical model must be complete and the weakest link determines its quality. The benefits of modeling are several, such as (Smith et al, 2007).

- Models require users to organize and sometimes quantify information and, in the process, often indicate areas where additional information is needed.
- Models provide a systematic approach to problem-solving.
- Models increase understanding of the problem.
- Models entail users to be very distinct about objectives.
- Models serve as a consistent tool for analysis.
- Models provide a standardized format for analyzing a problem.

Verbal modeling is thus the process of establishing and reproducing interrelationships between entities of a system in a model. It can be argued that modeling has an intrinsic value because of all the benefits models provide in an industrial development environment.

MODELING AND SIMULATION OF AIRCRAFT SYSTEMS

This section presents an overview of the M&S work for the aircraft systems. Described methods and tools are also applicable to passenger aircraft, automotive industry, and other complex products. The aircraft systems comprise fuel, ECS, hydraulic, and auxiliary power systems and also landing gear. Aircraft systems have many modeling challenges such as less compressible fluids and compressible air that give stiff differential equations, gravitational force effects, saturation and nonlinear cavitations. It is also a complex system of integrated systems that require models with integrated system software. Dynamic models based on physical differential equations have generally been used. Changes in tools have opened up for new possibilities for more advanced and more complete system simulations. The section also gives a detailed description of the simulation method called hosted simulation. The example used for hosted simulation demonstration purposes is a UAV (Unmanned Aerial Aircraft) fuel system.

Simulation of Mathematical Models

If a model is a simplified representation of a real-world system, a simulation is an execution of a model using input data to extract information from the model. A simulation is an experiment performed on a model, Figure 5. The simulation can be a real-life or hypothetical situation. By analogy with an experiment in the real world on a system that needs atmospheric conditions and delivers measurement

Figure 3. The aircraft system design process
Source: Auta et al, 2013

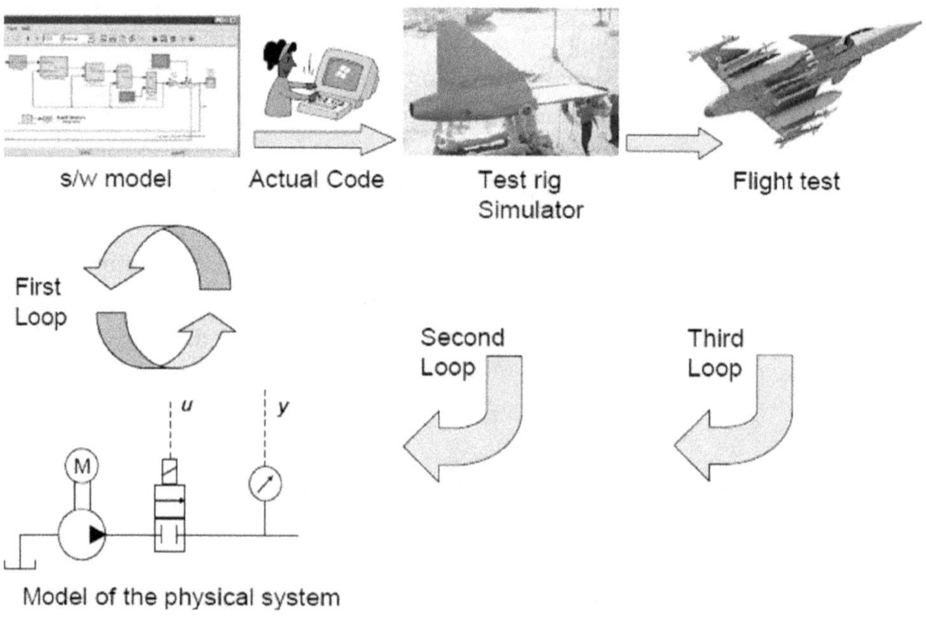

data, the model also needs inputs and delivers simulation data. The simulation enables the prediction of the behavior of the system from a set of parameters and initial conditions.

There are many reasons to simulate instead of setting up an experiment on a system in the real world (Chen et al, 2009). Some main reasons are:

- It is too expensive to perform experiments on real systems, e.g. extensive drop tank separation analysis.
- It is too dangerous. The pilot can practice a dangerous maneuver before performing it in the plane.
- The system may not yet exist, i.e. the model will act as a prototype that is evaluated and tested.
- Other positive features that simulation provide are:
- Variables not accessible in the real system can be observed in a simulation.
- It is easy to use and modify models and to change parameters and perform new simulations. With system design optimization many variants can be evaluated.
- The time scale of the system may be extended or shortened. For example, a pressure peak can be observed in detail or a flight of several hours can be simulated in minutes.
- The most important data flows in a model during simulation are:
- Parameters are constant during a simulation but can be changed in-between.
- Constants are not accessible for the simulation user.
- Variables are quantities that can vary with time.
- Inputs are variables that affect the model.
- For DAE/ODE/PDE models, states are an internal variable whose time derivative appears in the model and must, therefore, be initialized at simulation start.
- Outputs are variables that are observed.

The definitions of the terms vary depending on the domain. A promising standard, Functional Mockup Interface (FMI), has recently been developed, which can lead to a term alignment and simplified model integration between different technology domains and tools.

The mathematical models can be classified and hierarchized, (ARINC, 2006; Chen et al, 2005), in several different ways. The focus in the hierarchical classification below is on typical fluid system simulation models, which often are dynamic, non-linear, and have continuous time and continuous states. The model categories in the dashed rectangle can be seen as subset of the model categories outside it.

Figure 4. Simulation of nanostructured domains in a polymer blend where the interfaces are controlled at the nanoscale
Source: S. C. Glotzer, 1995

In contrast to a static model, the output of a dynamic model can also be a function of the history of the models' inputs and not only a function of the current inputs. For example, the pressure in the aircraft fuel tank is a function of previous flight conditions. The pressure will be different if the aircraft is climbing, diving, or in level flight just before the observed time. Dynamic models are typically represented with differential equations. Setting the time derivative of the states to zero in a dynamic model will result in a static model. Strictly speaking, there are only non-linear systems in the real world but many systems are often modeled by linear models to simplify the M&S. Mathematical models describing the fluid system are rarely linear.

Nonlinear models can be linearized in a working point for further analysis as if it were a linear model. The sampling of an analog signal corresponds to a conversion of "Continuous time" to "Discrete time".

Some other simulation output classifications are:

- Deterministic vs. probabilistic models. The simulation output of a deterministic model with a distinct set of input, model parameters, and model initialization will not differ from one simulation to another. Probabilistic models also include the probability distribution of the models' inputs and parameters. Extensive simulation of such models will result in the outputs of the model being given a probability distribution. Typical uncertain parameters are fluid properties and equipment performance degradation due to wear/aging. A good representation of the probability distribution of the inputs and parameters in the model reflects probable measurement data distribution from aircraft or rig.
- Quantitative vs. qualitative models. A mathematical model of an aircraft system is a quantitative model. A qualitative model could be a definition of quality criteria, ref [43], for a mathematical model for an aircraft system.
- Event-based (state machines).

A heterogeneous model is a model that consists of more than one model category, for example from a fuel system model with a continuous time equipment model connected to a discrete time control model.

Modeling and Simulation Tools

This chapter describes some aspects of modeling and simulation tools. Computational design is a growing field whose development is coupled with the rapid improvement of computers performance and their computational capability. There is no clear precision of the term computational design and it is interpreted differently in different engineering domains due to its broad implications. However, computational design methods are characterized by operating on computer models in different ways in order to extract information, (Gavel et al, 2003).

Historically has each technical domain has developed its own tools. This is a natural consequence of the requirements for a tool is just as high and specific to the product complexity in itself. This is in order to achieve an efficient development environment. The increased complexity of the designed systems has put higher requirements on the tool's capability to resemble the system and still have a high level of abstraction in combination with a user-friendly graphical interface (GUI). Models easily become unstructured and complex without an appropriate tool. There

Figure 5. A screenshot of a molecular dynamics simulation of water molecules moving in a carbon nanotube using the 3D Molecular Dynamics Simulator of MW
Source: Giese et al, 2010

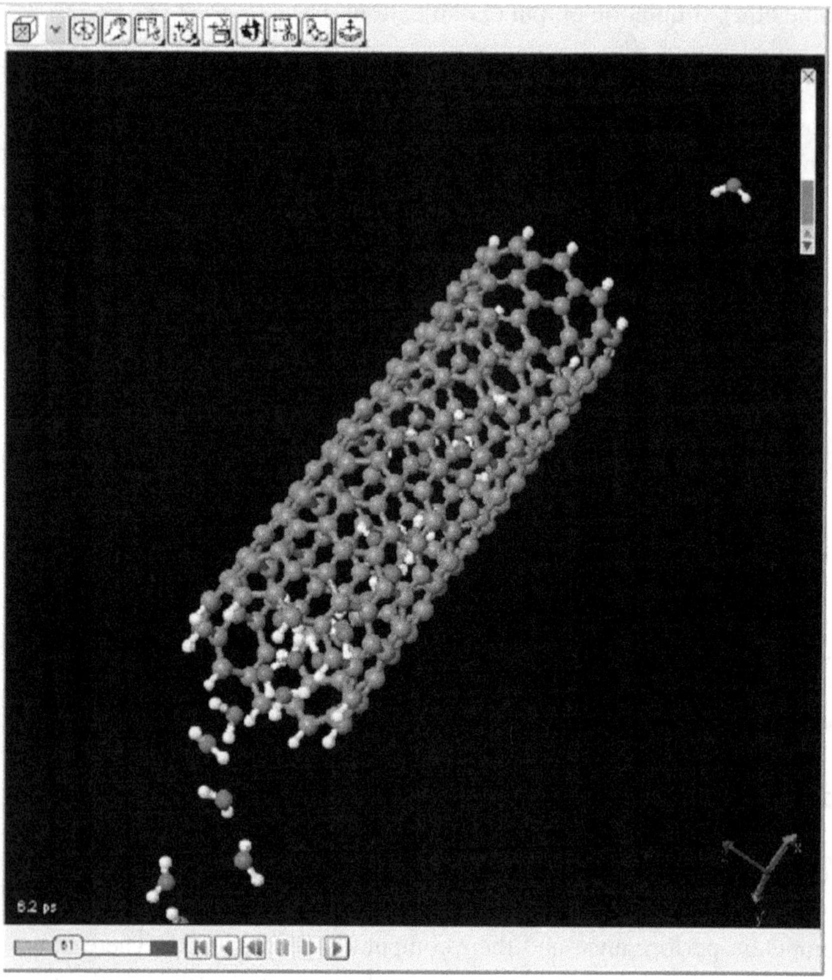

are several modeling-and-simulation tools on the market today that have a GUI that gives a good overview of the model. These are most often of the drag-and-drop type; this makes the modeling tool easy to use, thus minimizing the number of errors,, (Gavel et al, 2003).

Causality is the property of cause and effect in the system and it is an important condition for the choice of modeling technique and tool. For information flow or models of sensors or an Electronic Control Unit (ECU), the system's inputs and outputs decide the causality. For physical systems with mass and energy flow, the causality is a question of modeling techniques/tools. In non-causal models, the causality is not clearly stated, so the simulation tool has to sort the equations from model to the simulation code. When designing a causal model of a typical energy concerted system, the modeler has to choose what is considered to be a component's input and output. The bond graph modeling technique, (Paynter, 1961), is a method that aids transformation from non-causal to causal models. The bond graph is an energy-based graphical technique for building mathematical models of dynamic systems. Thus, there are basically two representations, the signal flow/port approach using block diagrams suited to casual parts of the system and the power

Figure 6. A screenshot of a quantum dynamics simulation of a scanning tunneling microscope that is an important tool in nanoscience

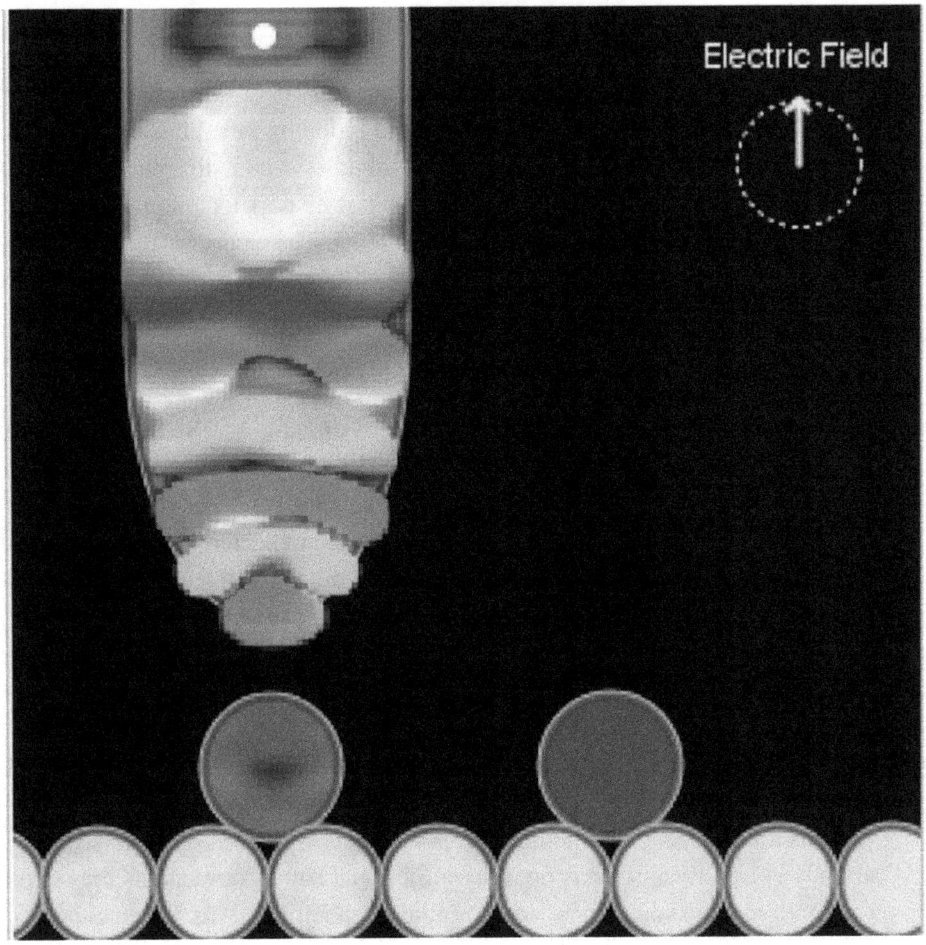

**For a more accurate representation see the electronic version.*

port approach, suitable for the non-causal parts. The chosen approach should be based on the dominating causality characteristics of the system but is sometimes an outcome of the tool available. The signal port approach clearly shows all variable couplings in the system. This is very useful for systems analysis and is therefore suitable for representing control systems and systems connected to them. However, a drawback is that the model may become complex and difficult to overview. In this case, power port modeling is more appropriate. In power port modeling, there have two directional nodes that contain the transfer of several variables. The power port is more compact and closely matches the real physical connection that by nature is bi-directional.

Two examples of tools/languages for aircraft system development are:

- MATLAB language is an object-oriented language for complex physical systems modeling. It is acceptable for multi-domain modeling, for example, in aerospace industries modeling of mechatronic systems. Such systems are composed of electrical, mechanical, and control systems as well as hydraulic subsystems. MATLAB uses equation-based modeling; the physical behavior of a model is described by differential, algebraic and discrete equations. MATLAB models are causal and use the power port technique. An M&S environment is needed to solve actual problems. The environment provides a customizable set of block libraries that let users design and simulate. There are both commercial and free M&S environments for MATLAB. Dymola, OpenMATLAB, and SimulationX, are well known commercial MATLAB tools.
- Simulink, is also an environment for multi-domain M&S. Simulink is a product from the tool vendor Mathworks. The biggest difference compared to MATLAB tools is that the models will be causal and use the signal flow technique. A causal block diagram is made up of connected operation blocks. The connections stand for signals, which are propagated from one block to the next. Blocks can be purely algebraic or may involve some notion of time such as delay or integration. A product called Simscape, has existed for some years that extends Simulink with tools for power port modeling of systems such as mechanical, electrical and, hydraulic.
- Most simulation packages come with predefined libraries with a set of numbers of equations representing physical components. One of the major reasons for choosing a particular tool is whether a suitable component library already exists. Even so, it is not uncommon that some components may need to be tailored and added in order to simulate a specific system. When choosing a library, it is important to know what levels of accuracy and bandwidth the components in the library are valid.

Modeling and Simulation History

This brief description of modeling and simulation history begins in the late 1970s as the availability of computers and computer-based calculation capacity began to increase for the engineers. In the late 1970s, Schlesinger (Sargent et al, 2007) defined the M&S activities, such as verification and validation and their relationship, and this has been improved upon by Sargent (Chatterjee et al, 1988) with the real world and simulation-world relationship with its analogies. Before the 1980s, the modeling of heavy aircraft system models was susceptible due to difficulties in visualizing and modifying the model. In the 1980s, the era as we know it today began, with tools that have user-friendly graphical user interfaces with features like "drag and drop" of block components and the power port (based on bond graph technique) concept or at least appears to be power port to the user. Boeing's EASY5 is one such example of an early M&S tool.

In the 1990s, the co-simulation between tools suits a common feature of commercial tools and heterogeneous simulation increased and the first version of the multi-domain modeling language MATLAB was released in 1984. The language has come to be used in both industry/academia and in different physical domains. In the 2000s, it has become common to generate code from models directly to the product or for hosted simulation, paper (Akaike et al, 1974). A new, mature and M&S-friendly design organization changed the complete fundamentals of the relationship of M&S and the design process. This change resulted in a new way of working, Model-Based Systems Engineering (MBSE), with a model-centric development approach.

Model-Based Systems Engineering

The International Council on Systems Engineering (INCOSE) defines Systems Engineering (SE) as "Systems Engineering is an interdisciplinary approach and means to enable the realization of successful systems. SE with a model based approach is called MBSE. MBSE uses "modeling" of the system in, for example, the design of complex systems and of its requirements to facilitate a more efficient design process. There are several definitions of how models are used in system development depending on when in the design process it is used and in which technical domain, for example:

- MBSE (Model Based Systems Engineering)
- MBD (Model Based Design)
- MBD (Model Based Development)
- MBD (Model Based Definition)
- MBE (Model Based Engineering)

- MDA (Model Driven Architecture)
- MDD (Model Driven Design)
- MDE (Model Driven Engineering)

As M&S grows, the above definitions are exchanged between domains and are used in wider/slightly different senses. Some of the definitions are principles and some are the means to perform the principles, e.g. the principle of MBSE is used to perform MBD, (ARINC, 2006) Some of the major benefits, Figure, of a model-centric approach are:

Documentation. The model is the only information carrier. Documentation and code are by-products and are generated based on the model. A model is also easier to share and communicate with people compared to a text, hence a visualization (model) is worth a thousand words and is unambiguous.

- Design by simulation. Executable models increase understanding of the behavior of the model and lead to better design quality.

Figure 7. System Build model example of system software
Source: ARINC, 2006

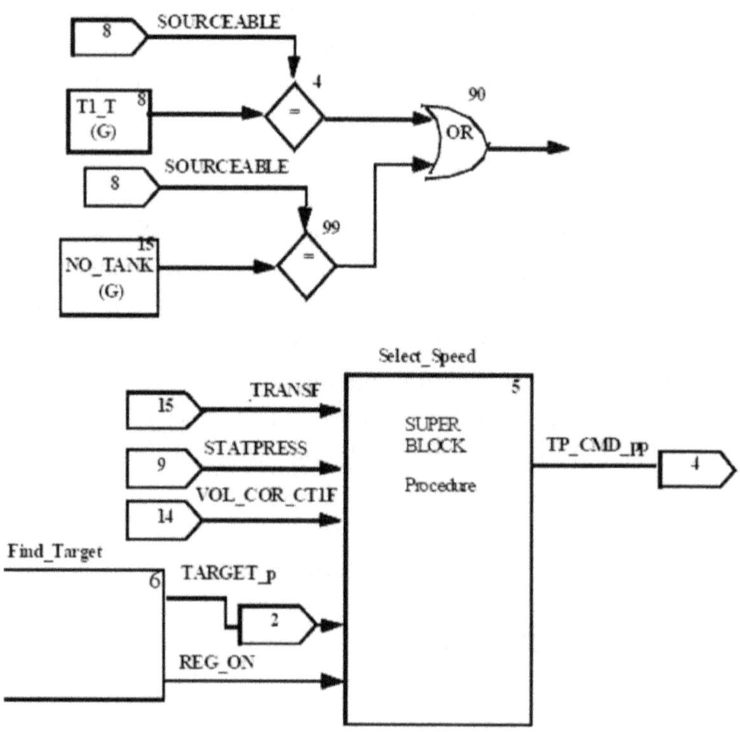

- Test and verification. Continuous tests and verifications can be made on a model.
- Auto-generated code decreases the number of actors involved and maintains consistency
- between specifying model and implementable code and its documents.

Some other benefits are:

- Early identification of requirements.
- Reuse of models.

MBSE is implemented by placing the model in the center, which leads to the following fundamentals:

- The interaction between developers and the various tasks is accomplished via the model.
- The model will have more than one purpose and be needed by more than one person.

The Model-Based Design Process

Engineering design is a technique to solve problems where a set of often doubtful objectives have to be balanced without violating a set of constraints. By employing modern modeling, simulation, and optimization techniques, vast improvements can be achieved in all parts of the design process. A great deal of Chapter has been done in the field of engineering design and has led to different design processes and methods. They all describe a phase-type process of different granularity with phases such as Concept Design, Preliminary Design, Specification, Prototype Development, Redesign, Detail Design, and Production. Ullman (1992), speaks of the design paradox, where very little is known about the design problem at the beginning but we have full design freedom. As time in the design process proceeds, knowledge about the problem is gained but the design freedom is lost due to the design decisions made during the process. To further stress the importance of the early phases of the design process, it is here that most of the cost is committed.

Computer models are a vital tool when evaluating the performance of the concept proposals. Most often, the modeling activities begin with simple stationary models of the systems, perhaps in MS Excel or equivalent, (Gavel, 2003). As the system goes into embodiment design, the software is modeled in a separate model with signal port technique, which is then co-simulated or host-simulated with the physical system model. However, most designs are actually modifications of already existing

Figure 8. Atomistic simulation of a nanostructure
Source: Andersson et al, 2009

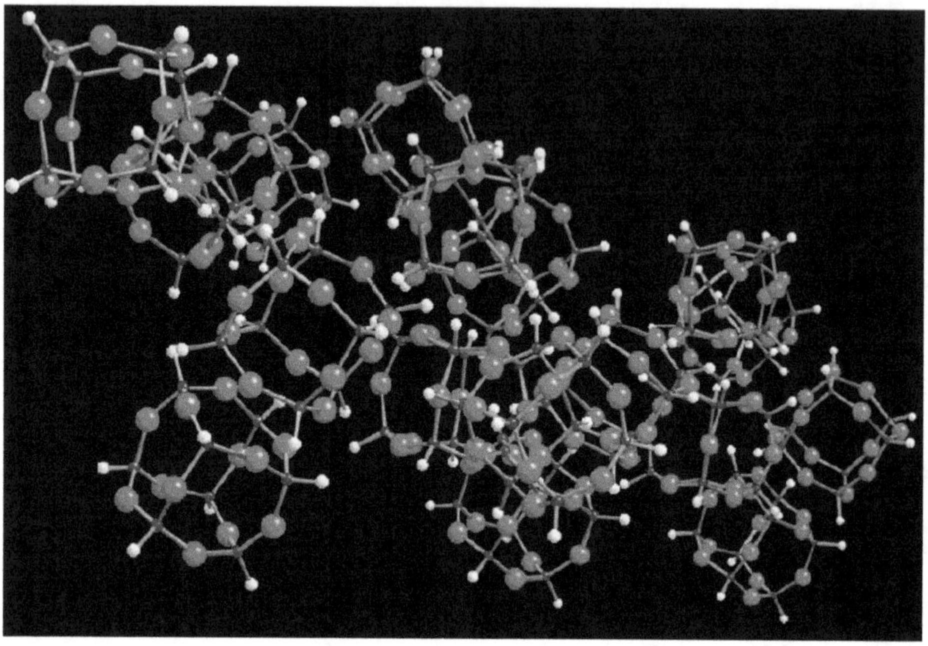

Figure 9. A typical complete system model, such as the fuel system model, may be divided into three major model
Source: Scholtz et al, 2002

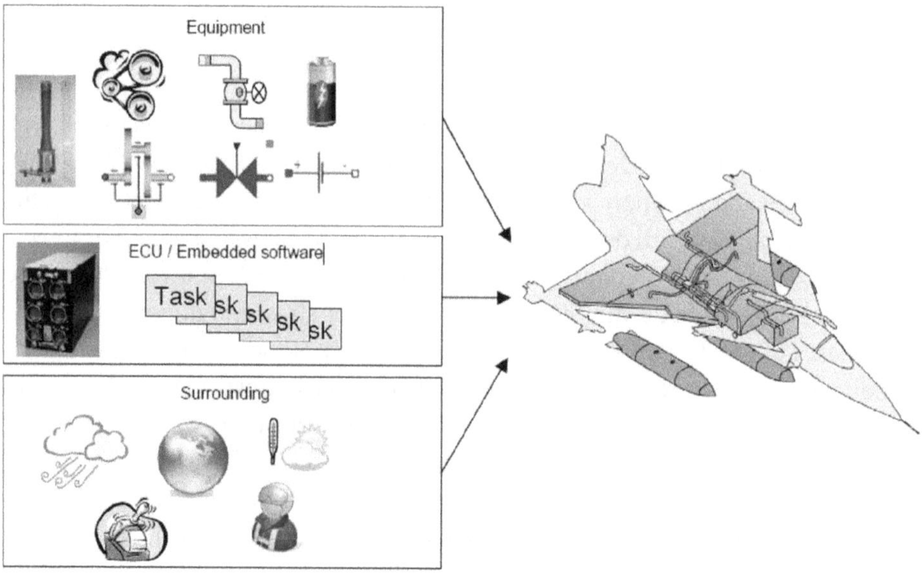

systems, if so, all the required models are already present in the early stages of the design process and relatively easy to modify for reuse. This also applies to other test facilities, such as test rigs, simulators, and test a/c, with the exception that these are not always so easy to modify.

Aircraft System Simulation and Simulators

Some models can be executed together in a so-called simulator. If the simulated system is an aircraft, then it is called a flight simulator. Simulation is an important activity for system level analysis/verification with the main objective being to reduce risk and cost. At component or system level, desktop simulations in the modeling tool or in specific mid-scale-simulation software/environments, add understanding and confidence to the system's design. For a whole aircraft system, this activity is characterized as large-scale-simulation with specific prerequisites. Some definitions related to simulation are provided below, (ARINC, 2006).

Mid-scale simulation: The activity performed when some simulation models of aircraft subsystems, developed with different modeling techniques, are integrated into a larger model, complex enough not to be simulatable in a desktop M&S tool.

Figure 10. Design concept for the TTR, showing internal components of the control system, motors and gearbox
Source: Gavel, 2003

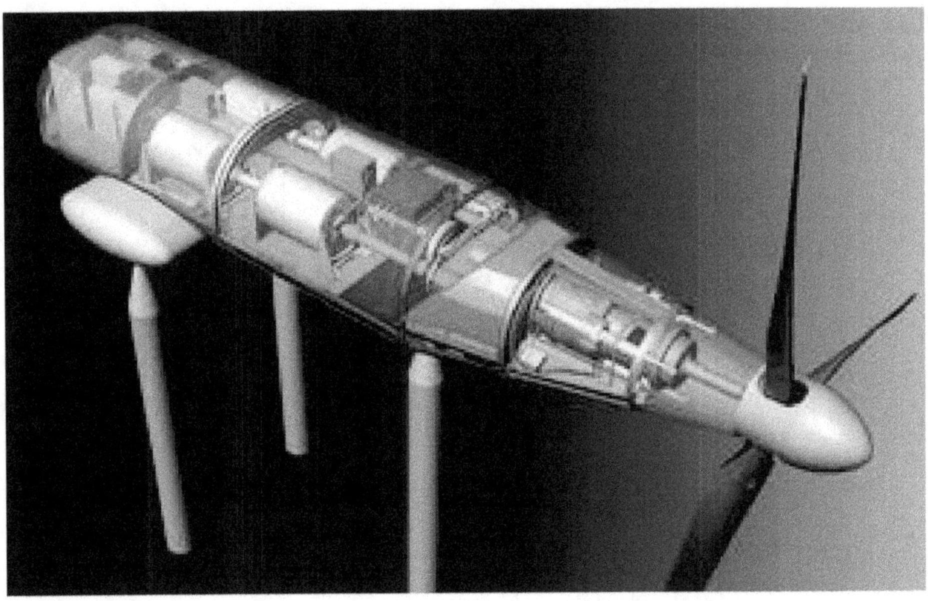

Large-scale simulation: The activity performed when several simulation models of the aircraft subsystems are integrated and specific arrangements for performance or interoperability exist. Examples of such arrangements are real-time execution, "pilot-in-the-loop simulation" or "hardware-in-the-loop simulation" (HILS) configurations.

Verification and Validation

Verification and validation (V&V) are two terms that are often mixed up. The general definitions of V&V are:

1. **Verification:** Did I build the thing right?
2. **Validation:** Did I build the right thing?

Verification tasks are often independent of context and can often be objectively answered with a yes or a no, e.g. formal verification. With M&S tools, such Dymola and Simulink, the number of computer programming and implementation errors has been reduced. Model verification then primarily ensures that an error-free simulation language has been used, that the simulation language has been properly implemented on the computer, and the model programmed correctly in the simulation language. Model validation, on the other hand, cannot be performed in early development phases like the concept phase due to the requirement of system experiment data. However,

Figure 11. Flow diagram of the simulation method
Source: Backlund, 2000

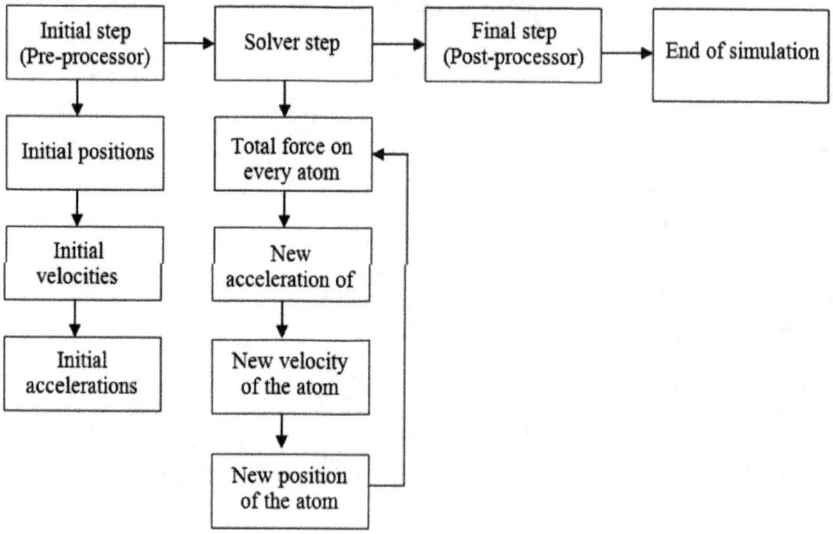

sensitivity analyses can still be made that point out model component parameters that have a strong influence on the simulation result or compare the assumed uncertainty data influence with similar model accuracy experience. Later in the development process, when more knowledge and measurement data is available, model validation with measurement data can begin. Model validation is context dependent, e.g. for a specific simulation task a model can be validated in parts of the flight envelope for some model outputs. For another simulation task with its context, a complete other model validation status can exist. In (Karlsson et al, 2010) a broader aspect and on a higher level has been taken concerning M&S result credibility. Eight factors have been defined as a five-level assessment of credibility for each factor, Verification Validation, Input Pedigree Results, Uncertainty, Results, Robustness, Use History M&S, Management, People, Qualifications. The approach clearly demonstrates the large number of factors that affect a model's credibility, which inevitably means that in the case of large models it is an extremely time-consuming task to make them credible.

The Gripen Aircraft System Models

Aircraft systems facing several modeling challenges such as mixing fuel and air in fuel systems, modeling gravitational force effects, and environmental dependency and flight envelope of the ECU. The aircraft systems' models are complex aircraft system models that include both the equipment and the software controlling and monitoring each system. Sub-models for aircraft system simulation can be organized into the following major categories.

Figure 12. Sensitivity analysis process
Source: Andersson, 2009

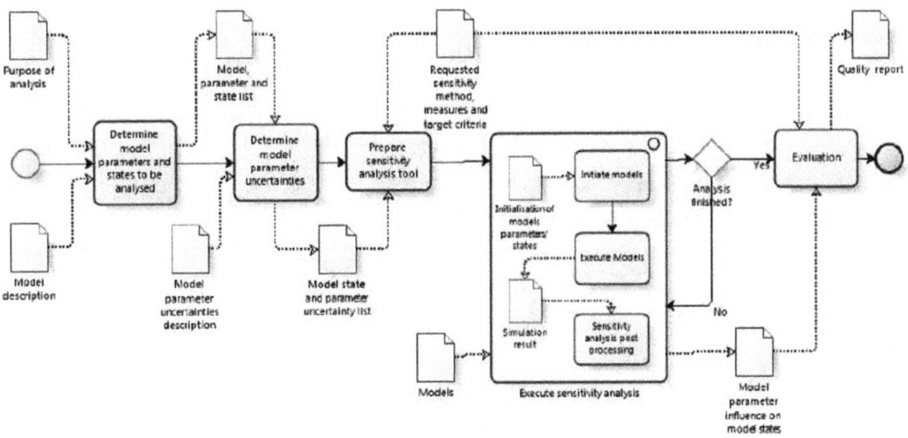

- The equipment model, with pumps, valves, and pipes for performance evaluation and dimensioning. The models are usually based on physical differential equations. Black-box models can be used for some equipment of minor interest such as sensors. Tables can be used for highly nonlinear equipment such as compressors and turbines
- The embedded software model, for control and monitoring of the equipment. The system software model can be hosted as a sub-routine and be generated automatically from the software model for facilitating the integrating of new code in the case of the redesign. The effects of the control units' hardware are rarely modeled.
- The environmental models with input data to the two categories above. The pilot, flight case and the atmosphere will be found here.

APPLICATIONS

MATLAB applications in behavior analysis of systems consisting of SWCNTs through molecular dynamics simulation. This chapter presents the applications of MATLAB in calculations for nano-systems, which mostly contain CNTs. The molecular dynamic simulation and analytical studies of CNTs. In this chapter the authors study the behavior of single-walled CNTs and functionalized CNTs (FCNTs)

Figure 13. Complete aircraft's vehicle systems
Source: Hayden, 2010

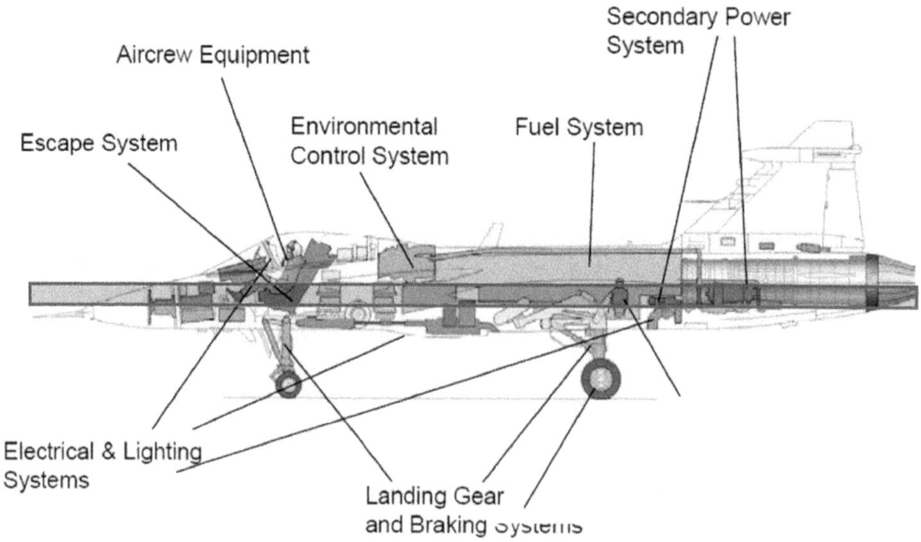

Figure 14. Example of computational modeling of CNT
Source: Scholtz et al, 2002

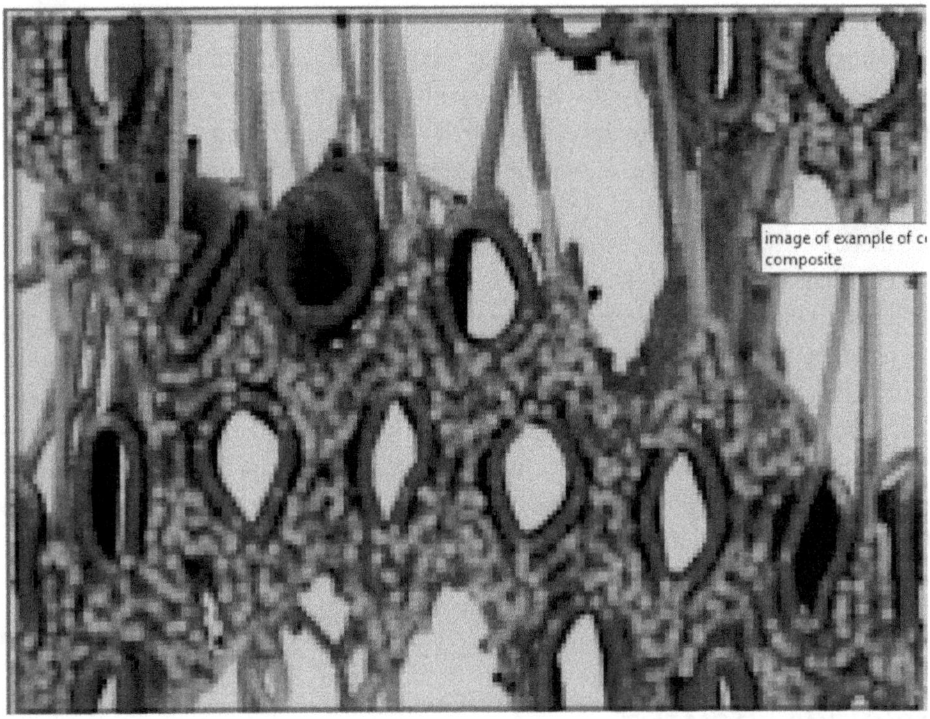

with four functional groups in water using the molecular dynamic simulation method. Glutamine, as a long chain functional group, and carboxyl, as a short chain functional group, have been used as functional groups in FCNTs. Four functional groups in each FCNT were localized at two positions: (i) all four functional groups were in the sidewalls of the nanotube and (ii) two functional groups were at the ends and two functional groups were in the sidewalls of the nanotube.

CONCLUSION

This chapter shows the modern design requirements with the help of simulation to assure the properties of aircraft like faster, miniature, highly maneuverable, self-healing, light weight warrant for materials, intelligence guided, smart, eco-friendly, with extraordinary mechanical and multifunctional properties in aviation sector.

MATLAB promotes reuse of models, which means that methods and measures for model development and validation, for example, should be used in several life cycle phases.

Using MATLAB results in inevitable model handovers, due to the long life of models. People from many disciplines, with and without M&S experience, use and make design decisions based on M&S outcomes. This means that methods and model quality measures should be comprehensible and transparent, in order be accepted in an organization. Measures should also be directly used as feedback in the model's improvement process.

The data used for validation, e.g. uncertainty data and its context, should be included in the model or strongly connected to it. This speeds up the process of the user gaining an insight into the model validity. By making the uncertainties visible to the user, the necessary and continuous model improvement process is supported.

An understanding of the adequate modeling level and of a model's future requirements improves the quality of what M&S contributes to a development project. There is a balance between too simple a model and too detailed a model. Too simple a model will not have the capacity to reflect the system or meet future, tougher model requirements while too detailed a model will, for example, suffer from model handling errors and cost more than necessary.

ACKNOWLEDGMENT

I would like to give thanks to my Parents Mr. Ramesh Chandra Lal Das and Mrs. Mira Devi and My brothers Mr. Dilip Das and Mr. Pradeep Kumar with my nephew D. Amay.

REFERENCES

Akaike, H. (1974). A new look at the statistical model identification. *IEEE Transactions on Automatic Control*, *19*(6), 716–723. doi:10.1109/TAC.1974.1100705

Andersson, H. (2009). *Aircraft Systems Modeling: Model Based Systems Engineering in Avionics Design and Aircraft Simulation*. Linköping Studies in Science and Technology, Thesis No. 1394.

Angulalakshmi, V. S., Tamilarasan, G., & Karthikeyan, S. (2014). Optimization of CVD synthesis conditions for the synthesis of multiwalled carbon nanotubes using response surface methodology. *J. Environ. Nanotechnol.*, *3*(2), 81–91. doi:10.13074/jent.2014.03.142066

ARINC 653. (2006). *Avionics Application Software Standard Interface*. ARINC 653, Specification, Part 1-3, Aeronautical Radio Incorporated, Annapolis, Maryland, USA.

Asghar, Raman, & Daud. (2015). Advanced oxidation processes for in-situ production of hydrogen peroxide/hydroxyl radical for textile wastewater treatment: A review. *J. of Clean. Produc.*, *87*, 826–838.

Auta, M., & Hameed, B. H. (2013). Coalesced chitosan activated carbon composite for batch and fixed-bed adsorption of cationic and anionic dyes. *Colloids and Surfaces. B, Biointerfaces*, *105*, 199–206. doi:10.1016/j.colsurfb.2012.12.021 PMID:23376092

Backlund, G. (2000). *The Effects of Modelling Requirements in Early Phases of Buyer-Supplier Relation*. Linköping Studies in Science and Technology, Thesis No. 812.

Balamurugan, J., Thanagamuthu, R., & Pandurangan, A. (2013). Effective synthesis of carbon nanotubes of high purity over Cr–Ni–SBA-15 and its application in high performance dye-sensitized solar cells. *RSC. Adv.*, *3*, 4321–4331.

Becker, M. C., Salvatore, P., & Zirpoli, F. (2005). The impact of virtual simulation tools on problem solving and new product development organization. *Research Policy*, *34*(9), 1305–1321. doi:10.1016/j.respol.2005.03.016

Biltgen, P. T. (2008). *Uncertainty quantification for capability-based systems-of-systems design*. 26th International Congress of the Aeronautical Sciences, Anchorage, AK.

Carloni, L. P., Passerone, R., Pinto, A., & Sangiovanni-Vincentelli, A. L. (2006). Languages and Tools for Hybrid Systems Design. *Foundations and Trends in Electronic Design Automation*, *1*(1-2), 1–193. doi:10.1561/1000000001

Cassandras, C. G., & Lafortune, S. (1999). *Introduction to Discrete Event Systems*. New York: Springer. doi:10.1007/978-1-4757-4070-7

Chatterjee, S., & Hadi, A. S. (1988). *Sensitivity analysis in linear regression*. New York: Wiley. doi:10.1002/9780470316764

Chen, W., & Jin, R. (2005). Analytical Variance-Based Global Sensitivity Analysis in Simulation-Based Design Under Uncertainty. *Journal of Mechanical Design*, *127*. . doi:10.1115/1.1904642

Cullen, A. C., & Frey, H. C. (1999). *Probabilistic Techniques in Exposure Assessment*. New York: Plenum Press.

Fabrycky, W. J., & Blanchard, B. S. (1991). *Life-Cycle Cost, and Economic Analysis*. Englewood Cliffs, NJ: Prentice Hall.

Fritzson, P. (2004). *Principles of Object-Oriented Modeling and Simulation with MATLAB 2.1*. New York: Wiley-IEEE Press.

Gains, B. (1979). Qua Vadis. *General Systems Yearbook.*, *24*, 1–9.

Gavel, H. (2007). *On aircraft fuel systems: conceptual design and modeling*. Linköping Studies in Science and Technology. Thesis No. 1067.

Gavel, H., & Andersson, J. (2003). *Using Optimization as a Tool in Fuel System Conceptual Design*. SAE World Aviation Congress and Display 2003, Paper No 2003-01-3052.

Gavel, H., Krus, P., Andersson, J., & Johansson, B. (2005). *Probabilistic design in the conceptual phase of an aircraft fuel system*. 7th AIAA Non-Deterministic Design Forum, paper No AIAA- 2005-2219, Austin, TX.

Giese, T., Oehler, B., & Sielemann, M. (2010). *A systematic approach to optimize conventional environmental control architectures*. Hamburg, Germany: Deutscher Luft- und Raumfahrtkongress.

Graf, S. (2008). OMEGA: Correct development of real time and embedded systems. *Software & Systems Modeling*, *7*(2), 127–130. doi:10.100710270-007-0077-5

Karlsson, L. (2010). *Simulator testing Basic course*. Linköping: SAAB AB.

Krus, P. (2000). *Post optimal system analysis using aggregated design impact matrix*. 2000 ASME International Design Engineering Technical Conferences & Computers and Information in Engineering Conference, Baltimore, MD.

Krus, P. (2008). *Computational tools for aircraft system analysis and optimization*. 26th International Congress of the Aeronautical Sciences, Anchorage, AK.

Kuhn, R. A. (2008). Multi level approach for aircraft electrical systems design. *6th International MATLAB Conference*, Bielefeld, Germany.

Larsson, E. (2007). *Modelling of a fuel system in MATLAB – applied to an unmanned aircraft* (Master thesis). Dept of Electrical Engineering, Linköping University.

Larsson, J., & Krus, P. (2003). Stability analysis of coupled simulation. *ASME International Mechanical Engineering Congress*, 861-868.

Lind, I., & Andersson, H. (2011). Model Based Systems Engineering for Aircraft Systems – How does MATLAB Based Tools Fit? *8th International MATLAB Conference*, Dresden, Germany.

Liscouet-Hanke, S. (2008). *A model-based methodology for integrated preliminary sizing and analysis of aircraft power system architectures*. Institut National des Sciences Appliquées de Toulouse.

Ljung, L., & Glad, T. (1991). *Modellbygge och simulering*. Studentlitteratur.

Malak, R. J., Jr. (2005). *A Framework for Validating Reusable Behavioral Models in Engineering Design* (Master's Thesis). G.W. Woodruff School of Mechanical Engineering, Georgia Institute of Technology, Atlanta, GA.

Malak, R. J., & Predis, C. J. J. (2007). Validating behavioral models for reuse. *Research in Engineering Design*, *2007*(18), 111–128. doi:10.100700163-007-0031-0

Mavris, D. N., & DeLaurentis, D. A. (2000). A probabilistic approach for examining aircraft concept feasibility and viability. *Aircraft Design*, *3*(2), 79–101. doi:10.1016/S1369-8869(00)00008-2

Muessig, P. R., Laack, D. R., & Wrobleski, J. J. (2000). An integrated approach to evaluating simulation credibility. *Proceedings of the 2000 Summer Computer Simulation Conference*.

NASA-STD-7009. Standard for models and simulations, National Aeronautics and Space Administration Washington, DC 20546-0001, November 7th, 2008, available online at http://standards.nasa.gov

National Science Foundation. (1996). Research Opportunities in Engineering Design. NSF Strategic Planning Workshop Final Report (NSF Grant DMI-9521590).

Neelamkavil, F. (1987). *Computer Simulation, and Modeling*. John Wiley & Sons Inc.

Oberkampf & DeLand. (2002). Error and uncertainty in modeling and simulation. Reliability Engineering & System Safety, 75(3), 333-357.

Paynter, H. M. (1961). *Analysis and Design of Engineering Systems*. MIT Press.

Potts, C. (1993). Software Engineering Research Revisited. *IEEE Software*, *10*(5), 19–28. doi:10.1109/52.232392

RESCENDO (Collaborative and Robust Engineering using Simulation Capability Enabling Next Design Optimizations), a research project funded by the European Union Seventh Framework Programme (FP7/2007-2013). Deliverable "D5.4.1 BDA Quality Laboratory State-of-the-Art". Release 1.0 2009. Courtesy of EADS.

Sargent, R. G. (2001). Some approaches and paradigms for verifying and validating simulation models. *Winter Simulation Conference, WSC'01*, 2, 106-114. 10.1109/WSC.2001.977251

Sargent, R. G. (2007). Verification and validation of simulation models. *2007 Winter Simulation Conference*, 124-37. 10.1109/WSC.2007.4419595

Schlesinger, S. (1979). Terminology for model credibility. *Simulation, 32*(3), 103-104.

Scholtz, D. (2002). Aircraft Systems – Reliability, mass Power and Costs. *European Workshop on Aircraft Design Education.*

Sinha, R., Liang, V., Paredis, C., & Khosla, P. (2001). A survey of modeling and simulation methods for design of engineering systems. *Journal of Computing and Information Science in Engineering*, *1*(1), 84–91. doi:10.1115/1.1344877

Smith, P., Prabhu, S., & Friedman, J. (2007). Best Practices for Establishing a Model-Based Design Culture, 2007-01-0777. *SAE World Congress & Exhibition*, Detroit, MI. 10.4271/2007-01-0777

Sobol, I. M. (1993). Sensitivity Analysis for Nonlinear Mathematical Models. *Mathematical Modeling & Computational Experiment*, *1*, 407–414.

Stevenson, W. J. (2002). Operations Management (7th ed.). Boston: McGraw-Hill/Irwin.

Ullman, D. (1992). *The Mechanical Design Process*. New York: McGraw-Hill Inc.

Wu, B., Zhu, D., Zhang, S., Lin, W., Wu, G., & Pan, B. (2015). The photochemistry of carbon nanotubes and its impact on the photo-degradation of dye pollutants in aqueous solutions. *Journal of Colloid and Interface Science*, *439*, 98–104. doi:10.1016/j.jcis.2014.10.020 PMID:25463180

Compilation of References

Abbas, S. M., Dixit, A. K., Chatterjee, R., & Goel, T. C. (2005). Complex permittivity and microwave absorption properties of $BaTiO_3$- polyaniline composite. *Materials Science and Engineering B*, *123*(2), 167–171. doi:10.1016/j.mseb.2005.07.018

Abdalla, M., Dean, D., Robinson, P., & Nyairo, E. (2008). Cure behavior of epoxy/MWCNT nanocomposites: The effect of nanotube surface modification. *Polymer*, *49*(16), 3310–3317. doi:10.1016/j.polymer.2008.05.016

Abu-Nada, E., & Oztop, H. F. (2009). Effects of Inclination Angle on Natural Convection in Enclosures Filled With Cu–Water Nano-Fluid. *International Journal of Heat and Fluid Flow*, *30*(4), 669–678. doi:10.1016/j.ijheatfluidflow.2009.02.001

Acocella, M. R., Corcione, C. E., Giuri, A., Maggio, M., Maffezzoli, A., & Guerra, G. (2016). Graphene oxide as a catalyst for ring opening reactions in amine crosslinking of epoxy resins. *RSC Advances*, *6*(28), 23858–23866. doi:10.1039/C6RA00485G

Afshari, M., Kotek, R., Tonelli, A. E., & Jung, D.-W. (2007). Producing polyamide nanofibers by electrospinning. In P. J. Brown & K. Stevens (Eds.), *Nanofibers and Nanotechnology in Textiles* (pp. 71–89). Woodhead Publishing. doi:10.1533/9781845693732.1.71

Agareva, N. A., Aleksandrov, A. P., Smirnova, L. A., & Bityurin, N. M. (2009). Synthesis of block polymethylmethacrylate containing precursor for photoinduced formation of gold nanoparticles, *Perspektivnye Materialy*, *1*, 5-12.

Agarwal, S., Wendorff, J. H., & Greiner, A. (2008). Use of electrospinning technique for biomedical applications. *Polymer*, *49*(26), 5603–5621. doi:10.1016/j.polymer.2008.09.014

Ahmed, J. U., Gotoha, Y., Ohkoshia, Y., Nagura, M., Endo, R., & Hara, T. (2005). Hydration in a new semiaromatic polyamide observed by humidity-controlled dynamic viscoelastometry and X-ray diffraction. *Journal of Polymer Science*, *43*(13), 1640–1648. doi:10.1002/polb.20446

Ahmed, S., Ikram, S., Kanchi, S., & Bisetty, K. (2018). *Advancements and Potential Prospects of Polymer/Metal Oxide Nanocomposites: From Laboratory Synthesis to Commercialization. In Biocomposites* (pp. 47–84). Pan Stanford.

Ajayan, P. M. (1999). Nanotubes from Carbon. *Chemical Reviews*, *99*(7), 1787–1800. doi:10.1021/cr970102g PMID:11849010

Akaike, H. (1974). A new look at the statistical model identification. *IEEE Transactions on Automatic Control*, *19*(6), 716–723. doi:10.1109/TAC.1974.1100705

Alegaonkar, A. P., & Alegaonkar, P. S. (2018). Nano-Carbon/Polymer Composites for Electromagnetic Shielding, Structural Mechanical and Field Emission Applications. *Thermoset Composites: Preparation. Properties and Applications*, *38*, 128.

Al-Ghamdi, A. A., Al-Hartomy, O. A., Al-Salamy, F., Al-Ghamdi, A. A., El-Mossalamy, E. H., Abdel Daiem, A. M., & El-Tantawy, F. (2012). Novel electromagnetic interference shielding effectiveness in the microwave band of magnetic nitrile butadiene rubber/magnetite nanocomposites. *Journal of Applied Polymer Science*, *125*(4), 2604–2613. doi:10.1002/app.36371

Ali, A. A., & El-Hamid, M. A. (2006). Electro-spinning optimization for precursor carbon nanofibers. *Composites. Part A, Applied Science and Manufacturing*, *37*(10), 1681–1687. doi:10.1016/j.compositesa.2005.10.008

Alizadeh, T., Nayeri, S., & Mirzaee, S. (2018). A high performance potentiometric sensor for lactic acid determination based on molecularly imprinted polymer/MWCNTs/PVC nanocomposite film covered carbon rod electrode. *Talanta*. PMID:30348364

Allaoui, A., Bai, S., Cheng, H.-M., & Bai, J. (2002). Mechanical and electrical properties of a MWNT/epoxy composite. *Composites Science and Technology*, *62*(15), 1993–1998. doi:10.1016/S0266-3538(02)00129-X

Allen, M. J., Tung, V. C., & Kaner, R. B. (2010). Honeycomb Carbon: A Review of Graphene. *Chemical Reviews*, *110*(1), 132–145. doi:10.1021/cr900070d PMID:19610631

Al-Saleh, M. H., Gelves, G., & Sundararaj, U. (2011). Copper nanowire/polystyrene nanocomposites: Lower percolation threshold and higher EMI shielding. *Compos Part A*, *42*(1), 92–97. doi:10.1016/j.compositesa.2010.10.003

Al-Saleh, M. H., & Sundararaj, U. (2009). A review of vapor grown carbon nanofiber/polymer conductive composites. *Carbon*, *47*(1), 2–22. doi:10.1016/j.carbon.2008.09.039

Al-Saleh, M. H., & Sundararaj, U. (2011). Review of the mechanical properties of carbon nanofiber/polymer composites. *Composites. Part A, Applied Science and Manufacturing*, *42*(12), 2126–2142. doi:10.1016/j.compositesa.2011.08.005

Al-Shamani, A. N., Yazdi, M. H., Alghoul, M. A., Abed, A. M., Ruslan, M. H., Mat, S., & Sopian, K. (2014). Nanofluids for Improved Efficiency in Cooling Solar Collectors: A Review. *Renewable & Sustainable Energy Reviews*, *38*, 348–367. doi:10.1016/j.rser.2014.05.041

Altan, M., & Uysal, A. (2016). An Experimental Study on Mechanical Behavior of Nanographene/Epoxy Nanocomposites. *Advances in Polymer Technology*.

Alzina, C., Mija, A., Vincent, L., & Sbirrazzuoli, N. (2012). Effects of incorporation of organically modified montmorillonite on the reaction mechanism of epoxy/amine cure. *The Journal of Physical Chemistry B, 116*(19), 5786–5794. doi:10.1021/jp3021722

Alzina, C., Sbirrazzuoli, N., & Mija, A. (2010). Hybrid nanocomposites: Advanced nonlinear method for calculating key kinetic parameters of complex cure kinetics. *The Journal of Physical Chemistry B, 114*(39), 12480–12487. doi:10.1021/jp1040629

Alzina, C., Sbirrazzuoli, N., & Mija, A. (2011). Epoxy-amine based nanocomposites reinforced by silica nanoparticles. Relationships between morphologic aspects, cure kinetics, and thermal properties. *The Journal of Physical Chemistry C, 115*(46), 22789–22795. doi:10.1021/jp2066259

American Chemistry Council – Plastic Division. (2014). *Plastics and Polymer Composites: Technology Roadmap for Automotive Markets*. Author.

Amir, W. W., Jumahat, A., & Mahmud, J. (2015). Effect of nanoclay content on flexural properties of glass fiber reinforced polymer (GFRP) composite. *J. Teknol, 76*, 31–35.

Andersson, H. (2009). *Aircraft Systems Modeling: Model Based Systems Engineering in Avionics Design and Aircraft Simulation*. Linköping Studies in Science and Technology, Thesis No. 1394.

Andrei, G., Dima, D., & Andrei, L. (2006). Lightweight magnetic composites for aircraft applications. *Journal of Optoelectronics and Advanced Materials, 8*(2), 726.

Angulalakshmi, V. S., Tamilarasan, G., & Karthikeyan, S. (2014). Optimization of CVD synthesis conditions for the synthesis of multiwalled carbon nanotubes using response surface methodology. *J. Environ. Nanotechnol., 3*(2), 81–91. doi:10.13074/jent.2014.03.142066

Araby, S., Meng, Q., Zhang, L., Kang, H., Majewski, P., Tang, Y., & Ma, J. (2014). Electrically and thermally conductive elastomer/graphene nanocomposites by solution mixing. *Polymer, 55*(1), 201–210. doi:10.1016/j.polymer.2013.11.032

Arena, M., Vertuccio, L., Barra, G., Viscardi, M., & Guadagno, L. (2018). Dynamic performance of self-sensing epoxy resin for aerospace structures. *AIP Conference Proceedings*. 10.1063/1.5045928

ARINC 653. (2006). *Avionics Application Software Standard Interface*. ARINC 653, Specification, Part 1-3, Aeronautical Radio Incorporated, Annapolis, Maryland, USA.

Arshad, S. N., Naraghi, M., & Chasiotis, I. (2011). Strong carbon nanofibers from electrospun polyacrylonitrile. *Carbon, 49*(5), 1710–1719. doi:10.1016/j.carbon.2010.12.056

Asghar, Raman, & Daud. (2015). Advanced oxidation processes for in-situ production of hydrogen peroxide/hydroxyl radical for textile wastewater treatment: A review. *J. of Clean. Produc., 87*, 826–838.

Ashby, M. F., Ferreira, P. J., & Schodek, D. L. (2009). Material Property Charts and their Uses. In Nanomaterials, Nanotechnologies and Design (pp. 147-176). Elsevier. doi:10.1016/B978-0-7506-8149-0.00007-6

Ashby, M. F. (1992). *Materials Selection in Mechanical Design*. Pergamon Press.

Askeland, D. R. (1996). *The Science and Engineering of Materials*. London: Chapman and Hall.

Ates, M., Eker, A. A., & Eker, B. (2017). Carbon nanotube-based nanocomposites and their applications. *Journal of Adhesion Science and Technology*, *31*(18), 1977–1997. doi:10.1080/01694243.2017.1295625

Atif, R., Shyha, I., & Inam, F. (2016). Mechanical, thermal, and electrical properties of graphene-epoxy nanocomposites—A review. *Polymers*, *8*(8), 281. doi:10.3390/polym8080281

Atif, R., Shyha, I., & Inam, F. (2017). Modeling and experimentation of multi-layered nanostructured graphene-epoxy nanocomposites for enhanced thermal and mechanical properties. *Journal of Composite Materials*, *51*(2), 209–220. doi:10.1177/0021998316640060

Augustin, T. (2018). *Structural health monitoring of carbon fiber reinforced polymers and carbon nanotube modified adhesive joints via electrical resistance measurement*. Technische Universität Hamburg.

Auta, M., & Hameed, B. H. (2013). Coalesced chitosan activated carbon composite for batch and fixed-bed adsorption of cationic and anionic dyes. *Colloids and Surfaces. B, Biointerfaces*, *105*, 199–206. doi:10.1016/j.colsurfb.2012.12.021 PMID:23376092

Azeez, A. A., Rhee, K. Y., Park, S. J., & Hui, D. (2013). Epoxy clay nanocomposites–processing, properties and applications: A review. *Composites. Part B, Engineering*, *45*(1), 308–320. doi:10.1016/j.compositesb.2012.04.012

Backlund, G. (2000). *The Effects of Modelling Requirements in Early Phases of Buyer-Supplier Relation*. Linköping Studies in Science and Technology, Thesis No. 812.

Balakrishnan, S., & Raghavan, D. (2003). Chemically functionalized clay epoxy nanocomposites for aerospace applications. In *Technical Proceedings of the 2003 Nanotechnology Conference and Trade Show* (*Vol. 3*, pp. 250-253). Academic Press.

Balamurugan, J., Thanagamuthu, R., & Pandurangan, A. (2013). Effective synthesis of carbon nanotubes of high purity over Cr–Ni–SBA-15 and its application in high performance dye-sensitized solar cells. *RSC. Adv.*, *3*, 4321–4331.

Baldissera, P., & Delprete, C. (2008). Deep Cryogenic Treatment: A Bibliographic Review. *The Open Mechanical Engineering Journal*, *2*(1), 1–11. doi:10.2174/1874155X00802010001

Baller, J., Becker, N., Ziehmer, M., Thomassey, M., Zielinski, B., Müller, U., & Sanctuary, R. (2009). Interactions between silica nanoparticles and an epoxy resin before and during network formation. *Polymer*, *50*(14), 3211–3219. doi:10.1016/j.polymer.2009.05.020

Baller, J., Thomassey, M., Ziehmer, M., & Sanctuary, R. (2011). The catalytic influence of alumina nanoparticles on epoxy curing. *Thermochimica Acta*, *517*(1–2), 34–39. doi:10.1016/j.tca.2011.01.029

Bal, S., & Samal, S. S. (2007). Carbon nanotube reinforced polymer composites—A state of the art. *Bulletin of Materials Science*, *30*(4), 379–386. doi:10.100712034-007-0061-2

Bardaji, M., Uznanski, P., Amiens, C., Chaudret, B., & Laguna, A. (2002). Aurophilic complexes as gold atom sources in organic media. *Chemical Communications*, (6): 598–599.

Bashir, Z. (1991). A critical review of the stabilisation of polyacrylonitrile. *Carbon*, *29*(8), 1081–1090. doi:10.1016/0008-6223(91)90024-D

Baur, J., & Silverman, E. (2007). Challenges and Opportunities in Multifunctional Nanocomposite Structures for Aerospace Applications. *MRS Bulletin*, *32*(4), 328–334. doi:10.1557/mrs2007.231

Becker, M. C., Salvatore, P., & Zirpoli, F. (2005). The impact of virtual simulation tools on problem solving and new product development organization. *Research Policy*, *34*(9), 1305–1321. doi:10.1016/j.respol.2005.03.016

Becker, O., Simon, P. G., & Dusek, K. (2005). Epoxy layered silicate nanocomposites. *Advances in Polymer Science*, *179*, 29–82. doi:10.1007/b107204

Bekyarova, E., Thostenson, E., Yu, A., Kim, H., Gao, J., Tang, J., ... Haddon, R. (2007). Multiscale carbon nanotube– carbon fiber reinforcement for advanced epoxy composites. *Langmuir*, *23*(7), 3970–3974. doi:10.1021/la062743p PMID:17326671

Bernard, O., Ulm, F. J., & Lemarchand, E. (2003). A Multiscale Micromechanics-Hydration Model for the Early-Age Elastic Properties of Cement-Based Materials. *Cement and Concrete Research*, *33*(9), 1293–1309. doi:10.1016/S0008-8846(03)00039-5

Bethune, D. S., Klang, C. H., de Vries, M. S., Gorman, G., Savoy, R., Vazquez, J., & Beyers, R. (1993). Cobalt-catalysed growth of carbon nanotubes with single-atomic-layer walls. *Nature*, *363*(6430), 605–607. doi:10.1038/363605a0

Beyer, G. (2002). Carbon nanotubes as flame retardants for polymers. *Fire and Materials*, *26*(6), 291–293. doi:10.1002/fam.805

Bharadwaj, R. K. (2001). Modeling the barrier properties of polymer-layered silicate nanocomposites. *Macromolecules*, *34*(26), 9189–9192. doi:10.1021/ma010780b

Bhardwaj, N., & Kundu, S. C. (2010). Electrospinning: A fascinating fiber fabrication technique. *Biotechnology Advances*, *28*(3), 325–347. doi:10.1016/j.biotechadv.2010.01.004 PMID:20100560

Bhattacharyya, A. R., Bose, S., Kulkarni, A. R., Pötschke, P., Häußler, L., Fischer, D., & Jehnichen, D. (2007). Styrene maleic anhydride copolymer mediated dispersion of single wall carbon nanotubes in polyamide 12: Crystallization behavior and morphology. *Journal of Applied Polymer Science*, *106*(1), 345–353. doi:10.1002/app.26680

Bhushan, B., & Gupta, B. K. (1991). *Handbook of Tribology*. McGraw Hill.

Bhuvaneswari, C. M., Kale, S. S., Gouda, G., Jayapal, P., & Tamilmani, K. (2017). Elastomers and Adhesives for Aerospace Applications. In *Aerospace Materials and Material Technologies* (pp. 563–586). Singapore: Springer. doi:10.1007/978-981-10-2134-3_26

Bhuyan, M. S. A., Uddin, M. N., Islam, M. M., Bipasha, F. A., & Hossain, S. S. (2016). Synthesis of graphene. *International Nano Letters*, *6*(2), 65–83. doi:10.100740089-015-0176-1

Bierwagen, G. P., & Tallman, D. E. (2001). Choice and measurement of crucial aircraft coatings system properties. *Progress in Organic Coatings*, *41*(4), 201–216. doi:10.1016/S0300-9440(01)00131-X

Bigley, C., & Greenwood, P. (2003). Using Silica to Control Bleed and Segregation in Self-Compacting Concrete. *Concrete (London)*, 43–45.

Bilgen, E., & Muftuoglu, A. (2008). Natural Convection in an Open Square Cavity with Slots. *International Communications in Heat and Mass Transfer*, *35*(8), 896–900. doi:10.1016/j.icheatmasstransfer.2008.05.001

Biltgen, P. T. (2008). *Uncertainty quantification for capability-based systems-of-systems design*. 26th International Congress of the Aeronautical Sciences, Anchorage, AK.

Bizdoaca, E., Spasova, M., Farle, M., Hilgendorff, M., & Caruso, F. (2002). Magnetically directed self-assembly of submicron spheres with a Fe_3O_4 nanoparticle shell. *Journal of Magnetism and Magnetic Materials*, *240*(1-3), 44–46. doi:10.1016/S0304-8853(01)00724-7

Bogdanova, L. M., Kuzub, L. I., Dzhavadyan, E. A., Torbov, V. I., Dremova, N. N., & Pomogailo, A. D. (2014). Mechanical properties of epoxy composites based on silver nanoparticles formed in situ, *Journal of Polymer Science. Ser. A*, *56*(3), 304–310.

Bogdanova, L. M., Kuzub, L. I., Spirin, M. G., Dzhardimalieva, G. I., & Irzhak, V. I. (2015). Features of the synthesis of silver nanoparticles in a curing epoxy matrix. *Vestnik KGTU*, *18*(16), 10–15.

Bogdanova, L. M., Shershnev, V. A., Spirin, M. G., Irzhak, V. I., & Dzhardimalieva, G. I. (in press). Evolution of silver nanoparticles synthesized in situ in a polycondensation epoxy matrix, *Russian. The Journal of Physical Chemistry A*.

Bogdanova, L., Kuzub, L., Dzhavadjan, E., Rabenok, E., Novikov, G., & Pomogailo, A. (2012). In situ synthesis and properties of epoxy nanocomposites. *Macromolecular Symposia*, *317–318*(1), 117–122. doi:10.1002/masy.201100099

Böger, L., Wichmann, M. H., Meyer, L. O., & Schulte, K. (2008). Load and health monitoring in glass fibre reinforced composites with an electrically conductive nanocomposite epoxy matrix. *Composites Science and Technology*, *68*(7-8), 1886–1894. doi:10.1016/j.compscitech.2008.01.001

Bognitzki, M., Czado, W., Frese, T., Schaper, A., Hellwig, M., Steinhart, M., ... Wendorff, J. H. (2001). Nanostructured Fibers via Electrospinning. *Advanced Materials*, *13*(1), 70–72. doi:10.1002/1521-4095(200101)13:1<70::AID-ADMA70>3.0.CO;2-H

Bogue, R. (2011). Nanocomposites: A review of technology and applications. *Assembly Automation*, *31*(2), 106–112. doi:10.1108/01445151111117683

Bourbigot, S., Bras, M. L., Dabrowski, F., Gilman, J. W., & Kashiwagi, T. (2000). PA-6 clay nanocomposite hybrid as char forming agent in intumescent formulations. *Fire and Materials*, *24*(4), 201–208. doi:10.1002/1099-1018(200007/08)24:4<201::AID-FAM739>3.0.CO;2-D

Breuer, O., & Sundararaj, U. (2004). Big returns from small fibers: A review of polymer/carbon nanotube composites. *Polymer Composites*, *25*(6), 630–645. doi:10.1002/pc.20058

Brinkman, H. C. (1952). The Viscosity of Concentrated Suspensions and Solutions. *The Journal of Chemical Physics*, *20*(4), 571. doi:10.1063/1.1700493

Britz, D. A., & Khlobystov, A. N. (2006). Noncovalent interactions of molecules with single walled carbon nanotubes. *Chemical Society Reviews*, *35*(7), 637–659. doi:10.1039/b507451g

Cai, Z.-Q., Movva, S., Chiou, N.-R., Guerra, D., Hioe, Y., Castro, J. M., & Lee, L. J. (2010). Effect of polyaniline surface modification of carbon nanofibers on cure kinetics of epoxy resin. *Journal of Applied Polymer Science*, *118*(4), 2328–2335.

Callister, W. D. (2000). *Materials Science and Engineering: An Introduction*. John Wiley and Sons.

Cao, G., & Wang, Y. (2011). *Nanostructures and Nanomaterials: Synthesis, Properties, and Applications*. World Scientific. doi:10.1142/7885

Carloni, L. P., Passerone, R., Pinto, A., & Sangiovanni-Vincentelli, A. L. (2006). Languages and Tools for Hybrid Systems Design. *Foundations and Trends in Electronic Design Automation*, *1*(1-2), 1–193. doi:10.1561/1000000001

Cassandras, C. G., & Lafortune, S. (1999). *Introduction to Discrete Event Systems*. New York: Springer. doi:10.1007/978-1-4757-4070-7

Cha, J., Chiashi, S., Inoue, T., Einarsson, E., Shiomi, J., & Maruyama, S. (2018). Fabrication of uniform vertically-aligned carbon nanotube–polymer composite thin films by capillary flow intrusion. *Japanese Journal of Applied Physics*, *57*(11), 115101. doi:10.7567/JJAP.57.115101

Chang, T.-E., Kisliuk, A., Rhodes, S., Brittain, W., & Sokolov, A. (2006). Conductivity and mechanical properties of well-dispersed single-wall carbon nanotube/polystyrene composite. *Polymer*, *47*(22), 7740–7746. doi:10.1016/j.polymer.2006.09.013

Charan, S., Singh, N., Khanna, P. K., & Patil, K. R. (2006). Direct synthesis of nanocrystalline silver from the reaction between silver carboxylates and n-trioctylphosphine. *Journal of Nanoscience and Nanotechnology*, *6*(7), 2095–2102. doi:10.1166/jnn.2006.339

Chatterjee, S., & Hadi, A. S. (1988). *Sensitivity analysis in linear regression*. New York: Wiley. doi:10.1002/9780470316764

Chen, S. (2018). *Polymer Based Nanocomposites as Multifunctional Structure for Space Radiation Shielding: A Study of Nanomaterial Fabrications and Evaluations*. Academic Press.

Chen, W., & Jin, R. (2005). Analytical Variance-Based Global Sensitivity Analysis in Simulation-Based Design Under Uncertainty. *Journal of Mechanical Design, 127*. . doi:10.1115/1.1904642

Chen, T. K., Tien, Y. I., & Wei, K. H. (2000). Synthesis and characterization of novel segmented polyurethane/clay nanocomposites. *Polymer*, *41*(4), 1345–1353. doi:10.1016/S0032-3861(99)00280-3

Chen, X., Wang, J., Lin, M., Zhong, W., Feng, T., Chen, X., ... Xue, F. (2008). Mechanical and thermal properties of epoxy nanocomposites reinforced with amino-functionalized multi-walled carbon nanotubes. *Materials Science and Engineering A*, *492*(1-2), 236–242. doi:10.1016/j.msea.2008.04.044

Chen, X., Wei, S., Gunesoglu, C., Zhu, J., Southworth, C. S., Sun, L. Y., ... Guo, Z. (2010). Electrospun Magnetic Fibrillar Polystyrene Nanocomposites Reinforced with Nickel Nanoparticles. *Macromolecular Chemistry and Physics*, *211*(16), 1775–1783. doi:10.1002/macp.201000153

Chernin, I. Z., Smekhov, F. M., & Zherdev, Yu. V. (1982). *Epoxy polymers and compositions*. Moscow: Khimiya. (in Russian)

Chiscan, O., Dumitru, I., Postolache, P., Tura, V., & Stancu, A. (2012). Electrospun PVC/Fe_3O_4 composite nanofibers for microwave absorption applications. *Materials Letters*, *68*, 251–254. doi:10.1016/j.matlet.2011.10.084

Chisholm, N., Mahfuz, H., Rangari, V. K., Ashfaq, A., & Jeelani, S. (2005). Fabrication and mechanical characterization of carbon/SiC-epoxy nanocomposites. *Composite Structures*, *67*(1), 115–124. doi:10.1016/j.compstruct.2004.01.010

Choi, W. J., Powell, R. L., & Kim, D. S. (2009). Curing behavior and properties of epoxy nanocomposites with amine functionalized multiwall carbon nanotubes. *Polymer Composites*, *30*(4), 415–421. doi:10.1002/pc.20571

Choi, W., Lahiri, I., Seelaboyina, R., & Kang, Y. S. (2010). Synthesis of Graphene and Its Applications: A Review. *Critical Reviews in Solid State and Material Sciences*, *35*(1), 52–71. doi:10.1080/10408430903505036

Choi, Y. K., Sugimoto, K. I., Song, S. M., & Endo, M. (2005). Mechanical and thermal properties of vapor-grown carbon nanofiber and polycarbonate composite sheets. *Materials Letters*, *59*(27), 3514–3520. doi:10.1016/j.matlet.2005.05.082

Chopra, S., Batthula, S., Deshmukh, K., & Peshwe, D. R. (2017c). Tribological Behaviour of Multi-Walled Carbon Nanotubes (MWCNT) Filled Polybutylene Terephthalate (PBT) Nanocomposites. *Transactions of the Indian Institute of Metals*, *70*(3), 801–807. doi:10.100712666-017-1055-4

Chopra, S., Deshmukh, K. A., Deshmukh, A., Gogte, C. L., & Peshwe, D. (2016). The Influence of Multi-Walled Carbon Nanotubes on the Nucleation, Crystallization and Tensile Properties of PA-6/MWCNT Composites. *Journal of Basic and Applied Research International*, *19*(4), 247–252.

Chopra, S., Deshmukh, K. A., Deshmukh, A., Gogte, C. L., & Peshwe, D. (2018b). Prediction, evaluation and mechanism governing interphase strength in tensile fractured PA-6/MWCNT nanocomposites. *Composites. Part A, Applied Science and Manufacturing*, *112*, 255–262. doi:10.1016/j.compositesa.2018.06.012

Chopra, S., Deshmukh, K. A., Deshmukh, A., & Peshwe, D. R. (2018a). Inflorescence type morphology and mirror–mist–hackle pattern in tensile fractograph of MWCNT/PBT nanocomposites. *International Journal of Materials Research*, *109*(6), 561–568. doi:10.3139/146.111630

Chopra, S., Deshmukh, K. A., & Peshwe, D. R. (2017a). *Cryogenic Treatment of Cubic Boron Nitride (CBN) Cutting Inserts*. Lambert Academic Publishing.

Chopra, S., Deshmukh, K. A., & Peshwe, D. R. (2017b). Theoretical prediction of interfacial properties of PBT/CNT nanocomposites and its experimental evaluation. *Mechanics of Materials*, *109*, 11–17. doi:10.1016/j.mechmat.2017.03.012

Chou, T. W., Gao, L., Thostenson, E. T., Zhang, Z., & Byun, J. H. (2010). An assessment of the science and technology of carbon-nanotube based fibers and composites. *Composites Science and Technology*, *70*(1), 1–19. doi:10.1016/j.compscitech.2009.10.004

Chua, C. K., & Pumera, M. (2015). Carbocatalysis: The state of "Metal-Free" catalysis. *Chemistry (Weinheim an der Bergstrasse, Germany)*, *21*(36), 12550–12562. doi:10.1002/chem.201501383

Chung, D. (2001). Electromagnetic interference shielding effectiveness of carbon materials. *Carbon*, *39*(2), 279–285. doi:10.1016/S0008-6223(00)00184-6

Chung, D. D. L. (1994). *Carbon Fiber Composites*. Butterworth-Heinemann. doi:10.1016/B978-0-08-050073-7.50009-9

Chung, D. D. L. (2000). Materials for electromagnetic interference shielding. *Journal of Materials Engineering and Performance*, *9*(3), 350–354. doi:10.1361/105994900770346042

Cientifica, C. M. P. (2002). *The Nanotechnology Opportunity Report*. Retrieved from www.cmp-cientifica.com/

Ci, L., Song, L., Jariwala, D., Elías, A. L., Gao, W., Terrones, M., & Ajayan, P. M. (2009). Graphene Shape Control by Multistage Cutting and Transfer. *Advanced Materials*, *21*(44), 4487–4491. doi:10.1002/adma.200900942

Cividanes, L. S. N., Simonetti, E. A., Moraes, M. B., Fernandes, F. W., & Thim, G. P. (2014). Influence of carbon nanotubes on epoxy resin cure reaction using different techniques: A comprehensive review. *Polymer Engineering and Science*, *54*(11), 2461–2469. doi:10.1002/pen.23775

Cividanes, L. S., Brunelli, D. D., Antunes, E. F., Corat, E. J., Sakane, K. K., & Thim, G. P. (2013). Cure study of epoxy resin reinforced with multiwalled carbon nanotubes by Raman and luminescence spectroscopy. *Journal of Applied Polymer Science*, *127*(1), 544–553. doi:10.1002/app.37815

Clary, D. R., & Mills, G. (2011). Photochemical generation of nanometer-sized Cu particles in octane. *The Journal of Physical Chemistry C, 115*(30), 14656–14663. doi:10.1021/jp2040136

Clary, D. R., Nabil, M., Sedeh, M. M., El-Hasadi, Y., & Mills, G. (2012). Photochemical generation of Ag, Pd, and Pt particles in octane. *The Journal of Physical Chemistry C, 116*(16), 9243–9250. doi:10.1021/jp210051j

Colaneri, N. F., & Shacklette, L. W. (1992). EMI shielding measurements of conductive polymer blends. *IEEE Transactions on Instrumentation and Measurement, 41*(2), 291–297. doi:10.1109/19.137363

Coleman, J. N., Khan, U., Blau, W. J., & Gun'ko, Y. K. (2006). Small but Strong: A review of the mechanical properties of carbon nanotube-polymer composites. *Carbon, 44*(9), 1624–1652. doi:10.1016/j.carbon.2006.02.038

Coleman, J. N., Khan, U., & Gun'ko, Y. K. (2006). Mechanical reinforcement of polymers using carbon nanotubes. *Advanced Materials, 18*(6), 689–706. doi:10.1002/adma.200501851

Cooper, C. A., Ravich, D., Lips, D., Mayer, J., & Wagner, H. D. (2002). Distribution and alignment of carbon nanotubes and nanofibrils in a polymer matrix. *Composites Science and Technology, 62*(7), 1105–1112. doi:10.1016/S0266-3538(02)00056-8

Cortés, P., Fraga, I., Calventus, Y., Román, F., Hutchinson, J. M., & Ferrando, F. (2014). A new epoxy-based layered silicate nanocomposite using a hyperbranched polymer: Study of the curing reaction and nanostructure development. *Materials (Basel), 7*(3), 1830–1849. doi:10.3390/ma7031830

Cortés, P., Lozano, K., Barrera, E. V., & Bonilla-Rios, J. (2003). Effects of nanofiber treatments on the properties of vapor-grown carbon fiber reinforced polymer composites. *Journal of Applied Polymer Science, 89*(9), 2527–2534. doi:10.1002/app.12309

Costa, A. P. D., Botelho, E. C., Costa, M. L., Narita, N. E., & Tarpani, J. R. (2012). A review of welding technologies for thermoplastic composites in aerospace applications. *Journal of Aerospace Technology and Management, 4*(3), 255–265. doi:10.5028/jatm.2012.040303912

Cox, H. L. (1952). The elasticity and strength of paper and other fibrous materials. *British Journal of Applied Physics, 3*(3), 72–79. doi:10.1088/0508-3443/3/3/302

Cui, Y., Kumar, S., Kona, B. R., & van Houcke, D. (2015). Gas barrier properties of polymer/clay nanocomposites. *RSC Advances, 5*(78), 63669–63690. doi:10.1039/C5RA10333A

Cullen, A. C., & Frey, H. C. (1999). *Probabilistic Techniques in Exposure Assessment*. New York: Plenum Press.

D'yachkov, P. N. (2006). *Carbon nanotubes. Structure, properties, application*. Moscow: Binom. (in Russian)

Da Silva, L. V., Pezzin, S. H., Rezende, M. C., & Amico, S. C. (2016). Glass fiber/carbon nanotubes/epoxy three component composites as radar absorbing materials. *Polymer Composites*, *37*(8), 2277–2284. doi:10.1002/pc.23405

Dahoud, M. S., Abdel-Rahem, R. A., Farrag, E. A., Ibrahim, S., Ayesh, A. S., & Alshoaibi, A. (2018). Rheological and electrical properties of multiwalled carbon nanotubes–polyvinyl alcohol nanocomposites treated at different pH conditions. *Journal of Thermoplastic Composite Materials*.

Dai, H. (2001). Nanotube Growth and Characterization. In M. S. Dresselhaus, G. Dresselhaus, & P. Avouris (Eds.), *Carbon Nanotubes* (pp. 29–53). Springer Berlin Heidelberg; doi:10.1007/3-540-39947-X_3

Dai, H. (2002). Carbon nanotubes: Opportunities and challenges. *Surface Science*, *500*(1-3), 218–241. doi:10.1016/S0039-6028(01)01558-8

Dang, Z.-M., Yuan, J.-K., Zha, J.-W., Zhou, T., Li, S.-T., & Hu, G.-H. (2012). Fundamentals, processes and applications of high-permittivity polymer–matrix composites. *Progress in Materials Science*, *57*(4), 660–723. doi:10.1016/j.pmatsci.2011.08.001

Darwish, M. S., Bakry, A., Kolek, O., Martinová, L., & Stibor, I. (2019). Electrospun functionalized magnetic polyamide 6 composite nanofiber: Fabrication and stabilization. *Polymer Composites*, *40*, 296–303. doi:10.1002/pc.24647

Dasari, A., Yu, Z. Z., & Mai, Y. W. (2009). Fundamental aspects and recent progress on wear/scratch damage in polymer nanocomposites. *Materials Science and Engineering R Reports*, *63*(2), 31–80. doi:10.1016/j.mser.2008.10.001

Das, N. C., Liu, Y., Yang, K., Peng, W., Maiti, S., & Wang, H. (2009). Single-walled carbon nanotube/poly(methyl methacrylate) composites for electromagnetic interference shielding. *Polymer Engineering and Science*, *49*(8), 1627–1634. doi:10.1002/pen.21384

De Volder, M. F., Tawfick, S. H., Baughman, R. H., & Hart, A. J. (2013). Carbon nanotubes: present and future commercial applications. *Science, 339*(6119), 535-539.

De Volder, M. F. L., Tawfick, S. H., Baughman, R. H., & Hart, A. J. (2013). Carbon Nanotubes: Present and Future Commercial Applications. *Science*, *339*(6119), 535–539. doi:10.1126cience.1222453 PMID:23372006

DeBerry, D. W. (1985). Modification of the electrochemical and corrosion behavior of stainless steels with an electroactive coating. *Journal of the Electrochemical Society*, *132*(5), 1022–1026. doi:10.1149/1.2114008

Deepa, B., Abraham, E., & Cherian, B. M. (2011). Cellulose-based bio- and nanocomposites. *Bioresource Technology*, *102*, 1988–1997. doi:10.1016/j.biortech.2010.09.030 PMID:20926289

Deguchi, R., Nishio, T., & Okada, A. (1992). *U.S. Patent No. 5,102,948*. Washington, DC: U.S. Patent and Trademark Office.

Deitzel, J. M., Kleinmeyer, J., Harris, D., & Beck Tan, N. C. (2001). The effect of processing variables on the morphology of electrospun nanofibers and textiles. *Polymer*, *42*(1), 261–272. doi:10.1016/S0032-3861(00)00250-0

Delhaes, P. (2013). *Carbon Based Solids and Materials*. John Wiley & Sons. doi:10.1002/9781118557617

Deshmukh, G. S., Peshwe, D. R., Pathak, S. U., & Ekhe, J. D. (2011). Evaluation of mechanical and thermal properties of Poly(butylene terephthalate) (PBT) composites reinforced with wollastonite. *Transactions of the Indian Institute of Metals*, *64*(1-2), 127–132. doi:10.100712666-011-0025-5

Dinca, I., Ban, C., Stefan, A., & Pelin, G. (2012). Nanocomposites as advanced materials for aerospace industry. *INCAS Bulletin*, *4*(4), 73–83.

Domingues, D., Logakis, E., & Skordos, A. (2012). The use of an electric field in the preparation of glass fibre/epoxy composites containing carbon nanotubes. *Carbon*, *50*(7), 2493–2503. doi:10.1016/j.carbon.2012.01.072

Doshi, J., & Reneker, D. H. (1995). Electrospinning process and applications of electrospun fibers. *Journal of Electrostatics*, *35*(2), 151–160. doi:10.1016/0304-3886(95)00041-8

Dresselhaus, M. S., & Avouris, P. (2001). Introduction to Carbon Materials Research. In M. S. Dresselhaus, G. Dresselhaus, & P. Avouris (Eds.), *Carbon Nanotubes* (pp. 1–9). Springer Berlin Heidelberg; doi:10.1007/3-540-39947-X_1

Dresselhaus, M. S., Dresselhaus, G., & Avouris, P. (Eds.). (2001). *Carbon Nanotubes: Synthesis, Structure, Properties, and Applications*. Berlin, Heidelberg: Springer-Verlag. doi:10.1007/3-540-39947-X

Duc, N. D., Nguyen, P. D., Cuong, N. H., Van Sy, N., & Khoa, N. D. (2018). An analytical approach on nonlinear mechanical and thermal post-buckling of nanocomposite double-curved shallow shells reinforced by carbon nanotubes. *Proceedings of the Institution of Mechanical Engineers, Part C: Journal of Mechanical Engineering Science*.

Du, J., & Cheng, H.-M. (2012). The Fabrication, Properties, and Uses of Graphene/Polymer Composites. *Macromolecular Chemistry and Physics*, *213*(10–11), 1060–1077. doi:10.1002/macp.201200029

Dumazet-Bonnamour, I., & Perchec, P. (2000). Colloidal dispersion of magnetite nanoparticles via in situ preparation with sodium polyoxyalkylene di-phosphonates. *Carbon*, *45*(8), 1614–1621.

Ebbesen, T. W., & Ajayan, P. M. (1992). Large-scale synthesis of carbon nanotubes. *Nature*, *358*(6383), 220–222. doi:10.1038/358220a0

Edie, D. D. (1998). The effect of processing on the structure and properties of carbon fibers. *Carbon*, *36*(4), 345–362. doi:10.1016/S0008-6223(97)00185-1

Editorial. (2015). Graphene opens up to new applications. *Nature Nanotechnology*, *10*(5), 381. doi:10.1038/nnano.2015.110

Einstein, A. (1906). Eine Neue bestimmung der molekuldimensionen. *Annalen Der Physik, Leipzig*, *19*(2), 289–306. doi:10.1002/andp.19063240204

Einstein, A. (1956). *Investigation on Theory of Brownian Motion*. Dover.

Ellis, B. (1993). *Chemistry and technology of epoxy resins*. Springer. doi:10.1007/978-94-011-2932-9

Endo, M., Kim, Y. A., Hayashi, T., Yanagisawa, T., Muramatsu, H., Ezaka, M., ... Dresselhaus, M. S. (2003). Microstructural changes induced in "stacked cup" carbon nanofibers by heat treatment. *Carbon*, *41*(10), 1941–1947. doi:10.1016/S0008-6223(03)00171-4

Erdem, R. (2016). Evaluation of electromagnetic shielding effectiveness of multi-axial fabrics and their reinforced PES composites. *Bulletin of Materials Science*, *39*(4), 963–970. doi:10.100712034-016-1234-7

Esmizadeh, E., Yousefi, A., & Naderi, G. (2015). Effect of type and aspect ratio of different carbon nanotubes on cure behavior of epoxy-based nanocomposite. *Iranian Polymer Journal*, *24*(1), 1–12. doi:10.100713726-014-0281-4

Ewels, C., Glerup, M., Krstic, V., Basiu, V. A., & Basiuk, E. V. (2007). Nitrogen and boron doping in carbon nanotubes. *Doped Nanomaterials and Nanodevices, 3*. Retrieved from http://ewels.info/science/publications/papers/2007.NitrogenChapter.pdf

Fabrycky, W. J., & Blanchard, B. S. (1991). *Life-Cycle Cost, and Economic Analysis*. Englewood Cliffs, NJ: Prentice Hall.

Fahma, F., Iwamoto, S., Hori, N., Iwata, T., & Takemura, A. (2011). Effect of pre-acid-hydrolysis treatment on morphology and properties of cellulose nanowhiskers from coconut husk. *Cellulose (London, England)*, *18*(2), 443–450. doi:10.100710570-010-9480-0

Fakirov, S. (Ed.). (2005). *Handbook of Thermoplastic Polyesters: Homopolymers, Copolymers, Blends and Composites*. Wiley.

Faraji, S., Yardim, M. F., Can, D. S., & Sarac, A. S. (2017). Characterization of polyacrylonitrile, poly(acrylonitrile-co-vinyl acetate), and poly(acrylonitrile-co-itaconic acid) based activated carbon nanofibers. *Journal of Applied Polymer Science, 134*(2). doi:10.1002/app.44381

Faraji, M., Yamini, Y., & Rezaee, M. (2010). Magnetic nanoparticles: Synthesis, stabilization, functionalization, characterization, and applications. *Journal of the Indian Chemical Society*, *7*(1), 1–37.

Farhani, F., & Niaki, K. S. (2008). *Cryogenic treatment of metals- a review of experimental studies, underlying mechanism and process equipment. 4th BSME-ASME International Conference on Thermal Engineering*, Dhaka, Bangladesh.

Farré, M., Gajda-Schrantz, K., Kantiani, L., & Barceló, D. (2009). Ecotoxicity and analysis of nanomaterials in the aquatic environment. *Analytical and Bioanalytical Chemistry*, *393*(1), 81–95. doi:10.100700216-008-2458-1 PMID:18987850

Federation of Indian Chambers of Commerce and Industry (FICCI). (2017). Plastic Industry for Infrastructure. *3rd National Conference on Sustainable Infrastructure with Plastics.*

Feng, L., Xie, N., & Zhong, J. (2014). Carbon nanofibers and their composites: A review of synthesizing, properties and applications. *Materials (Basel), 7*(5), 3919–3945. doi:10.3390/ma7053919 PMID:28788657

Ferdosian, F., Ebrahimi, M., & Jannesari, A. (2013). Curing kinetics of solid epoxy/DDM/nanoclay: Isoconversional models versus fitting model. *Thermochimica Acta, 568*, 67–73. doi:10.1016/j.tca.2013.06.001

Fiedler, B., Gojny, F. H., Wichmann, M. H., Bauhofer, W., & Schulte, K. (2004). Can carbon nanotubes be used to sense damage in composites? *Annales de chimie, 29*(6), 81–94. doi:10.3166/acsm.29.6.81-94

Finegan, I. C., & Tibbetts, G. G. (2001). Electrical conductivity of vapor-grown carbon fiber/thermoplastic composites. *Journal of Materials Research, 16*(6), 1668–1674. doi:10.1557/JMR.2001.0231

Finney, E. E., & Finke, R. G. (2008). Nanocluster nucleation and growth kinetic and mechanistic studies: A review emphasizing transition-metal nanoclusters. *Journal of Colloid and Interface Science, 317*(2), 351–374. doi:10.1016/j.jcis.2007.05.092

Fletcher, A., Gupta, M. C., Dudley, K. L., & Vedeler, E. (2010). Elastomer foam nanocomposites for electromagnetic dissipation and shielding applications. *Composites Science and Technology, 70*(6), 953–958. doi:10.1016/j.compscitech.2010.02.011

Frenkel', Ya. I. (1975). *Kineticheskaya teoriya zhidkostei [Kinetic Theory of Liquids]*. Leningrad: Nauka.

Fritzson, P. (2004). *Principles of Object-Oriented Modeling and Simulation with MATLAB 2.1*. New York: Wiley-IEEE Press.

Fu, X., Ramos, M., Al-Jumaily, A. M., Meshkinzar, A., & Huang, X. (n.d.). Stretchable strain sensor facilely fabricated based on multi-wall carbon nanotube composites with excellent performance. *Journal of Materials Science*, 1-11.

Fu, X., & Chung, D. D. L. (1996). Submicron carbon filament cement-matrix composites for electromagnetic interference shielding. *Cement and Concrete Research, 26*(10), 1467–1472. doi:10.1016/0008-8846(96)00146-9

Gains, B. (1979). Qua Vadis. *General Systems Yearbook., 24*, 1–9.

Galpaya, D. G. D., Fernando, J. F. S., Rintoul, L., Motta, N., Waclawik, E. R., Yan, C., & George, G. A. (2015). The effect of graphene oxide and its oxidized debris on the cure chemistry and interphase structure of epoxy nanocomposites. *Polymer, 71*, 122–134. doi:10.1016/j.polymer.2015.06.054

Ganguli, R. (2013). Optimal design of composite structures: A historical review. *Journal of the Indian Institute of Science, 93*(4), 557–570.

Ganzhorn, M., Vijayaraghavan, A., Dehm, S., Hennrich, F., Green, A. A., Fichtner, M., ... Kappes, M. M. (2011). Hydrogen sensing with diameter-and chirality-sorted carbon nanotubes. *ACS Nano*, *5*(3), 1670–1676. doi:10.1021/nn101992g PMID:21341751

Gao, H., Liu, L., Luo, Y., & Jia, D. (2011). In-situ preparation of epoxy/silver nanocomposites by thermal decomposition of silver–imidazole complex. *Materials Letters*, *65*(23–24), 3529–3532. doi:10.1016/j.matlet.2011.07.086

Garboczi, E. J., & Bentz, D. P. (1997). Analytical formulas for Interfacial Transition Zone Properties. *Journal of Advanced Cement Based Materials*, *6*(3-4), 99–108. doi:10.1016/S1065-7355(97)90016-X

García, E., Wardle, B., & John Hart, A. (2008). Joining prepreg composite interfaces with aligned carbon nanotubes. *Composites. Part A, Applied Science and Manufacturing*, *39*(6), 1065–1070. doi:10.1016/j.compositesa.2008.03.011

Gavel, H. (2007). *On aircraft fuel systems: conceptual design and modeling*. Linköping Studies in Science and Technology. Thesis No. 1067.

Gavel, H., & Andersson, J. (2003). *Using Optimization as a Tool in Fuel System Conceptual Design*. SAE World Aviation Congress and Display 2003, Paper No 2003-01-3052.

Gavel, H., Krus, P., Andersson, J., & Johansson, B. (2005). *Probabilistic design in the conceptual phase of an aircraft fuel system.* 7th AIAA Non-Deterministic Design Forum, paper No AIAA-2005-2219, Austin, TX.

Gay, D., & Hoa, S. V. (2007). *Composite materials: design and applications*. CRC Press. doi:10.1201/9781420045208

Geim, A. K., & Novoselov, K. S. (2009). The rise of graphene. In *Nanoscience and Technology* (Vol. 1–0, pp. 11–19). Co-Published with Macmillan Publishers Ltd. doi:10.1142/9789814287005_0002

Gelves, G. A., Al-Saleh, M. H., & Sundararaj, U. (2011). Highly electrically conductive and high performance EMI shielding nanowire/polymer nanocomposites by miscible mixing and precipitation. *Journal of Materials Chemistry*, *21*(3), 829–836. doi:10.1039/C0JM02546A

George, K., Panda, B. P., Mohanty, S., & Nayak, S. K. (2018). Recent developments in elastomeric heat shielding materials for solid rocket motor casing application for future perspective. *Polymers for Advanced Technologies*, *29*(1), 8–21. doi:10.1002/pat.4101

Gergin, İ., Ismar, E., & Sarac, A. S. (2017). Oxidative stabilization of polyacrylonitrile nanofibers and carbon nanofibers containing graphene oxide (GO): a spectroscopic and electrochemical study., Oxidative stabilization of polyacrylonitrile nanofibers and carbon nanofibers containing graphene oxide (GO): a spectroscopic and electrochemical study. *Beilstein Journal of Nanotechnology*, *8*(16), 1616–1628. doi:10.3762/bjnano.8.161, 10.3762/bjnano.8.161

Ghaemy, M., Amini Nasab, S. M., & Barghamadi, M. (2007). Nonisothermal cure kinetics of diglycidylether of bisphenol-A/amine system reinforced with nanosilica particles. *Journal of Applied Polymer Science*, *104*(6), 3855–3863. doi:10.1002/app.25994

Ghaffari, M., Ehsani, M., Vandalvand, M., Avazverdi, E., Askari, A., & Goudarzi, A. (2015). Studying the effect of micro- and nano-sized ZnO particles on the curing kinetic of epoxy/polyaminoamide system. *Progress in Organic Coatings*, *89*, 277–283. doi:10.1016/j.porgcoat.2015.08.016

Ghanbari, K., & Ehsani, M. (2018). Change in surface properties of polydimethylsiloxane/polyurethane/carbon nanotubes elastomeric coatings. *Progress in Organic Coatings*, *125*, 393–402. doi:10.1016/j.porgcoat.2018.09.022

Giese, T., Oehler, B., & Sielemann, M. (2010). *A systematic approach to optimize conventional environmental control architectures*. Hamburg, Germany: Deutscher Luft- und Raumfahrtkongress.

Gilbert, E. N., Hayes, B. S., & Seferis, J. C. (2003). Nano-alumina modified epoxy based film adhesives. *Polymer Engineering and Science*, *43*(5), 1096–1104. doi:10.1002/pen.10093

Gilman, J. W., Jackson, C. L., Morgan, A. B., Harris, R., Manias, E., Giannelis, E. P., ... Phillips, S. H. (2000). Flammability properties of polymer– layered-silicate nanocomposites. Polypropylene and polystyrene nanocomposites. *Chemistry of Materials*, *12*(7), 1866–1873. doi:10.1021/cm0001760

Gkikas, G., Sioulas, D., Lekatou, A., Barkoula, N. M., & Paipetis, A. S. (2012). Enhanced bonded aircraft repair using nano-modified adhesives. *Materials & Design*, *41*, 394–402. doi:10.1016/j.matdes.2012.04.052

Gobi, N., Vijayakumar, D., Keles, O., & Erogbogbo, F. (2017). Infusion of Graphene Quantum Dots to Create Stronger, Tougher, and Brighter Polymer Composites. *ACS Omega*, *2*(8), 4356–4362. doi:10.1021/acsomega.6b00517

Gohardani, A. S., Doulgeris, G., & Singh, R. (2011). Challenges of future aircraft propulsion: A review of distributed propulsion technology and its potential application for the all electric commercial aircraft. *Progress in Aerospace Sciences*, *47*(5), 369–391. doi:10.1016/j.paerosci.2010.09.001

Gohardani, O., Elola, M. C., & Elizetxea, C. (2014). Potential and prospective implementation of carbon nanotubes on next generation aircraft and space vehicles: A review of current and expected applications in aerospace sciences. *Progress in Aerospace Sciences*, *70*, 42–68. doi:10.1016/j.paerosci.2014.05.002

Gojny, F. H., Wichmann, M. H., Fiedler, B., & Schulte, K. (2005). Influence of different carbon nanotubes on the mechanical properties of epoxy matrix composites–a comparative study. *Composites Science and Technology*, *65*(15-16), 2300–2313. doi:10.1016/j.compscitech.2005.04.021

Gonza'lez, M., Pozuelo, J., & Baselga, J. (2018). Electromagnetic Shielding Materials in GHz Range. *Chemical Record (New York, N.Y.)*, *18*, 1–11. PMID:29380939

Gouda, M., Hebeish, A. A., & Aljaafari, A. I. (2016). New route for development of electromagnetic shielding based on cellulosic nanofibers. *Journal of Industrial Textiles*, *46*(8), 1598–1615. doi:10.1177/1528083715627166

Govindaraj, B., & Sarojadevi, M. (2018). Microwave-assisted synthesis of nanocomposites from polyimides chemically cross-linked with functionalized carbon nanotubes for aerospace applications. *Polymers for Advanced Technologies*, *29*(6), 1718–1726. doi:10.1002/pat.4275

Gowariker, V. R., Viswanathan, N. V., & Sreedhar, J. (1986). *Polymer Science*. New Age International Publishers.

Grachev, V. P., Kondrashov, S. V., Akatenkov, R. V., Aleksashin, V. N., Deev, I. S., Anoshkin, I. V., ... Irzhak, V. I. (2014). Modification of epoxy polymers by small additives of multiwall carbon nanotube. *Polymer Science, Series A*, *56*(3), 330–336. doi:10.1134/S0965545X14030079

Graf, S. (2008). OMEGA: Correct development of real time and embedded systems. *Software & Systems Modeling*, *7*(2), 127–130. doi:10.100710270-007-0077-5

Grayfer, E. D., Makotchenko, V. G., Nazarov, A. S., Kim, S. J., & Fedorov, V. E. (2011). Graphene: Chemical approaches to the synthesis and modification. *Russian Chemical Reviews*, *80*(8), 751–04. doi:10.1070/RC2011v080n08ABEH004181

Grimes, C. A. (1994). EMI shielding characteristics of permalloymultilayer thin films. In *IEEE Aerospace Applications Conf. Proc*, IEEE.

Guadagno, L., De Vivo, B., Di Bartolomeo, A., Lamberti, P., Sorrentino, A., Tucci, V., ... Vittoria, V. (2011). Effect of functionalization on the thermo-mechanical and electrical behavior of multi-wall carbon nanotube/epoxy composites. *Carbon*, *49*(6), 1919–1930. doi:10.1016/j.carbon.2011.01.017

Gugliuzza, A., & Drioli, E. (2013). A review on membrane engineering for innovation in wearable fabrics and protective textiles. *Journal of Membrane Science*, *446*, 350–375. doi:10.1016/j.memsci.2013.07.014

Gu, H., Tadakamalla, S., Huang, Y., Colorado, H. A., Luo, Z., Haldolaarachchige, N., ... Guo, Z. (2012). Polyaniline stabilized magnetite nanoparticle reinforced epoxy nanocomposites. *ACS Applied Materials & Interfaces*, *4*(10), 5613–5624. doi:10.1021/am301529t

Guo, T., Nikolaev, P., Rinzler, A. G., Tomanek, D., Colbert, D. T., & Smalley, R. E. (1995). Self-assembly of tubular fullerenes. *Journal of Physical Chemistry*, *99*(27), 10694–10697. doi:10.1021/j100027a002

Gupta, T. K., Singh, B. P., Dhakate, S. R., Singh, V. N., & Mathur, R. B. (2013). Improved nanoindentation and microwave shielding properties of modified MWCNT reinforced polyurethane composites. *Journal of Materials Chemistry. A, Materials for Energy and Sustainability*, *1*(32), 9138. doi:10.1039/c3ta11611e

Gur'eva, L. L., Tkachuk, A. I., Kuzub, L. I., Estrina, G. A., Knerel'man, E. I., Khodos, I. I., & Rozenberg, B. A. (2013). Synthesis of silver nanoparticles with polystyrylcarboxylate ligands, *Journal of Polymer Science. Ser. B*, *55*(3–4), 139–146.

Gu, S. Y., Ren, J., & Vancso, G. J. (2005). Process optimization and empirical modeling for electrospun polyacrylonitrile (PAN) nanofiber precursor of carbon nanofibers. *European Polymer Journal*, *41*(11), 2559–2568. doi:10.1016/j.eurpolymj.2005.05.008

Gusev, A. I. (2007). *Nanomaterials, nanostructures, nanotechnologies*. Moscow: Fizmatlit. (in Russian)

Guzman de Villoria, R., Ydrefos, L., Hallander, P., Ishiguro, K., Noedin, P., & Wardle, B. (2012). Aligned carbon nanotube reinforcement of aerospace carbon fiber composites: substructural strength evaluation for aerostructure applications. In *Proceedings of the 53rd AIAA/ASME/ASCE/AHS/ASC structures, structural dynamics and materials conference*. American Institutes of Aeronautics and Astronautics. 10.2514/6.2012-1566

Guzmán de Villoria, R., Hart, A. J., & Wardle, B. L. (2011). Continuous high-yield production of vertically aligned carbon nanotubes on 2D and 3D substrates. *ACS Nano*, *5*(6), 4850–4857. doi:10.1021/nn2008645 PMID:21591620

Hajdu, J. (1997). New challenges for electroless plating technologies. *Trans Inst Metal Finish*, *75*(1), B7–B10.

Hajibeygi, M., Shabanian, M., Omidi-Ghallemohamadi, M., & Khonakdar, H. A. (2017). Optical, thermal and combustion properties of self-colored polyamide nanocomposites reinforced with azo dye surface modified ZnO nanoparticles. *Applied Surface Science*, *416*, 628–638. doi:10.1016/j.apsusc.2017.04.139

Hamilton, R. L., & Crosser, O. K. (1962). Thermal Conductivity of Heterogeneous Two-Component System. *Industrial & Engineering Chemistry Fundamentals*, *1*(3), 187–191. doi:10.1021/i160003a005

Han, B., Cheng, A., Ji, G., Wu, S., & Shen, J. (2004). Effect of organophilic montmorillonite on polyurethane/montmorillonite nanocomposites. *Journal of Applied Polymer Science*, *91*(4), 2536–2542. doi:10.1002/app.13398

Han, J. H., Zhang, H., Chen, M. J., Wang, D., Liu, Q., Wu, Q. L., & Zhang, Z. (2015). The combination of carbon nanotube buckypaper and insulating adhesive for lightning strike protection of the carbon fiber/epoxy laminates. *Carbon*, *94*, 101–113. doi:10.1016/j.carbon.2015.06.026

Haque, A., Hossain, F., Dean, D., & Shamsuzzoha, M. (2002). *s2-glass/vinyl ester polymer nanocomposites: manufacturing, structures, thermal and mechanical properties*. Paper presented at the 17th Annual Technical Conference, American Society for Composites.

Heimann, M., Wirts-Ruetters, M., Boehme, B., & Wolter, K.-J. (2008). *Investigations of carbon nanotubes epoxy composites for electronics packaging*. Paper presented at the Electronic Components and Technology Conference, 2008. ECTC 2008. 58th. 10.1109/ECTC.2008.4550214

Higgins, B. A., & Brittain, W. J. (2005). Polycarbonate carbon nanofiber composites. *European Polymer Journal*, *41*(5), 889–893. doi:10.1016/j.eurpolymj.2004.11.040

Hirsch, J., & Al-Samman, T. (2013). Superior light metals by texture engineering: Optimized aluminum and magnesium alloys for automotive applications. *Acta Materialia*, *61*(3), 818–843. doi:10.1016/j.actamat.2012.10.044

Hoang, A. S. (2011). Electrical conductivity and electromagnetic interference shielding characteristics of multiwalled carbon nanotube filled polyurethane composite films. *Adv. Nat. Sci. Nanosci. Nanotechnol.*, *2*(2), 025007. doi:10.1088/2043-6262/2/2/025007

Hoffmann, D., Landfester, K., & Antonietti, M. (2001). Encapsulation of magnetite in polymer particles via the miniemulsion polymerization process. *Magnetohydrodynamics*, *37*(3), 217–221.

Holzinger, M., Vostrowsky, O., Hirsch, A., Hennrich, F., Kappes, M., Weiss, R., & Jellen, F. (2001). Sidewall Functionalization of Carbon Nanotubes This work was supported by the European Union under the 5th Framework Research Training Network 1999, HPRNT 1999-00011 FUNCARS. *Angewandte Chemie International Edition*, *40*(21), 4002–4005. doi:10.1002/1521-3773(20011105)40:21<4002::AID-ANIE4002>3.0.CO;2-8 PMID:12404474

Hong, S. G., & Tsai, J. S. (2001). The adsorption and curing behavior of the epoxy/amidoamine system in the presence of metal oxides. *Journal of Thermal Analysis and Calorimetry*, *63*(1), 31–46. doi:10.1023/A:1010119900035

Huang, G., Wang, S., Song, P. A., Wu, C., Chen, S., & Wang, X. (2014). Combination effect of carbon nanotubes with graphene on intumescent flame-retardant polypropylene nanocomposites. *Composites. Part A, Applied Science and Manufacturing*, *59*, 18–25. doi:10.1016/j.compositesa.2013.12.010

Huang, X., Xie, L., Yang, K., Wu, C., Jiang, P., Li, S., ... Tanaka, T. (2014). Role of interface in highly filled epoxy/$BaTiO_3$ nanocomposites. Part I-correlation between nanoparticle surface chemistry and nanocomposite dielectric property. *IEEE Transactions on Dielectrics and Electrical Insulation*, *21*(2), 467–479. doi:10.1109/TDEI.2013.004165

Huang, Z. M., He, C. L., Yang, A. Z., Zhang, Y., Han, X.-J., Yin, J., & Wu, Q. (2006). Encapsulating drugs in biodegradable ultrafine fibers through co-axial electrospinning. *Journal of Biomedical Materials Research. Part A*, *77*(1), 169–179. doi:10.1002/jbm.a.30564 PMID:16392131

Huang, Z. P., Xu, J. W., Ren, Z. F., Wang, J. H., Siegal, M. P., & Provencio, P. N. (1998). Growth of highly oriented carbon nanotubes by plasma-enhanced hot filament chemical vapor deposition. *Applied Physics Letters*, *73*(26), 3845–3847. doi:10.1063/1.122912

Huang, Z.-M., Zhang, Y.-Z., Kotaki, M., & Ramakrishna, S. (2003). A review on polymer nanofibers by electrospinning and their applications in nanocomposites. *Composites Science and Technology*, *63*(15), 2223–2253. doi:10.1016/S0266-3538(03)00178-7

Huiping, C. J. H. L. T. (2003). Interlaminar properties of carbon/phenolic composite modified with nano-carbon black. *Fiber Reinforced Plastics/Composite, 4.*

Hu, M., Gao, J., Dong, Y., Li, K., Shan, G., Yang, S., & Li, R. K.-Y. (2012). Flexible transparent PES/silver nanowires/PET sandwichstructured film for high efficiency electromagnetic interference shielding. *Langmuir*, *28*(18), 7101–7106. doi:10.1021/la300720y PMID:22533864

Hummers, W. S. Jr, & Offeman, R. E. (1958). Preparation of Graphitic Oxide. *Journal of the American Chemical Society*, *80*(6), 1339–1339. doi:10.1021/ja01539a017

Hussain, F., Hojjati, M., Okamoto, M., & Gorga, R. E. (2006). Polymer-matrix nanocomposites, processing, manufacturing, and application: An overview. *Journal of Composite Materials*, *40*(17), 1511–1575. doi:10.1177/0021998306067321

Hussain, S. T., Abbas, F., Kausar, A., & Khan, M. R. (2013). New polyaniline/polypyrrole/polythiophene and functionalized multiwalled carbon nanotube-based nanocomposites: Layer-by-layer in situ polymerization. *High Performance Polymers*, *25*(1), 70–78. doi:10.1177/0954008312456048

Hussein, A., & Kim, B. (2018). Graphene/polymer nanocomposites: The active role of the matrix in stiffening mechanics. *Composite Structures*, *202*, 170–181. doi:10.1016/j.compstruct.2018.01.023

Hutchinson, J. M., Shiravand, F., & Calventus, Y. (2013). Intra- and extra-gallery reactions in tri-functional epoxy polymer layered silicate nanocomposites. *Journal of Applied Polymer Science*, *128*(5), 2961–2970. doi:10.1002/app.38452

Hutchinson, J. M., Shiravand, F., Calventus, Y., Fernández-Francos, X., & Ramis, X. (2014). Highly exfoliated nanostructure in trifunctional epoxy/clay nanocomposites using boron trifluoride as initiator. *Journal of Applied Polymer Science*, *131*(6), 40020–40029. doi:10.1002/app.40020

Iijima, S. (1991). Helical Microtubules of Graphitic Carbon. *Nature*, *354*(6348), 56–58. doi:10.1038/354056a0

Indumathi, J., Bijwe, J., & Ghosh, A. K. (1999a). *Cryotreatment - an effective technique to enhance the abrasive wear performance of high performance engineering polymers and composites. International Conference on Polyimides*, Newark, NJ.

Indumathi, J., Bijwe, J., Ghosh, A. K., Fahim, M., & Krishnaraj, N. (1999b). Wear of cryo-treated engineering polymers and composites. *Wear*, *225-229*, 343–353. doi:10.1016/S0043-1648(99)00063-0

Irzhak, T.F., & Irzhak, V.I. (2017). Epoxy nanocomposites. *Polymer Sciences, A*, *59*(6), 791-825.

Irzhak, V. I. (2011). Epoxide composite materials with carbon nanotubes. *Russian Chemical Reviews*, *80*(8), 787–806. doi:10.1070/RC2011v080n08ABEH004196

Irzhak, V. I. (2016). The mechanisms of the formation of metal-containing nanoparticles. *Review Journal of Chemistry*, *6*(4), 370–404. doi:10.1134/S2079978016040026

Irzhak, V. I. (2017). Digestive ripening of nanoparticles, *Russian. The Journal of Physical Chemistry A*, *91*(8), 1502–1506.

Irzhak, V. I., Rozenberg, B. A., & Enikolopyan, N. S. (1979). *Network polymers. Synthesis, structure, properties*. Moscow: Nauka. (in Russian)

Ismar, E., Karazehir, T., Ates, M., & Sarac, A. S. (2018). Electrospun carbon nanofiber web electrode: Supercapacitor behavior in various electrolytes. *Journal of Applied Polymer Science*, *135*(4), 45723. doi:10.1002/app.45723

Ismar, E., & Sarac, A. S. (2016). Synthesis and characterization of poly (acrylonitrile-co-acrylic acid) as precursor of carbon nanofibers. *Polymers for Advanced Technologies*, *27*(10), 1383–1388. doi:10.1002/pat.3807

Ismar, E., & Sarac, A. S. (2018). Oxidation of polyacrylonitrile nanofiber webs as a precursor for carbon nanofiber: Aligned and non-aligned nanofibers. *Polymer Bulletin*, *75*(2), 485–499. doi:10.100700289-017-2043-x

Ivankovic, M., Brnardic, I., Ivankovic, H., & Mencer, H. J. (2006). DSC study of the cure kinetics during nanocomposite formation: Epoxy/poly(oxypropylene) diamine/organically modified montmorillonite system. *Journal of Applied Polymer Science*, *99*(2), 550–557. doi:10.1002/app.22488

Ivanovskii, A. L. (2012). Graphene-based and graphene-like materials. *Russian Chemical Reviews*, *81*(7), 571–605. doi:10.1070/RC2012v081n07ABEH004302

Iwahori, Y., Ishiwata, S., Sumizawa, T., & Ishikawa, T. (2005). Mechanical properties improvements in two-phase and three-phase composites using carbon nano-fiber dispersed resin. *Composites. Part A, Applied Science and Manufacturing*, *36*(10), 1430–1439. doi:10.1016/j.compositesa.2004.11.017

Jagtap, S. B., Rao, V. S., Barman, S., & Ratna, D. (2015). Nanocomposites based on epoxy resin and organoclay functionalized with a reactive modifier having structural similarity with the curing agent. *Polymer*, *63*, 41–71. doi:10.1016/j.polymer.2015.02.038

Jakubinek, M. B., Ashrafi, B., Zhang, Y., Martinez-Rubi, Y., Kingston, C. T., Johnston, A., & Simard, B. (2015). Single-walled carbon nanotube–epoxy composites for structural and conductive aerospace adhesives. *Composites. Part B, Engineering*, *69*, 87–93. doi:10.1016/j.compositesb.2014.09.022

Jang, J. S., Varischetti, J., Lee, G. W., & Suhr, J. (2011). Experimental and analytical investigation of mechanical damping and CTE of both SiO2 particle and carbon nanofiber reinforced hybrid epoxy composites. *Composites. Part A, Applied Science and Manufacturing*, *42*(1), 98–103. doi:10.1016/j.compositesa.2010.10.008

Jarosz, P., Schauerman, C., Alvarenga, J., Moses, B., Mastrangelo, T., Raffaelle, R., ... Landi, B. (2011). Carbon nanotube wires and cables: Near-term applications and future perspectives. *Nanoscale*, *3*(11), 4542–4553. doi:10.1039/c1nr10814j PMID:21984338

Jen, M. H. R., Tseng, Y. C., & Wu, C. H. (2005). Manufacturing and mechanical response of nanocomposite laminates. *Composites Science and Technology*, *65*(5), 775–779. doi:10.1016/j.compscitech.2004.10.010

Jeon, H., Park, J., & Shon, M. (2013). Corrosion protection by epoxy coating containing multi-walled carbon nanotubes. *Journal of Industrial and Engineering Chemistry*, *19*(3), 849–853. doi:10.1016/j.jiec.2012.10.030

Jia, J., Duan, Y. Y., & Wang, S. H. (2007). Preparation and characterization of antibacterial silver containing nanofibers for wound dressing applications. *J US-China Med Sci*, *22*, 452–454.

Jiang, K., Schadler, L. S., Siegel, R. W., Zhang, X., Zhang, H., & Terrones, M. (2004). *Protein immobilization on carbon nanotubes via a two-step process of diimide-activated amidation.* Academic Press. doi:10.1039/B310359E

Jin, Y., & Boyd, R. H. (1998). Subglass chain dynamics and relaxation in polyethylene: A molecular dynamics simulation study. *The Journal of Chemical Physics*, *108*(23), 9912–9923. doi:10.1063/1.476430

John, B., Nair, C. R., & Ninan, K. N. (2010). Effect of nanoclay on the mechanical, dynamic mechanical and thermal properties of cyanate ester syntactic foams. *Materials Science and Engineering A*, *527*(21-22), 5435–5443. doi:10.1016/j.msea.2010.05.016

John, B., Reghunadhan Nair, C. P., Mathew, D., & Ninan, K. N. (2008). Foam sandwich composites with cyanate ester based syntactic foam as core and carbon-cyanate ester as skin: Processing and properties. *Journal of Applied Polymer Science*, *110*(3), 1366–1374. doi:10.1002/app.28658

Jorio, A., Kauppinen, E., & Hassanien, A. (2007). *Carbon-nanotube metrology. In Carbon Nanotubes* (pp. 63–100). Springer.

Joseph, N., Varghese, J., & Sebastian, M. T. (2017). In situ polymerized polyaniline nanofiber-based functional cotton and nylon fabrics as millimeter-wave absorbers. *Polymer Journal*, *49*(4), 391–399. doi:10.1038/pj.2016.121

Joshi, M., & Chatterjee, U. (2016). Polymer nanocomposite: An advanced material for aerospace applications. Advanced Composite Materials for Aerospace Engineering, 241-264.

Jouni, M., Faure-Vincent, J., Fedorko, P., Djurado, D., Boiteux, G., & Massardier, V. (2014). Charge carrier transport and low electrical percolation threshold in multiwalled carbon nanotube polymer nanocomposites. *Carbon*, *76*, 10–18. doi:10.1016/j.carbon.2014.04.031

Journet, C., Maser, W. K., Bernier, P., Loiseau, A., de la Chapelle, M. L., Lefrant, S., ... Fischer, J. E. (1997). Large-scale production of single-walled carbon nanotubes by the electric-arc technique. *Nature*, *388*(6644), 756–758. doi:10.1038/41972

Kabir, E., Kumar, V., Kim, K. H., Yip, A. G. K., & Sohn, J. R. (2018). Environmental Impacts of Nanomaterials. *Journal of Environmental Management*, *225*, 261–271. doi:10.1016/j.jenvman.2018.07.087 PMID:30096714

Kablov, E. N., Kondrashov, S. V., & Yurkov, G. Yu. (2013). Prospects of using carbonaceous nanoparticles in binders for polymer composites. *Nanotechnologies in Russia*, *8*(3-4), 163–185. doi:10.1134/S1995078013020080

Kadhim, N., Mei, Y., Wang, Y., Li, Y., Meng, F., Jiang, M., & Zhou, Z. (2018). Remarkable Improvement in the Mechanical Properties of Epoxy Composites Achieved by a Small Amount of Modified Helical Carbon Nanotubes. *Polymers*, *10*(10), 1103. doi:10.3390/polym10101103

Kalia, S. (2010). Cryogenic Processing: A Study of Materials at Low Temperatures. *Journal of Low Temperature Physics*, *158*(5-6), 934–945. doi:10.100710909-009-0058-x

Kang, T. S., Lee, S. W., Joo, J., & Lee, J. Y. (2005). Electrically conducting polypyrrole fibers spun by electrospinning. *Synthetic Metals*, *153*(1-3), 61–64. doi:10.1016/j.synthmet.2005.07.135

Karasinski, E. N., Da Luz, M. G., Lepienski, C. M., & Coelho, L. A. F. (2013). Nanostructured coating based on epoxy/metal oxides: Kinetic curing and mechanical properties. *Thermochimica Acta*, *569*, 167–211. doi:10.1016/j.tca.2013.07.015

Karlsson, L. (2010). *Simulator testing Basic course*. Linköping: SAAB AB.

Kasagi, T., Tsutaoka, T., & Hatakeyama, K. (1999). Particle size effect on the complex permeability for perm-alloy composite materials. *IEEE Transactions on Magnetics*, *35*(5), 3424–3426. doi:10.1109/20.800545

Kaseem, M., Hamad, K., & Ko, Y. G. (2016). Fabrication and materials properties of polystyrene/carbon nanotube (PS/CNT) composites: A review. *European Polymer Journal*, *79*, 36–62. doi:10.1016/j.eurpolymj.2016.04.011

Kashi, S., Gupta, R. K., Baum, T., Kao, N., & Bhattacharya, S. N. (2016). Morphology, electromagnetic properties and electromagnetic interference shielding performance of poly lactide/graphene nanoplatelet nanocomposites. *Materials & Design*, *95*, 119–126. doi:10.1016/j.matdes.2016.01.086

Kashiwagi, T., Du, F., Winey, K. I., Groth, K. M., Shields, J. R., Bellayer, S. P., ... Douglas, J. F. (2005). Flammability properties of polymer nanocomposites with single-walled carbon nanotubes: Effects of nanotube dispersion and concentration. *Polymer*, *46*(2), 471–481. doi:10.1016/j.polymer.2004.10.087

Kashiwagi, Y., Yamamoto, M., & Nakamoto, M. (2006). Facile size-regulated synthesis of silver nanoparticles by controlled thermolysis of silver alkylcarboxylates in the presence of alkylamines with different chain lengths. *Journal of Colloid and Interface Science*, *300*(1), 169–175. doi:10.1016/j.jcis.2006.03.041

Katunin, A., Krukiewicz, K., Turczyn, R., Sul, P., Łasica, A., & Bilewicz, M. (2017). Synthesis and characterization of the electrically conductive polymeric composite for lightning strike protection of aircraft structures. *Composite Structures*, *159*, 773–783. doi:10.1016/j.compstruct.2016.10.028

Kausar, A., & Hussain, S. T. (2013). Effect of multi-walled carbon nanotube reinforcement on the physical properties of poly (thiourea-azo-ether)-based nanocomposites. *Journal of Plastic Film & Sheeting*, *29*(4), 365–383. doi:10.1177/8756087913487003

Kausar, A., Rafique, I., & Muhammad, B. (2016). Review of applications of polymer/carbon nanotubes and epoxy/CNT composites. *Polymer-Plastics Technology and Engineering*, *55*(11), 1167–1191. doi:10.1080/03602559.2016.1163588

Kawashima, S., Chaouche, M., Corr, D. J., & Shah, S. P. (2013). Rate of thixotropic rebuilding of cement pastes modified with highly purified attapulgite clays. *Cement and Concrete Research*, *53*, 112–118. doi:10.1016/j.cemconres.2013.05.019

Keith, J. M., Janda, N. B., King, J. A., Perger, W. F., & Oxby, T. J. (2005). Shielding Effectiveness Density Theory for Carbon Fiber/Nylon 6,6 Composites. *Polymer Composites*, *26*(5), 671–678. doi:10.1002/pc.20139

Kelly, A., & Tyson, A. W. (1965). Tensile properties of fibre-reinforced metals: Copper/tungsten and copper/molybdenum. *Journal of the Mechanics and Physics of Solids*, *13*(6), 329–350. doi:10.1016/0022-5096(65)90035-9

Khanafer, K. M., Al-Amiri, A. M., & Pop, I. (2007). Numerical Simulation of Unsteady Mixed Convection in a Driven Cavity using an Externally Excited Sliding Lid. *European Journal of Mechanics - B/Fluids*, *26*(5), 669–687. doi:10.1016/j.euromechflu.2006.06.006

Khanafer, K., Vafai, L., & Lightstone, M. (2003). Buoyancy-Driven Heat Transfer Enhancement in a Two-Dimensional Enclosure Utilizing Nano-Fluids. *International Journal of Heat and Mass Transfer*, *46*(19), 3639–3653. doi:10.1016/S0017-9310(03)00156-X

Khanna, P. K., Kulkarni, D., & Beri, R. K. (2008). Synthesis and characterization of myristic acid capped silver nanoparticles. *Journal of Nanoparticle Research*, *10*(6), 1059–1062. doi:10.100711051-008-9366-3

Khan, S. D., Arora, M., Puri, C., Wahab, M. A., & Saini, P. (2014). Synthesis and characterization of acrylic resin/activated carbon composites. *Indian Journal of Pure and Applied Physics*, *52*(4), 251–254.

Khan, S. D., Arora, M., Wahab, M. A., & Saini, P. (2014). *Permittivity and Electromagnetic Interference Shielding Investigations of Activated Charcoal Loaded Acrylic Coating Compositions.* Journal of Polymers. doi:10.1155/2014/193058

Khan, S. U., Li, C. Y., Siddiqui, N. A., & Kim, J. K. (2011). Vibration damping characteristics of carbon fiber-reinforced composites containing multi-walled carbon nanotubes. *Composites Science and Technology*, *71*(12), 1486–1494. doi:10.1016/j.compscitech.2011.03.022

Khan, Z. U., Kausar, A., & Ullah, H. (2016). A review on composite papers of graphene oxide, carbon nanotube, polymer/GO, and polymer/CNT: Processing strategies, properties, and relevance. *Polymer-Plastics Technology and Engineering*, *55*(6), 559–581. doi:10.1080/03602559.2015.1098693

Khozin, V. G. (2004). *Reinforcement of epoxy polymers*. Kazan: Dom pechati. (in Russian)

Kim, B. K., Seo, J. W., & Jeong, H. M. (2003). Morphology and properties of waterborne polyurethane/clay nanocomposites. *European Polymer Journal*, *39*(1), 85–91. doi:10.1016/S0014-3057(02)00173-8

Kim, H. R., Kim, B. S., & Kim, I. S. (2012). Fabrication and EMI shielding effectiveness of Ag decorated highly porous poly (vinyl alcohol) /Fe_2O_3 nanofibrous composites. *Materials Chemistry and Physics*, *135*(2-3), 1024–1029. doi:10.1016/j.matchemphys.2012.06.008

Kim, H., Abdala, A. A., & Macosko, C. W. (2010). Graphene/Polymer Nanocomposites. *Macromolecules*, *43*(16), 6515–6530. doi:10.1021/ma100572e

Kim, J. Y. (2011). Poly(butylene terephthalate) Nanocomposites Containing Carbon Nanotube. In B. Reddy (Ed.), *Advances in Nanocomposites – Synthesis, Characterization and Industrial Applications* (pp. 707–726). InTech Open Publication.

Kim, M., Kim, Y., Lee, K. M., Jeong, S. Y., Lee, E., Baeck, S. H., & Shim, S. E. (2016). Electrochemical improvement due to alignment of carbon nanofibers fabricated by electrospinning as an electrode for supercapacitor. *Carbon*, *99*, 607–618. doi:10.1016/j.carbon.2015.12.068

Kimmel, W. D., & Gerke, D. D. (1995). Controlling EMI with cable shields. *Medical Device & Diagnostic Industry*, *17*(7), 112–115.

King, J. A., Klimek, D. R., Miskioglu, I., & Odegard, G. M. (2015). Mechanical properties of graphene nanoplatelet/epoxy composites. *Journal of Composite Materials*, *49*(6), 659–668. doi:10.1177/0021998314522674

Kiselev, N. A., Moravsky, A. P., Ormont, A. B., & Zakharov, D. N. (1999). SEM and HREM study of the internal structure of nanotube rich carbon arc cathodic deposits. *Carbon*, *37*(7), 1093–1103. doi:10.1016/S0008-6223(98)00309-1

Kishi, T., & Maekawa, K. (1997). Multi-Component Model for Heat of Hydration Heating of Blended Cement with Blast Furnace Slag and Fly Ash. *Concrete Library of JSCE*, *30*(12), 125–139.

Kodgire, P. V., Bhattacharyya, A. R., Bose, S., Gupta, N., Kulkarni, A. R., & Misra, A. (2006). Control of multiwall carbon nanotubes dispersion in polyamide6 matrix: An assessment through electrical conductivity. *Chemical Physics Letters*, *432*(4-6), 480–485. doi:10.1016/j.cplett.2006.10.088

Koenig, J. L. (1999). *Spectroscopy of polymers*. Elsevier. doi:10.1016/B978-044410031-3/50005-0

Kondrashov, S. V., Merkulova, Yu. I., Marakhovskii, P. S., D'yachkova, T. P., Shashkeev, K. A., Popkov, O. V., ... Yurkov, G. Yu. (2017). Degradation of physicomechanical properties of epoxy nanocomposites with carbon nanotubes upon heat and humidity aging. *Russian Journal of Applied Chemistry*, *90*(5), 788–796. doi:10.1134/S1070427217050202

Kong, D., & Park, C. E. (2003). Real time exfoliation behavior of clay layers in epoxy-clay nanocomposites real time exfoliation behavior of clay layers in epoxy-clay nanocomposites. *Chemistry of Materials*, *15*(2), 419–424. doi:10.1021/cm0205837

Kong, Q., Luo, Z., Wang, Y., & Wang, B. (2018). Fabrication of super-stretchable and electrical conductive membrane of spandex/multi-wall carbon nanotube/reduced graphene oxide composite. *Journal of Polymer Research*, *25*(11), 231. doi:10.100710965-018-1597-0

Kortaberria, G., Sangermano, M., & Mondragon, I. (2012). In situ synthetized silver/epoxy nanocomposites: Electrical characterization in terms of dielectric relaxation spectroscopy. *Macromolecular Symposia*, *321-322*(1), 112–117. doi:10.1002/masy.201251119

Kosti, S. (2014). Numerical Study of Heat Flux Boundary in Nanofluid-Filled Cavity. *Nanomaterials and Energy*, *3*(6), 193–205. doi:10.1680/nme.14.00017

Kosti, S., Das, M. K., & Saha, A. K. (2013). Buoyancy-Driven Flow and Heat Transfer in a Nano-Fluid-Filled Enclosure. *Nanomaterials and Energy*, *2*(4), 200–211. doi:10.1680/nme.13.00013

Kosti, S., & Malvi, C. S. (2018). Cumulative Influence of Nanoparticles on MMCs Time-Temperature History Curve. *Nanomaterials and Energy*, *7*(1), 1–10. doi:10.1680/jnaen.17.00011

Kosti, S., & Pathak, P. (2018). Genetic Algorithm Based Finite Difference Simulation of Solidification Process of MMC's. *Materials Today: Proceedings*, *5*(2), 8271–8279. doi:10.1016/j.matpr.2017.11.518

Kozlov, G., & Dolbin, I. (2018). Effect of a Nanofiller Structure on the Degree of Reinforcement of Polymer–Carbon Nanotube Nanocomposites with the Use of a Percolation Model. *Journal of Applied Mechanics and Technical Physics*, *59*(4), 765–769. doi:10.1134/S0021894418040259

Kroto, H. W., Heath, J. R., O'Brien, S. C., Curl, R. F., & Smalley, R. E. (1985). C 60: Buckminsterfullerene. *Nature*, *318*(6042), 162–163. doi:10.1038/318162a0

Krüger, A. (2010). *Carbon Materials and Nanotechnology*. John Wiley & Sons. doi:10.1002/9783527629602

Krus, P. (2000). *Post optimal system analysis using aggregated design impact matrix*. 2000 ASME International Design Engineering Technical Conferences & Computers and Information in Engineering Conference, Baltimore, MD.

Krus, P. (2008). *Computational tools for aircraft system analysis and optimization*. 26th International Congress of the Aeronautical Sciences, Anchorage, AK.

Kuan, C.-F., Chen, W.-J., Li, Y.-L., Chen, C.-H., Kuan, H.-C., & Chiang, C.-L. (2010). Flame retardance and thermal stability of carbon nanotube epoxy composite prepared from sol–gel method. *Journal of Physics and Chemistry of Solids*, *71*(4), 539–543. doi:10.1016/j.jpcs.2009.12.031

Kuhn, R. A. (2008). Multi level approach for aircraft electrical systems design. *6th International MATLAB Conference*, Bielefeld, Germany.

Kumar, A., Manju, P., Krishnan S, G., & Nayak, S. K. (2017b). Physical, Mechanical, Surface, Thermal and Morphological Studies of PLA-Sepiolite Bionanocomposites. *Journal of Mineral, Metal and Material Engineering, 3*, 103-111.

Kumar, M. S., Sharma, N., & Ray, B. C. (2008). Mechanical Behavior of Glass/Epoxy Composites at Liquid Nitrogen Temperature. *Journal of Reinforced Plastics and Composites, 27*(9), 937–944. doi:10.1177/0731684407085877

Kumar, S., Dang, T. D., Arnold, F. E., Bhattacharyya, A. R., Min, B. G., Zhang, X., ... Smalley, R. E. (2002). Synthesis, structure, and properties of PBO/SWNT Composites&. *Macromolecules, 35*(24), 9039–9043. doi:10.1021/ma0205055

Kumar, S., Doshi, H., Srinivasarao, M., Park, J. O., & Schiraldi, D. A. (2002). Fibers from polypropylene/nano carbon fiber composites. *Polymer, 43*(5), 1701–1703. doi:10.1016/S0032-3861(01)00744-3

Kumar, S., Samal, S. K., Mohanty, S., & Nayak, S. K. (2017a). Synthesis and Characterization of Nanoclay-Reinforced Trifunctional "Bioresin-Modified" Epoxy Blends Enhanced with Mechanical and Thermal Properties. *ChemistrySelect, 2*(35), 11445–11455. doi:10.1002lct.201702041

Kuzub, L. I., Bogdanova, L. M., Kurkin, T. S. & Buzin, P. V. (2010). Principal aspects of the synthesis of silver nanoparticles with carboxylate ligands. *Proceeding of Articles "Structure and Dynamics of Molecular Systems"*, 340-45.

Kuzub, L. I., Bogdanova, L. M., Kurkin, T. S., Torbov, V. I., Gur'eva, L. L., Rozenberg, B. A. & Buzin, P. V. (2009). Investigation of the regularities of synthesis of monodisperse silver nanoparticles with chemically bound organic ligands. *Proceeding of Articles "Structure and Dynamics of Molecular Systems"*, 134-39.

Kuzub, L. I., Gur'eva, L. L., Grishchuk, A. A., Estrina, G. A., Estrin, Ya. I., & Badamshina, E. R. (2015). Regularities of the formation of silver nanoparticles with oligostyrylcarboxylate ligands, *Journal of Polymer Science. Ser. B, 57*(6), 608–615.

Kuzumaki, T., Miyazawa, K., Ichinose, H., & Ito, K. (1998). Processing of Carbon Nanotube Reinforced Aluminium Composites'. *Journal of Materials Research, 13*(9), 2445–2449. doi:10.1557/JMR.1998.0340

Kwon, Y.-K., Lee, Y. H., Kim, S.-G., Jund, P., Tománek, D., & Smalley, R. E. (1997). Morphology and Stability of Growing Multiwall Carbon Nanotubes. *Physical Review Letters, 79*(11), 2065–2068. doi:10.1103/PhysRevLett.79.2065

Lagarkov, A. N., & Rozanov, K. N. (2009). High-frequency behavior of magnetic composites. *Journal of Magnetism and Magnetic Materials, 321*(14), 2082–2092. doi:10.1016/j.jmmm.2008.08.099

Lai, F. H., & Yang, Y. T. (2011). Lattice Boltzmann Simulation of Natural Convection Heat Transfer of Al2O3/Water Nano-Fluids in a Square Enclosure. *International Journal of Thermal Sciences, 50*(10), 1930–1941. doi:10.1016/j.ijthermalsci.2011.04.015

LaMer, V. K., & Dinegar, R. H. (1950). Theory, production and mechanism of formation of monodispersed hydrosols. *Journal of the American Chemical Society*, *72*(11), 4847–4854. doi:10.1021/ja01167a001

Lan, X., Liu, Y., Lv, H., Wang, X., Leng, J., & Du, S. (2009). Fiber reinforced shape-memory polymer composite and its application in a deployable hinge. *Smart Materials and Structures*, *18*(2), 024002. doi:10.1088/0964-1726/18/2/024002

Larson, C., Smith, J. R., & Armstrong, G. J. (2013). Current research on surface finishing and coatings for aerospace bodies and structures–a review. *Transactions of the IMF*, *91*(3), 120-132. 10.1179/0020296713Z.000000000102

Larsson, E. (2007). *Modelling of a fuel system in MATLAB – applied to an unmanned aircraft* (Master thesis). Dept of Electrical Engineering, Linköping University.

Larsson, J., & Krus, P. (2003). Stability analysis of coupled simulation. *ASME International Mechanical Engineering Congress*, 861-868.

Lee, J., & Ramakrishna, S. (2017). Carbon Nanotube Wires and Cables: Near-Term Applications and Future Perspectives. *Nanotechnology for Energy Sustainability*, 485-506.

Lee, B. O., Woo, W. J., & Kim, M.-S. (2001). EMI Shielding Effectiveness of Carbon Nanofiber Filled Poly(vinyl alcohol) Coating Materials. *Macromolecular Materials and Engineering*, *286*(2), 114–118. doi:10.1002/1439-2054(20010201)286:2<114::AID-MAME114>3.0.CO;2-8

Lee, B. O., Woo, W. J., Park, H. S., Hahm, H. S., Wu, J. P., & Kim, M. S. (2002). Influence of aspect ratio and skin effect on EMI shielding of coating materials fabricated with carbon nanofiber/PVDF. *Journal of Materials Science*, *37*(9), 1839–1843. doi:10.1023/A:1014970528482

Lee, C., Wei, X., Kysar, J. W., & Hone, J. (2008). Measurement of the Elastic Properties and Intrinsic Strength of Monolayer Graphene. *Science*, *321*(5887), 385–388. doi:10.1126cience.1157996 PMID:18635798

Lee, H. Y., Cruz, H., & Son, Y. (2018). Effects of incorporation of polyester on the electrical resistivity of polycarbonate/multi-walled carbon nanotube nanocomposite. *Journal of Composite Materials*.

Lee, H., & Neville, K. (1967). *Handbook of epoxy resins*. New York: McGraw-Hill.

Lee, J.-H., Lee, S.-H., Choi, C. J., Jang, S. P., & Choi, S. U. S. (2010). A Review of Thermal Conductivity Data, Mechanism and Models for Nanofluids. *International Journal of Micro-Nano Scale Transport*, *1*(4), 269–322. doi:10.1260/1759-3093.1.4.269

Lee, Y.-H., Kim, J.-S., Noh, J., Lee, I., Kim, H. J., Choi, S., ... Choi, J. W. (2013). Wearable Textile Battery Rechargeable by Solar Energy. *Nano Letters*, *13*(11), 5753–5761. doi:10.1021/nl403860k PMID:24164580

Leng, J., & Lau, A. K.-t. (2010). *Multifunctional polymer nanocomposites*. CRC Press. doi:10.1201/b10462

Leng, J., Wu, X., & Liu, Y. (2009). Infrared light-active shape memory polymer filled with nanocarbon particles. *Journal of Applied Polymer Science*, *114*(4), 2455–2460. doi:10.1002/app.30724

Levchik, S. V., & Weil, E. D. (2004). Thermal decomposition, combustion and flame-retardancy of epoxy resins—A review of the recent literature. *Polymer International*, *53*(12), 1901–1929. doi:10.1002/pi.1473

Liang, X., Chang, A. S. P., Zhang, Y., Harteneck, B. D., Choo, H., Olynick, D. L., & Cabrini, S. (2009). Electrostatic Force Assisted Exfoliation of Prepatterned Few-Layer Graphenes into Device Sites. *Nano Letters*, *9*(1), 467–472. doi:10.1021/nl803512z PMID:19072062

Liang, Y. L., & Pearson, R. A. (2010). The toughening mechanism in hybrid epoxy-silica-rubber nanocomposites (HESRNs). *Polymer*, *51*(21), 4880–4890. doi:10.1016/j.polymer.2010.08.052

Liang, Z., Gou, J., Zhang, C., Wang, B., & Kramer, L. (2004). Investigation of molecular interactions between (10, 10) single-walled nanotube and Epon 862 resin/DETDA curing agent molecules. *Materials Science and Engineering A*, *365*(1–2), 228–234. doi:10.1016/j.msea.2003.09.032

Li, D., & Xia, Y. (2004). Electrospinning of Nanofibers: Reinventing the Wheel? *Advanced Materials*, *16*(14), 1151–1170. doi:10.1002/adma.200400719

Lifshits, M., & Slezov, V. V. (1959). Kinetics of diffusive decomposition of supersaturated solid solutions. *Journal of Experimental and Theoretical Physics*, *35*(2), 479–492.

Li, J., Liu, H., Guo, J., Hu, Z., Wang, Z., Wang, B., ... Guo, Z. (2017). Flexible, conductive, porous, fibrillar polymer– gold nanocomposites with enhanced electromagnetic interference shielding and mechanical properties. *Journal of Materials Chemistry. C, Materials for Optical and Electronic Devices*, *5*(5), 1095–1105. doi:10.1039/C6TC04780G

Li, K. Y., Kuan, C. F., Kuan, H. C., Chen, C. H., Shen, M. Y., Yang, J. M., & Chiang, C. L. (2014). Preparation and properties of novel epoxy/graphene oxide nanosheets (GON) composites functionalized with flame retardant containing phosphorus and silicon. *Materials Chemistry and Physics*, *146*(3), 354–362. doi:10.1016/j.matchemphys.2014.03.037

Li, L., & Chung, D. D. L. (1991). Electrically conducting powder filled polyimidesiloxane. *Composites*, *22*(3), 211–218. doi:10.1016/0010-4361(91)90321-7

Li, L., & Chung, D. D. L. (1993). Effect of viscosity on the electrical properties of conducting thermoplastic composites made by compression molding of a powder mixture. *Polymer Composites*, *14*(6), 467–472. doi:10.1002/pc.750140604

Li, L., & Chung, D. D. L. (1994). Electrical and mechanical properties of electrically conductive polyethersulfone composites. *Composites*, *25*(3), 215–224. doi:10.1016/0010-4361(94)90019-1

Li, L., Zeng, Z., Zou, H., & Liang, M. (2015). Curing characteristics of an epoxy resin in the presence of functional graphite oxide with amine-rich surface. *Thermochimica Acta*, *614*, 76–102. doi:10.1016/j.tca.2015.06.006

Li, L., Zou, H., Liang, M., & Chen, Y. (2014). Study on the effect of poly(oxypropylene)diamine modified organic montmorillonite on curing kinetics of epoxy nanocomposites. *Thermochimica Acta*, *597*, 93–100. doi:10.1016/j.tca.2014.10.008

Lim, C.-S., Guzman, M., Schaefer, J., & Minaie, B. (2013). Fabrication and properties of dense thin films containing functionalized carbon nanofibers. *Thin Solid Films*, *534*, 111–115. doi:10.1016/j.tsf.2013.02.010

Li, N., Huang, Y., Du, F., He, X., Lin, X., Gao, H., ... Eklund, P. C. (2006). Electromagnetic Interference (EMI) shielding of single-walled carbon nanotube epoxy composites. *Nano Letters*, *6*(6), 1141–1145. doi:10.1021/nl0602589 PMID:16771569

Lin, C., & Chung, D. D. L. (2008). Nanoclay paste as a thermal interface material for smooth surfaces. *Journal of Electronic Materials*, *37*(11), 1698–1709. doi:10.100711664-008-0538-9

Lin, D., Tian, X., Wu, F., & Xing, B. (2010). Fate and transport of engineered nanomaterials in the environment. *Journal of Environmental Quality*, *39*(6), 1896–1908. doi:10.2134/jeq2009.0423 PMID:21284287

Lind, I., & Andersson, H. (2011). Model Based Systems Engineering for Aircraft Systems – How does MATLAB Based Tools Fit? *8th International MATLAB Conference*, Dresden, Germany.

Ling, J., Zhai, W., Feng, W., Shen, B., Zhang, J., & Zheng, W. G. (2013). Facile Preparation of Lightweight Microcellular Polyetherimide/Graphene Composite Foams for Electromagnetic Interference Shielding. *ACS Applied Materials & Interfaces*, *5*(7), 2677–2684. doi:10.1021/am303289m PMID:23465462

Lin, Y., Hill, D. E., Bentley, J., Allard, L. F., & Sun, Y. P. (2003). Characterization of Functionalized Single-Walled Carbon Nanotubes at Individual Nanotube-Thin Bundle Level. *The Journal of Physical Chemistry B*, *107*(38), 10453–10457. doi:10.1021/jp0306141

Liscouet-Hanke, S. (2008). *A model-based methodology for integrated preliminary sizing and analysis of aircraft power system architectures*. Institut National des Sciences Appliquées de Toulouse.

Liu, F., Wang, H., Xue, L., Fan, L., & Zhu, Z. (2008). Effect of microstructure on the mechanical properties of PAN-based carbon fibers during high-temperature graphitization. *Journal of Materials Science*, *43*(12), 4316–4322. doi:10.100710853-008-2633-y

Liu, J. M., & Yang, S. C. (1991). Novel colloidal polyaniline fibrils made by template guided chemical polymerization. *Journal of the Chemical Society. Chemical Communications*, (21): 1529–1531. doi:10.1039/c39910001529

Liu, W., Hoa, S. V., & Pugh, M. (2005). Fracture toughness and water uptake of high-performance epoxy/nanoclay nanocomposites. *Composites Science and Technology*, *65*(15-16), 2364–2373. doi:10.1016/j.compscitech.2005.06.007

Liu, Y., Du, H., Liu, L., & Leng, J. (2014). Shape memory polymers and their composites in aerospace applications: A review. *Smart Materials and Structures*, *23*(2), 023001. doi:10.1088/0964-1726/23/2/023001

Liu, Y., Teng, H., Hou, H., & You, T. (2009). Nonenzymatic glucose sensor based on renewable electrospun Ni nanoparticle-loaded carbon nanofiber paste electrode. *Biosensors & Bioelectronics*, *24*(11), 3329–3334. doi:10.1016/j.bios.2009.04.032 PMID:19450966

Li, W. Z., Xie, S. S., Qian, L. X., Chang, B. H., Zou, B. S., Zhou, W. Y., ... Wang, G. (1701–1703). ... Wang, G. (1996). Large-Scale Synthesis of Aligned Carbon Nanotubes. *Science*, *274*(5293), 1701–1703. doi:10.1126cience.274.5293.1701

Li, Y., Chen, C., Zhang, S., Ni, Y., & Huang, J. (2008). Electrical conductivity and electromagnetic interference shielding characteristics of multiwalled carbon nanotube filled polyacrylate composite films. *Applied Surface Science*, *254*(18), 5766–5771. doi:10.1016/j.apsusc.2008.03.077

Liying, Z., Lin, B. W., Kye, Y. S., & Ma, J. (2013). Effect of carbon nanofiber reinforcement on electromagnetic interference shielding effectiveness of syntactic foam. *Journal of Materials Science*, *48*(21), 7757–7763. doi:10.100710853-013-7597-x

Ljung, L., & Glad, T. (1991). *Modellbygge och simulering*. Studentlitteratur.

Llobet, E. (2013). Gas sensors using carbon nanomaterials: A review. *Sensors and Actuators. B, Chemical*, *179*, 32–45. doi:10.1016/j.snb.2012.11.014

Logakis, E., & Skordos, A. (2012). *Lightning strike performance of carbon nanotube loaded aerospace composites.* Paper presented at the 15th European Conference on Composite Materials.

Lonjon, A., Demont, P., Dantras, E., & Lacabanne, C. (2012). Electrical conductivity improvement of aeronautical carbon fiber reinforced polyepoxy composites by insertion of carbon nanotubes. *Journal of Non-Crystalline Solids*, *358*(15), 1859–1862. doi:10.1016/j.jnoncrysol.2012.05.038

Lourie, O., & Wagner, H. D. (1998). Evaluation of Young's modulus of carbon nanotubes by micro-Raman spectroscopy. *Journal of Materials Research*, *13*(9), 2418–2422. doi:10.1557/JMR.1998.0336

Lubineau, G., & Rahaman, A. (2012). A review of strategies for improving the degradation properties of laminated continuous-fiber/epoxy composites with carbon-based nanoreinforcements. *Carbon*, *50*(7), 2377–2395. doi:10.1016/j.carbon.2012.01.059

Lu, J., Moon, K. S., & Wong, C. P. (2008). Silver/polymer nanocomposite as a high-k polymer matrix for dielectric composites with improved dielectric performance. *Journal of Materials Chemistry*, *18*(40), 4821–4826. doi:10.1039/b807566b

Luo, J. J., & Daniel, I. M. (2003). Characterization and modeling of mechanical behavior of polymer/clay nanocomposites. *Composites Science and Technology*, *63*(11), 1607–1616. doi:10.1016/S0266-3538(03)00060-5

Lu, S., Qu, R., & Forcada, J. (2009). Preparation of Magnetic Polymeric Composite Nanoparticles by Seeded Emulsion Polymerization. *Materials Letters*, *63*(9-10), 770–772. doi:10.1016/j.matlet.2008.12.045

Lu, Y., & Chen, W. (2012). Sub-nanometre sized metal clusters: From synthetic challenges to the unique property discoveries. *Chemical Society Reviews*, *41*(9), 3594–3624. doi:10.1039/c2cs15325d

Maharsia, R., Gupta, N., & Jerro, H. D. (2006). Investigation of flexural strength properties of rubber and nanoclay reinforced hybrid syntactic foams. *Materials Science and Engineering A*, *417*(1-2), 249–258. doi:10.1016/j.msea.2005.10.063

Mahmood, N., Islam, M., Hameed, A., & Saeed, S. (2013). Polyamide 6/Multiwalled Carbon Nanotubes Nanocomposites with Modified Morphology and Thermal Properties. *Polymers*, *5*(4), 1380–1391. doi:10.3390/polym5041380

Mahmood, N., Islam, M., Hameed, A., Saeed, S., & Khan, A. K. (2013). Polyamide-6-based composites reinforced with pristine or functionalized multi-walled carbon nanotubes produced using melt extrusion technique. *Journal of Composite Materials*, *48*(10), 1197–1207. doi:10.1177/0021998313484779

Mahrholz, T., Mosch, J., Ro¨stermundt, D., Riedel, U., & Herbeck, L. (2003). *Materials for Aerospace Applications*. 20th AAAF Colloquium, Paris, France.

Malak, R. J., Jr. (2005). *A Framework for Validating Reusable Behavioral Models in Engineering Design* (Master's Thesis). G.W. Woodruff School of Mechanical Engineering, Georgia Institute of Technology, Atlanta, GA.

Malak, R. J., & Predis, C. J. J. (2007). Validating behavioral models for reuse. *Research in Engineering Design*, *2007*(18), 111–128. doi:10.100700163-007-0031-0

Mallick, P. K. (2007). *Fiber-reinforced composites: materials, manufacturing, and design*. CRC Press. doi:10.1201/9781420005981

Manchado, M. L., Valentini, L., Biagiotti, J., & Kenny, J. (2005). Thermal and mechanical properties of single-walled carbon nanotubes–polypropylene composites prepared by melt processing. *Carbon*, *43*(7), 1499–1505. doi:10.1016/j.carbon.2005.01.031

Mandich, N. V. (1994). EMI shielding by electroless planting of ABSplastics. *Plating and Surface Finishing*, *81*(10), 60–63.

Ma, P.-C., Siddiqui, N. A., Marom, G., & Kim, J.-K. (2010). Dispersion and functionalization of carbon nanotubes for polymer-based nanocomposites: A review. *Composites. Part A, Applied Science and Manufacturing*, *41*(10), 1345–1367. doi:10.1016/j.compositesa.2010.07.003

Ma, Q., Yu, W., Dong, X., Wang, J., Liu, G., & Xu, J. (2012). Electrospinning preparation and properties of Fe_3O_4/Eu(BA)$_3$ phen /PVP magnetic-photoluminescent bifunctional composite nanofibers. *Journal of Nanoparticle Research*, *14*(10), 1203–1209. doi:10.100711051-012-1203-z

Markstein, H. W. (1995). Effective shielding defeats EMI. Electronic. *Packaging & Production, 35*(2), 4.

Masciangioli, T., & Zhang, W. X. (2003). Environment Technologies at the Nano-scale. *Environmental Science & Technology, 37*(5), 102A–109A. doi:10.1021/es0323998 PMID:12666906

Matějka, L., Dukh, O., & Kolařík, J. (2000). Reinforcement of crosslinked rubbery epoxies by in-situ formed silica. *Polymer, 41*(4), 1449–1459. doi:10.1016/S0032-3861(99)00317-1

Mauro, M., Acocella, M. R., Corcione, C. E., Maffezzoli, A., & Guerra, G. (2014). Catalytic activity of graphite-based nanofillers on cure reaction of epoxy resins. *Polymer, 55*(22), 5612–5615. doi:10.1016/j.polymer.2014.09.019

Mavris, D. N., & DeLaurentis, D. A. (2000). A probabilistic approach for examining aircraft concept feasibility and viability. *Aircraft Design, 3*(2), 79–101. doi:10.1016/S1369-8869(00)00008-2

Maxwell, J. C. (1873). *A Treatise on Electricity and Magnetism* (1st ed.; Vol. 1). Clarendoon Press.

McCombs, J. B., Golland, D. I., & Mayer, G. (1974). The effects of crystallite orientation and size on the strength of polycrystals. *Materials Science and Engineering, 15*(2-3), 275–282. doi:10.1016/0025-5416(74)90061-5

McDonough, J. R., Choi, J. W., Yang, Y., La Mantia, F., Zhang, Y., & Cui, Y. (2009). Carbon nanofiber supercapacitors with large areal capacitances. *Applied Physics Letters, 95*(24), 243109. doi:10.1063/1.3273864

Media Innovation Studio. (2018, August 7). Retrieved August 7, 2018, from http://mediainnovationstudio.org/graphene-uav-first-for-uclan-and-university-of-manchester/

Meenakshi, K. S., & Sudhan, E. P. J. (2016). Development of novel TGDDM epoxy nanocomposites for aerospace and high performance applications – Study of their thermal and electrical behavior. *Arabian Journal of Chemistry, 9*(1), 79–85. doi:10.1016/j.arabjc.2011.05.020

Meincke, O., Kaempfer, D., Weickmann, H., Friedrich, C., Vathauer, M., & Warth, H. (2004). Mechanical properties and electrical conductivity of carbon-nanotube filled polyamide-6 and its blends with acrylonitrile/butadiene/styrene. *Polymer, 45*(3), 739–748. doi:10.1016/j.polymer.2003.12.013

Meng, H., Sui, G. X., Xie, G. Y., & Yang, R. (2009). Friction and wear behavior of carbon nanotubes reinforced polyamide 6 composites under dry sliding and water lubricated condition. *Composites Science and Technology, 69*(5), 606–611. doi:10.1016/j.compscitech.2008.12.004

Messersmith, P. B., & Giannelis, E. P. (1995). Synthesis and barrier properties of poly (ε-caprolactone)-layered silicate nanocomposites. *Journal of Polymer Science. Part A, Polymer Chemistry, 33*(7), 1047–1057. doi:10.1002/pola.1995.080330707

Micheli, D., Apollo, C., Pastore, R., & Marchetti, M. (2010). X-Band microwave characterization of carbon-based nanocomposite material, absorption capability comparison and RAS design simulation. *Composites Science and Technology*, *70*(2), 400–409. doi:10.1016/j.compscitech.2009.11.015

Mihajlovic, M., Mihajlovic, M., Dankers, P. Y., Masereeuw, R., & Sijbesma, R. P. (2018). Carbon Nanotube Reinforced Supramolecular Hydrogels for Bioapplications. *Macromolecular Bioscience*, 1800173. PMID:30085403

Min, C., Shen, X., Shi, Z., Chen, L., & Xu, Z. (2010). The electrical properties and conducting mechanisms of carbon nanotube/polymer nanocomposites: A review. *Polymer-Plastics Technology and Engineering*, *49*(12), 1172–1181. doi:10.1080/03602559.2010.496405

Ming, P., Song, Z., Gong, S., Zhang, Y., Duan, J., Zhang, Q., ... Cheng, Q. (2015). Nacre-inspired integrated nanocomposites with fire retardant properties by graphene oxide and montmorillonite. *Journal of Materials Chemistry. A, Materials for Energy and Sustainability*, *3*(42), 21194–21200. doi:10.1039/C5TA05742F

Mirabedini, S. M., & Kiamanesh, A. (2013). The effect of micro and nano-sized particles on mechanical and adhesion properties of a clear polyester powder coating. *Progress in Organic Coatings*, *76*(11), 1625–1632. doi:10.1016/j.porgcoat.2013.07.009

Miyagawa, H., Rich, M. J., & Drzal, L. T. (2006). Thermo-physical properties of epoxy nanocomposites reinforced by carbon nanotubes and vapor grown carbon fibers. *Thermochimica Acta*, *442*(1), 67–73. doi:10.1016/j.tca.2006.01.016

Miyauchi, M., Simmons, T. J., Miao, J., Gagner, J. E., Shriver, Z. H., Aich, U., ... Linhardt, R. J. (2011). Electrospun polyvinylpyrrolidone fibers with high concentrations of ferromagnetic and superparamagnetic nanoparticles. *ACS Applied Materials & Interfaces*, *3*(6), 1958–1964. doi:10.1021/am200187x PMID:21561090

Monetta, T., Acquesta, A., & Bellucci, F. (2015). Graphene/epoxy coating as multifunctional material for aircraft structures. *Aerospace*, *2*(3), 423–434. doi:10.3390/aerospace2030423

Moniruzzaman, M., & Winey, K. I. (2006). Polymer nanocomposites containing carbon nanotubes. *Macromolecules*, *39*(16), 5194–5205. doi:10.1021/ma060733p

Montserrat, S., Román, F. M., Hutchinson, J., & Campos, L. (2008). Analysis of the cure of epoxy based layered silicate nanocomposites: Reaction kinetics and nanostructure development. *Journal of Applied Polymer Science*, *108*(2), 923–938. doi:10.1002/app.27297

Moran, J. I., Alvarez, V. A., Cyras, V. P., & Vázquez, A. (2008). Extraction of cellulose and preparation of nanocellulose from sisal fibers. *Cellulose (London, England)*, *15*(1), 149–159. doi:10.100710570-007-9145-9

Morant, C., Andrey, J., Prieto, P., Mendiola, D., Sanz, J. M., & Elizalde, E. (2006). XPS characterization of nitrogen-doped carbon nanotubes. *Physica Status Solidi. A, Applications and Materials Science*, *203*(6), 1069–1075. doi:10.1002/pssa.200566110

Moravsky, A., Wexler, E., & Loutfy, R. (2004). Growth of Carbon Nanotubes by Arc Discharge and Laser Ablation. In *Carbon Nanotubes* (Vol. 1–0, pp. 65–97). CRC Press. doi:10.1201/9780203494936.ch3

Mordkovich, V. Z. (2003). Carbon Nanofibers: A New Ultrahigh-Strength Material for Chemical Technology. *Theoretical Foundations of Chemical Engineering*, *37*(5), 429–438. doi:10.1023/A:1026082323244

Morgan, N. (1991). *Corrosion inhibiting EMI/RFI shielding coating and method of its use.* EP Patent application EP19,900,314,408.

Morgan, P. (2005). *Carbon Fibers and Their Composites.* Taylor & Francis. doi:10.1201/9781420028744

Moriche, R., Jiménez-Suárez, A., Sánchez, M., Prolongo, S. G., & Ureña, A. (2018). Sensitivity, influence of the strain rate and reversibility of GNPs based multiscale composite materials for high sensitive strain sensors. *Composites Science and Technology*, *155*, 100–107. doi:10.1016/j.compscitech.2017.12.002

Motozuka, S., Tagaya, M., Hotta, Y., Morinaga, M., Ikoma, T., Honma, T., ... Tanaka, J. (2013). Mechanochemical Fabrication of Carbon Fiber/Nylon-6 Composites with Interfacial Bondings. *Industrial & Engineering Chemistry Research*, *52*(5), 2182–2189. doi:10.1021/ie301333n

Moustafa, A. F. (2017). Isothermal Reduction Process and Kinetic of Nanomaterials in Reducing Atmosphere: A review. *Journal of Analytical and Applied Pyrolysis*, *127*, 126–139. doi:10.1016/j.jaap.2017.08.015

Mubarak, M. M. (2009). *Shedding the Extra Pounds: Weight Reduction Strategies for Fuel Efficient Cars.* A Frost and Sullivan Market Survey report.

Muessig, P. R., Laack, D. R., & Wrobleski, J. J. (2000). An integrated approach to evaluating simulation credibility. *Proceedings of the 2000 Summer Computer Simulation Conference.*

Murr, L. E., Gaytan, S. M., Ramirez, D. A., Martinez, E., Hernandez, J., Amato, K. N., ... Wicker, R. B. (2012). Metal fabrication by additive manufacturing using laser and electron beam melting technologies. *Journal of Materials Science and Technology*, *28*(1), 1–14. doi:10.1016/S1005-0302(12)60016-4

Nagasawa, C., Kumagai, Y., Urabe, K., & Shinagawa, S. (1999). Electromagnetic shielding particleboard with nickel-plated wood particles. *Journal of Porous Materials*, *6*(3), 247–254. doi:10.1023/A:1009692232398

Nakamoto, M., Kashiwagi, Y., & Yamamoto, M. (2005). Synthesis and size regulation of gold nanoparticles by controlled thermolysis of ammonium gold(I) thiolate in the absence or presence of amines. *Inorganica Chimica Acta*, *358*(14), 4229–4236. doi:10.1016/j.ica.2005.03.037

Namsaeng, J., Punyodom, W., & Worajittiphon, P. (2018). Synergistic effect of welding electrospun fibers and MWCNT reinforcement on strength enhancement of PAN–PVC non-woven mats for water filtration. *Chemical Engineering Science.*

Nan, L. I., & Van der Meer, R. (1986). *Synergistic effect of metal flake and metal or metal coated fiber on EMI shielding effectiveness of thermoplastics*. US Patent 4,566,990.

NASA-STD-7009. Standard for models and simulations, National Aeronautics and Space Administration Washington, DC 20546-0001, November 7th, 2008, available online at http://standards.nasa.gov

Nasouri, K., & Shoushtari, A. M. (2017). Fabrication of magnetite nanoparticles/Polyvinylpyrrolidone composite nanofibers and their application as Electromagnetic interference shielding material. *Journal of Thermoplastic Composite Materials*, *31*(4), 431–446. doi:10.1177/0892705717704488

Nasouri, K., Shoushtari, A. M., & Mojtahedi, M. R. M. (2015). Thermodynamic studies on polyvinylpyrrolidone solution systems used for fabrication of electrospun nanostructures: Effects of the solvent. *Advances in Polymer Technology*, *34*(3), 21495–21502. doi:10.1002/adv.21495

Nasouri, K., Shoushtari, A. M., & Mojtahedi, M. R. M. (2016). Theoretical and experimental studies on EMI shielding mechanisms of multi-walled carbon nanotubes reinforced high performance composite nanofibers. *Journal of Polymer Research*, *23*(4), 71–78. doi:10.100710965-016-0943-3

National Science Foundation. (1996). Research Opportunities in Engineering Design. NSF Strategic Planning Workshop Final Report (NSF Grant DMI-9521590).

Navalon, S., Dhakshinamoorthy, A., Alvaro, M., & Garcia, H. (2014). Carbocatalysis by graphene-based materials. *Chemical Reviews*, *114*(12), 6179–6212. doi:10.1021/cr4007347

Navarro, E., Baun, A., Behra, R., Hartmann, N. B., Filser, J., Miao, A.-J., ... Sigg, L. (2008). Environmental behavior and ecotoxicity of engineered nanoparticles to algae, plants, and fungi. *Ecotoxicology (London, England)*, *17*(5), 372–386. doi:10.100710646-008-0214-0 PMID:18461442

Nayak, L., Khastgir, D., & Chaki, T. K. (2013). A mechanistic study on electromagnetic shielding effectiveness of polysulfone/carbon nanofibers nanocomposites. *Journal of Materials Science*, *48*(4), 1492–1502. doi:10.100710853-012-6904-2

Neelamkavil, F. (1987). *Computer Simulation, and Modeling*. John Wiley & Sons Inc.

Ni, Y., Chen, L., Teng, K., Shi, J., Qian, X., Xu, Z., ... Ma, M. (2015). Superior Mechanical Properties of Epoxy Composites Reinforced by 3D Interconnected Graphene Skeleton. *ACS Applied Materials & Interfaces*, *7*(21), 11583–11591. doi:10.1021/acsami.5b02552

Ni, Z. H., Wang, H. M., Kasim, J., Fan, H. M., Yu, T., Wu, Y. H., ... Shen, Z. X. (2007). Graphene Thickness Determination Using Reflection and Contrast Spectroscopy. *Nano Letters*, *7*(9), 2758–2763. doi:10.1021/nl071254m PMID:17655269

Ni, Z., Wang, Y., Yu, T., & Shen, Z. (2008). Raman spectroscopy and imaging of graphene. *Nano Research*, *1*(4), 273–291. doi:10.100712274-008-8036-1

Njuguna, J., & Pielichowski, K. (2003). Polymer nanocomposites for aerospace applications: Properties. *Advanced Engineering Materials*, *5*(11), 769–778. doi:10.1002/adem.200310101

Njuguna, J., & Pielichowski, K. (2004). Polymer Nanocomposites for Aerospace Applications: Fabrication. *Advanced Engineering Materials*, *6*(4), 193–203. doi:10.1002/adem.200305111

Njuguna, J., Pielichowski, K., & Fan, J. (2012). Polymer nanocomposites for aerospace applications. In F. Gao (Ed.), *Advances in Polymer Nanocomposites* (pp. 472–539). Woodhead Publishing. doi:10.1533/9780857096241.3.472

Nolan, P. E., Lynch, D. C., & Cutler, A. H. (1998). Carbon Deposition and Hydrocarbon Formation on Group VIII Metal Catalysts. *The Journal of Physical Chemistry B*, *102*(21), 4165–4175. doi:10.1021/jp980996o

Novikov, G. F., Rabenok, E. V., Bogdanova, L. M., & Irzhak, V. I. (2017a). Temperature dependence of direct current conductivity in Ag-ED20 nanocomposite films, *Russian. The Journal of Physical Chemistry A*, *91*(10), 1971–1975.

Novikov, G. F., Rabenok, E. V., Bogdanova, L. M., & Irzhak, V. I. (2017b). Dielectric properties of films of Ag-ED20 epoxy nanocomposite synthesized in situ. Temperature dependence of direct current conductivity. *Journal of Polymer Science: Part A, General Papers*, *59*(5), 741–750.

Novoselov, K. S., Geim, A. K., Morozov, S. V., Jiang, D., Zhang, Y., Dubonos, S. V., ... Firsov, A. A. (2004). Electric Field Effect in Atomically Thin Carbon Films. *Science*, *306*(5696), 666–669. doi:10.1126cience.1102896 PMID:15499015

O'connell, M. J. (2006). *Carbon nanotubes: properties and applications*. CRC Press. doi:10.1201/9781420004212

O'Donnell, S., Sprong, K., & Haltli, B. (2004). Potential impact of carbon nanotube reinforced polymer composite on commercial aircraft performance and economics. In *AIAA 4th Aviation Technology, Integration and Operations (ATIO) Forum* (p. 6402). AIAA. 10.2514/6.2004-6402

O'Donnell, S., Sprong, K., & Haltli, B. (2004). Potential impact of carbon nanotube reinforced polymer composite on commercial aircraft performance and economics. In *AIAA fourth aviation technology, intergration and operations (ATIO) forum*. Chicago, IL: AIAA. Project no. 02MSR055-F3.

O'Shea, P. (1998). How to meet the shielding needs of a 500-MHz PC. *Evaluation Engineer*, *37*(6), 40–46.

Oberkampf & DeLand. (2002). Error and uncertainty in modeling and simulation. Reliability Engineering & System Safety, 75(3), 333-357.

Obraztsov, A. N., Obraztsova, E. A., Tyurnina, A. V., & Zolotukhin, A. A. (2007). Chemical vapor deposition of thin graphite films of nanometer thickness. *Carbon*, *45*(10), 2017–2021. doi:10.1016/j.carbon.2007.05.028

Ogasawara, T., Ishida, Y., Ishikawa, T., & Yokota, R. (2004). Characterization of multi-walled carbon nanotube/phenylethynyl terminated polyimide composites. *Composites. Part A, Applied Science and Manufacturing*, *35*(1), 67–74. doi:10.1016/j.compositesa.2003.09.003

Ogasawara, T., Moon, S.-Y., Inoue, Y., & Shimamura, Y. (2011). Mechanical properties of aligned multi-walled carbon nanotube/epoxy composites processed using a hot-melt prepreg method. *Composites Science and Technology*, *71*(16), 1826–1833. doi:10.1016/j.compscitech.2011.08.009

Okabe, T., Takehara, T., Inose, K., Hirano, N., Nishikawa, M., & Uehara, T. (2013). Curing reaction of epoxy resin composed of mixed base resin and curing agent: Experiments and molecular simulation. *Polymer*, *54*(17), 4660–4668. doi:10.1016/j.polymer.2013.06.026

Olmedo, L., Hourquebie, P., & Jousse, F. (1997). *Handbook of Organic Conductive Molecules and Polymers* (Vol. 2). Chichester, UK: John Wiley & Sons.

Omrani, A., Rostami, A. A., Ravari, F., & Mashak, A. (2011). Curing behavior and structure of a novel nanocomposite from glycerol diglycidyl ether and 3,3-dimethylglutaric anhydride. *Thermochimica Acta*, *517*(1–2), 9–15. doi:10.1016/j.tca.2011.01.024

Otaegi, I., Aramburu, N., Müller, A., & Guerrica-Echevarría, G. (2018). Novel Biobased Polyamide 410/Polyamide 6/CNT Nanocomposites. *Polymers*, *10*(9), 986. doi:10.3390/polym10090986

Ott, H. W. (2009). *Electromagnetic Compatibility Engineering*. Hoboken, NJ: John Wiley & Sons. doi:10.1002/9780470508510

Paiva, L. B., de Morales, A. R., & Valenzuela Diaz, F. R. (2008). Organoclays: Properties, preparation and application. *Applied Clay Science*, *42*(1–2), 8–24. doi:10.1016/j.clay.2008.02.006

Pande, K. N., & Peshwe, D. R. (2015). *Effectiveness of Optimized Cryo-Treatment on Polymer Composites*. Patent no. 1724/MUM/2014 A.

Pande, K. N. (2012). *Effect of Cryogenic Treatment on Polymers, Composites and Blends*. Nagpur: Visvesvaraya National Institute of Technology.

Pande, K. N., Peshwe, D. R., & Kumar, A. (2012). Effect of the Cryogenic Treatment on Polyamide and Optimization of Its Parameters for the Enhancement of Wear Performance. *Transactions of the Indian Institute of Metals*, *65*(3), 313–319. doi:10.100712666-012-0135-8

Pande, S., Singh, B. P., Mathur, R. B., Dhami, T. L., Saini, P., & Dhawan, S. K. (2009). Improved Electromagnetic Interference Shielding Properties of MWCNT–PMMA Composites Using Layered Structures. *Nanoscale Research Letters*, *4*(4), 327–334. doi:10.100711671-008-9246-x PMID:20596500

Pandey, J. K., Ahn, S., Lee, C. S., Mohanty, A. K., & Misra, M. (2010). Recent advances in the application of natural fiber based composites. *Macromolecular Materials and Engineering*, *295*(11), 975–989. doi:10.1002/mame.201000095

Panels, J. E., Lee, J. W., Park, K. Y., Kang, S. Y., Marquez, M., Wiesner, U., & Joo, Y. L. (2008). Synthesis and characterization of magnetically active carbon nanofiber/iron oxide composites with hierarchical pore structure. *Nanotechnology*, *19*(45), 455612–455618. doi:10.1088/0957-4484/19/45/455612 PMID:21832789

Paquin, A. M. (1958). *Epoxydverbindungen und Epoxydharze*. Berlin: Springer. doi:10.1007/978-3-642-86616-6

Park, H. S., Yang, I. M., Wu, J. P., Kim, M.-S., Hahm, H.-S., Kim, S.-K., & Rhee, H.-W. (2001). Synthesis of siliconeacrylic resins and their applications to superweatherable coatings. *Journal of Applied Polymer Science*, *81*(7), 1614–1623. doi:10.1002/app.1592

Park, J. K., & Kim, D. S. (2014a). Effects of an aminosilane and a tetra-functional epoxy on the physical properties of di-functional epoxy/graphene nanoplatelets nanocomposites. *Polymer Engineering and Science*, *54*(4), 969–976. doi:10.1002/pen.23624

Park, J. K., & Kim, D. S. (2014b). Preparation and physical properties of an epoxy nanocomposite with amine-functionalized graphenes. *Polymer Engineering and Science*, *54*(5), 985–991. doi:10.1002/pen.23368

Park, S. J., & Kim, H. C. (2001). Thermal stability and toughening of epoxy resin with polysulfone resin. *Journal of Polymer Science. Part B, Polymer Physics*, *39*(1), 121–128. doi:10.1002/1099-0488(20010101)39:1<121::AID-POLB110>3.0.CO;2-N

Patton, R. D., Pittman, J. C. U. Jr, Wang, L., & Hill, J. R. (1999). Vapor grown carbon fiber composites with epoxy and poly(phenylene sulfide) matrices. *Composites. Part A, Applied Science and Manufacturing*, *30*(9), 1081–1091. doi:10.1016/S1359-835X(99)00018-4

Pavia, D. L., Lampman, G. M., Kriz, G. S., & Vyvyan, J. R. (2009). *Introduction to spectroscopy*. Cengage Learning.

Pavlidou, S., & Papaspyrides, C. D. (2008). A review on polymer–layered silicate nanocomposites. *Progress in Polymer Science*, *33*(12), 1119–1198. doi:10.1016/j.progpolymsci.2008.07.008

Paynter, H. M. (1961). *Analysis and Design of Engineering Systems*. MIT Press.

Pei, S., & Cheng, H.-M. (2012). The reduction of graphene oxide. *Carbon*, *50*(9), 3210–3228. doi:10.1016/j.carbon.2011.11.010

Pervin, F., Zhou, Y., Rangari, V. K., & Jeelani, S. (2005). Testing and evaluation on the thermal and mechanical properties of carbon nano fiber reinforced SC-15 epoxy. *Materials Science and Engineering A*, *405*(1), 246–253. doi:10.1016/j.msea.2005.06.012

Podsiadlo, P., Shim, B. S., & Kotov, N. A. (2009). Polymer/clay and polymer/carbon nanotube hybrid organic–inorganic multilayered composites made by sequential layering of nanometer scale films. *Coordination Chemistry Reviews*, *253*(23), 2835–2851. doi:10.1016/j.ccr.2009.09.004

Pogue, R. T., Ye, J., Klosterman, D. A., Glass, A. S., & Chartoff, R. P. (1998). Evaluating fiber–matrix interaction in polymer–matrix composites by inverse gas chromatography. *Composites. Part A, Applied Science and Manufacturing*, *29*(9), 1273–1281. doi:10.1016/S1359-835X(98)00074-8

Polte, J., Tuaev, X., Wuithschick, M., Fischer, A., Thuenemann, A. F., Rademann, K., ... Emmerling, F. (2012). Formation mechanism of colloidal silver nanoparticles: Analogies and differences to the growth of gold nanoparticles. *ACS Nano*, *6*(7), 5791–5802. doi:10.1021/nn301724z

Pomogailo, A. D., Rozenberg, A. S., & Dzhardimalieva, G. I. (2005). *Metal-polymer nanocomposites* (L. Nicolais & G. Carotenuto, Eds.). Hoboken, NJ: Wiley.

Popov, V. N. (2004). Carbon nanotubes: Properties and application. *Materials Science and Engineering R Reports*, *43*(3), 61–102. doi:10.1016/j.mser.2003.10.001

Pora, J. (2001). Composite materials in the airbus A380–from history to future. *Proceedings of ICCM13*. Available on: https://pdfs.semanticscholar.org/31ff/a71022ce6bb0549d1711facf792b6b171414.pdf

Potts, C. (1993). Software Engineering Research Revisited. *IEEE Software*, *10*(5), 19–28. doi:10.1109/52.232392

Prabakaran, K., Mohanty, S., & Nayak, S. K. (2014 a). Influence of surface modified nanoclay on electrochemical properties of PVDF-HFP composite electrolytes. *International Journal of Plastics Technology*, *18*(3), 349–361. doi:10.100712588-014-9089-5

Prabakaran, K., Mohanty, S., & Nayak, S. K. (2015). PEO/PVdF–HFP electrolytes for natural dye sensitized solar cell applications: Effect of modified nano-TiO 2 on electrochemical and photovoltaic performance. *Journal of Materials Science Materials in Electronics*, *26*(6), 3887–3897. doi:10.100710854-015-2916-4

Prabhakaran, R. (2011). Lightning Strikes on Metal and Composite Aircraft and their Mitigation. *Journal of Aerospace Sciences and Technologies*, *63*(3), 208–222.

Pritchard, R. (1964). The transparency of crystalline polymers. *Polymer Engineering and Science*, *4*(1), 66–71. doi:10.1002/pen.760040114

Purwanto, S., & Salim, M. (2018). *Magnetic properties α-Fe2O3 Filled MWNT/PolyVinyl Alcohol Composites Film.* Paper presented at the Journal of Physics: Conference Series. 10.1088/1742-6596/1091/1/012022

Pustkova, P., Hutchinson, J. M., Román, F., & Montserrat, S. (2009). Homopolymerization effects in polymer layered silicate nanocomposites based upon epoxy resin: Implications for exfoliation. *Journal of Applied Polymer Science*, *114*(2), 1040–1047. doi:10.1002/app.30697

Qiu, J., & Wang, S. (2010). Reaction kinetics of functionalized carbon nanotubes reinforced polymer composites. *Materials Chemistry and Physics*, *121*(1–2), 295–301. doi:10.1016/j.matchemphys.2010.01.039

Qiu, S. L., Wang, C. S., Wang, Y. T., Liu, C. G., Chen, X. Y., Xie, H. F., ... Cheng, R. S. (2011). Reaction kinetics of functionalized carbon nanotubes reinforced polymer composites. *Express Polymer Letters*, *5*(9), 809–818. doi:10.3144/expresspolymlett.2011.79

Qi, X., Jia, Z., & Yang, Y. (2013). Influence of the dispersion of nano titanium dioxide on the tribological performance of fabric self-lubricating liner. *Journal of Applied Polymer Science*, *130*(3), 2100–2105. doi:10.1002/app.39343

Ra, E. J., Raymundo-Piñero, E., Lee, Y. H., & Béguin, F. (2009). High power supercapacitors using polyacrylonitrile-based carbon nanofiber paper. *Carbon*, *47*(13), 2984–2992. doi:10.1016/j.carbon.2009.06.051

Rafiee, M. A., Rafiee, J., Wang, Z., Song, H., Yu, Z. Z., & Koratkar, N. (2009). Enhanced mechanical properties of nanocomposites at low graphene content. *ACS Nano*, *3*(12), 3884–3890. doi:10.1021/nn9010472 PMID:19957928

Rafique, I., Kausar, A., Anwar, Z., & Muhammad, B. (2016). Exploration of epoxy resins, hardening systems, and epoxy/carbon nanotube composite designed for high performance materials: A review. *Polymer-Plastics Technology and Engineering*, *55*(3), 312–333. doi:10.1080/03602559.2015.1070874

Rahaman, A., & Mohanty, A. (2014). Effect of carbon nanotubes on the curing and thermomechanical behavior of epoxy/carbon nanotubes composites. *Polymer Composites*, *35*(29), 441–449. doi:10.1002/pc.22679

Rahaman, M. S. A., Ismail, A. F., & Mustafa, A. (2007). A review of heat treatment on polyacrylonitrile fiber. *Polymer Degradation & Stability*, *92*(8), 1421–1432. doi:10.1016/j.polymdegradstab.2007.03.023

Raimondo, M., Guadagno, L., Speranza, V., Bonnaud, L., Dubois, P., & Lafdi, K. (2018). Multifunctional graphene/POSS epoxy resin tailored for aircraft lightning strike protection. *Composites. Part B, Engineering*, *140*, 44–56. doi:10.1016/j.compositesb.2017.12.015

Rakov, E. G. (2006). *Nanotubes and fullerenes*. Moscow: Universitetskaya kniga. (in Russian)

Rakov, E. G. (2013). Carbon nanotubes in new materials. *Russian Chemical Reviews*, *82*(1), 27–47. doi:10.1070/RC2013v082n01ABEH004227

Ramakrishna, S., Fujihara, K., Teo, W. E., Yong, T., Ma, Z., & Ramaseshan, R. (2006). Electrospun nanofibers: Solving global issues. *Materials Today*, *9*(3), 40–50. doi:10.1016/S1369-7021(06)71389-X

Ramanathan, T., Abdala, A. A., Stankovich, S., Dikin, D. A., Herrera-Alonso, M., Piner, R. D., ... Brinson, L. C. (2008). Functionalized graphene sheets for polymer nanocomposites. *Nature Nanotechnology*, *3*(6), 327–331. doi:10.1038/nnano.2008.96 PMID:18654541

Ramasubramaniam, R., Chen, J., & Liu, H. (2003). Homogeneous carbon nanotube/polymer composites for electrical applications. *Applied Physics Letters*, *83*(14), 2928–2930. doi:10.1063/1.1616976

Rana, S., Alagirusamy, R., & Joshi, M. (2011). Development of carbon nanofibre incorporated three phase carbon/epoxy composites with enhanced mechanical, electrical and thermal properties. *Composites. Part A, Applied Science and Manufacturing*, *42*(5), 439–445. doi:10.1016/j.compositesa.2010.12.018

Rao, M. D. (2003). Recent applications of viscoelastic damping for noise control in automobiles and commercial airplanes. *Journal of Sound and Vibration*, *262*(3), 457–474. doi:10.1016/S0022-460X(03)00106-8

Rathod, V. T., Kumar, J. S., & Jain, A. (2017). Polymer and ceramic nanocomposites for aerospace applications. *Applied Nanoscience*, *7*(8), 519–548. doi:10.100713204-017-0592-9

Ravichandran, K., Praseetha, P. K., Arun, T., & Gobalakrishnan, S. (2018). Synthesis of Nanocomposites. In Synthesis of Inorganic Nanomaterials (pp. 141-168). Academic Press. doi:10.1016/B978-0-08-101975-7.00006-3

Ray, B. C. (2005a). Loading Rate Effects on Mechanical Properties of Polymer Composites at Ultra-low Temperatures. *Journal of Applied Polymer Science*, 95–98.

Ray, B. C. (2005b). Thermal Shock and Thermal Fatigue on Delamination of Glass-fiber-reinforced Polymeric Composites. *Journal of Reinforced Plastics and Composites*, *24*(1), 111–116. doi:10.1177/0731684405042953

Ray, S. S., Yamada, K., Okamoto, M., & Ueda, K. (2003). New polylactide-layered silicate nanocomposites. 2. Concurrent improvements of material properties, biodegradability and melt rheology. *Polymer*, *44*(3), 857–866. doi:10.1016/S0032-3861(02)00818-2

Reina, A., Jia, X., Ho, J., Nezich, D., Son, H., Bulovic, V., ... Kong, J. (2009). Large Area, Few-Layer Graphene Films on Arbitrary Substrates by Chemical Vapor Deposition. *Nano Letters*, *9*(1), 30–35. doi:10.1021/nl801827v PMID:19046078

RESCENDO (Collaborative and Robust Engineering using Simulation Capability Enabling Next Design Optimizations), a research project funded by the European Union Seventh Framework Programme (FP7/2007-2013). Deliverable "D5.4.1 BDA Quality Laboratory State-of-the-Art". Release 1.0 2009. Courtesy of EADS.

Rettenbacher, A. S., Elliott, B., Hudson, J. S., Amirkhanian, A., & Echegoyen, L. (2006). Preparation and Functionalization of Multilayer Fullerenes (Carbon Nano-Onions). *Chemistry (Weinheim an der Bergstrasse, Germany)*, *12*(2), 376–387. doi:10.1002/chem.200500517 PMID:16189840

Rinzler, A. G., Liu, J., Dai, H., Nikolaev, P., Huffman, C. B., Rodríguez-Macías, F. J., ... Smalley, R. E. (1998). Large-scale purification of single-wall carbon nanotubes: Process, product, and characterization. *Applied Physics. A, Materials Science & Processing*, *67*(1), 29–37. doi:10.1007003390050734

Rodriguez, A. J., Guzman, M. E., Lim, C. S., & Minaie, B. (2011). Mechanical properties of carbon nanofiber/fiber-reinforced hierarchical polymer composites manufactured with multiscale-reinforcement fabrics. *Carbon*, *49*(3), 937–948. doi:10.1016/j.carbon.2010.10.057

Rohatagi, P. K. (1993). Metal Matrix Composites. *Defence Science Journal*, *43*(4), 323–349. doi:10.14429/dsj.43.4336

Román, F., Calventus, Y., Colomer, P., & Hutchinson, J. M. (2012). Effect of carbon nanotubes on the curing and thermomechanical behavior of epoxy/carbon nanotubes composites. *Thermochimica Acta*, *541*, 76–85.

Rosa, M. F., Medeiros, E. S., Malmonge, J. A., Gregorski, K. S., Wood, D. F., Mattoso, L. H. C., ... Imam, S. H. (2010). Cellulose nanowhiskers from coconut husk fibers: Effect of preparation conditions on their thermal and morphological behavior. *Carbohydrate Polymers*, *81*(1), 83–92. doi:10.1016/j.carbpol.2010.01.059

Rosca, I., & Hoa, S. (2011). Method for reducing contact resistivity of carbon nanotube-containing epoxy adhesives for aerospace applications. *Composites Science and Technology*, *71*(2), 95–100. doi:10.1016/j.compscitech.2010.10.016

Rosso, P., & Ye, L. (2007). Epoxy/silica nanocomposites: Nanoparticle-induced cure kinetics and microstructure. *Macromolecular Rapid Communications*, *28*(1), 121–126. doi:10.1002/marc.200600588

Roy, S., Petrova, R. S., & Mitra, S. (2018). Effect of carbon nanotube (CNT) functionalization in epoxy-CNT composites. *Nanotechnology Reviews*, *7*(6), 475–485. doi:10.1515/ntrev-2018-0068 PMID:30637182

Rozenberg, B. A. (1986). Kinetics, thermodynamics and mechanism of reactions of epoxy oligomers with amines. *Advances in Polymer Science*, *75*, 113–165. doi:10.1007/BFb0017916

Rozenberg, B. A., & Tenne, R. (2008). Polymer-assisted fabrication of nanoparticles and nanocomposites. *Progress in Polymer Science*, *33*(1), 40–112. doi:10.1016/j.progpolymsci.2007.07.004

Rubel, A., & Landis, F. (1969). Numerical Study of Natural Convection in a Vertical Rectangular Enclosure. *Physics of Fluids*, *12*, 208–213.

Ryu, S. H., Sin, J. H., & Shanmugharaj, A. M. (2014). Study on the effect of hexamethylene diamine functionalized graphene oxide on the curing kinetics of epoxy nanocomposites. *European Polymer Journal*, *52*, 88–97. doi:10.1016/j.eurpolymj.2013.12.014

Sadchikov, V. V., & Prudnikova, Z. G. (1997). Amorphous materials in electromagnetic shields. *STAL. Sciences et Techniques de l'Animal de Laboratoire*, *4*, 66–69.

Sadhu, S., & Bhowmick, A. K. (2005). Effect of nanoclay on the dynamic mechanical properties of styrene butadiene and acrylonitrile butadiene rubber vulcanizates. *Rubber Chemistry and Technology*, *78*(2), 321–335. doi:10.5254/1.3547886

Saini, P. (2013). Electrical properties and electromagnetic interference shielding response of electrically conducting thermosetting nanocomposites. In Thermoset Nanocomposites. Wiley-VCH Verlag GmbH & Co. KGaA. doi:10.1002/9783527659647.ch10

Saini, P., & Choudhary, V. (2013). Enhanced electromagnetic interference shielding effectiveness of polyaniline functionalized carbon nanotubes filled polystyrene. Journal of Nanoparticle Research, 15. doi:10.100711051-012-1415-2

Saini, P., & Arora, M. (2012). *Microwave absorption and EMI shielding behavior of nanocomposites based on intrinsically conducting polymers, graphene and carbon nanotubes. In New Polymers for Special Applications*. InTech.

Saini, P., Arora, M., Gupta, G., Gupta, B. K., Singh, V. N., & Choudhary, V. (2013). High permittivity polyaniline-barium titanate nanocomposites with excellent electro-magnetic interference shielding response. *Nanoscale*, *5*(10), 4330–4336. doi:10.1039/c3nr00634d PMID:23563991

Saini, P., Choudhary, V., Singh, B. P., Mathur, R. B., & Dhawan, S. K. (2009). Polyaniline MWCNT nanocomposites for microwave absorption and EMI shielding. *Materials Chemistry and Physics*, *113*(2-3), 919–926. doi:10.1016/j.matchemphys.2008.08.065

Saini, P., Choudhary, V., Singh, B. P., Mathur, R. B., & Dhawan, S. K. (2011). Enhanced microwave absorption behavior of polyaniline-CNT/polystyrene blend in 12.4-18.0 GHz range. *Synthetic Metals*, *161*(15-16), 1522–1526. doi:10.1016/j.synthmet.2011.04.033

Saini, P., Choudhary, V., Sood, K. N., & Dhawan, S. K. (2009). Electromagnetic interference shielding behavior of polyaniline/ graphite composites prepared by in situ emulsion pathway. *Journal of Applied Polymer Science*, *113*(5), 3146–3155. doi:10.1002/app.30183

Saini, P., Choudhary, V., Vijayan, N., & Kotnala, R. K. (2012). Improved electromagnetic interference shielding response of poly (aniline)-coated fabrics containing dielectric and magnetic nanoparticles. *The Journal of Physical Chemistry C*, *116*(24), 13403–13412. doi:10.1021/jp302131w

Saito, R., Dresselhaus, G., & Dresselhaus, M. S. (1998). *Physical properties of carbon nanotubes*. World Scientific. doi:10.1142/p080

Saito, Y., & Uemura, S. (2000). Field emission from carbon nanotubes and its application to electron sources. *Carbon*, *38*(2), 169–182. doi:10.1016/S0008-6223(99)00139-6

Sakai, K., Asano, N., Wada, Y., & Yoshikado, S. (2010). Composite electromagnetic wave absorber made of soft magnetic material and polystyrene resin and control of permeability and permittivity. *Journal of the European Ceramic Society*, *30*(2), 347–353. doi:10.1016/j.jeurceramsoc.2009.05.044

Saleh, M. H., & Sundararaj, U. (2013). X-band EMI shielding mechanisms and shielding effectiveness of high structure carbon black/polypropylene composites. *Journal of Physics. D, Applied Physics*, *46*, 1–7.

Sanctuary, R., Baller, J., Zielinski, B., Becker, N., Krüger, K., Philipp, U., ... Ziehmer, M. (2009). Influence of Al_2O_3 nanoparticles on the isothermal cure of an epoxy resin. *Journal of Physics Condensed Matter*, *21*(3), 035118–035126. doi:10.1088/0953-8984/21/3/035118

Sandler, J., Kirk, J., Kinloch, I., Shaffer, M., & Windle, A. (2003). Ultra-low electrical percolation threshold in carbon-nanotube-epoxy composites. *Polymer*, *44*(19), 5893–5899. doi:10.1016/S0032-3861(03)00539-1

Sandler, J., Shaffer, M., Prasse, T., Bauhofer, W., Schulte, K., & Windle, A. (1999). Development of a dispersion process for carbon nanotubes in an epoxy matrix and the resulting electrical properties. *Polymer*, *40*(21), 5967–5971. doi:10.1016/S0032-3861(99)00166-4

Sandler, J., Werner, P., Shaffer, M. S., Demchuk, V., Altstädt, V., & Windle, A. H. (2002). Carbon-nanofibre-reinforced poly (ether ether ketone) composites. *Composites. Part A, Applied Science and Manufacturing*, *33*(8), 1033–1039. doi:10.1016/S1359-835X(02)00084-2

Sang, B., Li, Z. W., Li, X. H., Yu, L. G., & Zhang, Z. J. (2016). Graphene-based flame retardants: A review. *Journal of Materials Science*, *51*(18), 8271–8295. doi:10.100710853-016-0124-0

Sangermano, M., Yagci, Y., & Rizza, G. (2007). In situ synthesis of silver-epoxy nanocomposites by photoinduced electron transfer and cationic polymerization processes. *Macromolecules*, *40*(25), 8827–8829. doi:10.1021/ma702051g

Sanghoon, K., Joon-Suk, O., & Myeong-Gi, K. (2014). Electromagnetic interference (EMI) transparent shielding of reduced graphene oxide (RGO) interleaved structure fabricated by electrophoretic deposition. *Appl Mater Interfaces*, *6*(20), 17647–17653. doi:10.1021/am503893v PMID:25238628

Santosh, C., Velmurugan, V., Jacob, G., Jeong, S. K., Grace, A. N., & Bhatnagar, A. (2016). Role of Nanomaterials in Water Treatment Applications: A Review. *Chemical Engineering Journal*, *306*, 1116–1137. doi:10.1016/j.cej.2016.08.053

Santra, A. K., Sen, S., & Chakraborty, N. (2008). Study of Heat Transfer Augmentation in a Differentially Heated Square Cavity using Copper–Water Nano-Fluid. *International Journal of Thermal Sciences*, *47*(9), 1113–1122. doi:10.1016/j.ijthermalsci.2007.10.005

Sarac, A. S. (2015). *Nanofibers of Conjugated Polymers*. Pan Stanford.

Sargent, R. G. (2001). Some approaches and paradigms for verifying and validating simulation models. *Winter Simulation Conference, WSC'01*, 2, 106-114. 10.1109/WSC.2001.977251

Sargent, R. G. (2007). Verification and validation of simulation models. *2007 Winter Simulation Conference*, 124-37. 10.1109/WSC.2007.4419595

Sarvi, A., & Sundararaj, U. (2014). Electrical permittivity and electrical conductivity of multiwall carbon nanotube-polyaniline (MWCNT-PANi) core-shell nanofibers and MWCNT-PANi/polystyrene composites. *Macromolecular Materials and Engineering*, *299*(8), 1013–1020. doi:10.1002/mame.201300406

Sattar, R., Kausar, A., & Siddiq, M. (2015). Advances in thermoplastic polyurethane composites reinforced with carbon nanotubes and carbon nanofibers: A review. *Journal of Plastic Film & Sheeting*, *31*(2), 186–224. doi:10.1177/8756087914535126

Sayuti, M., Erh, O. M., Sarhan, A. A., & Hamdi, M. (2014). Investigation on the morphology of the machined surface in end milling of aerospace AL6061-T6 for novel uses of SiO2 nanolubrication system. *Journal of Cleaner Production*, *66*, 655–663. doi:10.1016/j.jclepro.2013.11.058

Schaetz, A., Zeltner, M., & Stark, W. J. (2012). Carbon modifications and surfaces for catalytic organic transformations. *ACS Catalysis*, *2*(6), 1267–1284. doi:10.1021/cs300014k

Schartel, B., Bartholmai, M., & Knoll, U. (2006). Some comments on the main fire retardancy mechanisms in polymer nanocomposites. *Polymers for Advanced Technologies*, *17*(9-10), 772–777. doi:10.1002/pat.792

Schartel, B., Pötschke, P., Knoll, U., & Abdel-Goad, M. (2005). Fire behaviour of polyamide 6/multiwall carbon nanotube nanocomposites. *European Polymer Journal*, *41*(5), 1061–1070. doi:10.1016/j.eurpolymj.2004.11.023

Schlesinger, S. (1979). Terminology for model credibility. *Simulation, 32*(3), 103-104.

Scholtz, D. (2002). Aircraft Systems – Reliability, mass Power and Costs. *European Workshop on Aircraft Design Education.*

Schwartz, L., Garboczi, E., & Bentz, D. (1995). Interfacial Transport in Porous Media: Applications to DC Electrical Conductivity of Mortars. *Journal of Applied Physics*, *78*(10), 5898–5908. doi:10.1063/1.360591

Sebastian, J., Schehl, N., Bouchard, M., Boehle, M., Li, L., Lagounov, A., & Lafdi, K. (2014). Health monitoring of structural composites with embedded carbon nanotube coated glass fiber sensors. *Carbon*, *66*, 191–200. doi:10.1016/j.carbon.2013.08.058

Šehić, A., Vasiljević, J., Demšar, A., Leskovšek, M., Bukošek, V., Medved, J., ... Simončič, B. (2018). Polyamide 6 composite fibers with incorporated mixtures of melamine cyanurate, carbon nanotubes, and carbon black. *Journal of Applied Polymer Science.*

Sengupta, R., Chakraborty, S., Bandyopadhyay, S., Dasgupta, S., Mukhopadhyay, R., Auddy, K., & Deuri, A. S. (2007). A short review on rubber/clay nanocomposites with emphasis on mechanical properties. *Polymer Engineering and Science*, *47*(11), 1956–1974. doi:10.1002/pen.20921

Senokos, E. (n.d.). *Development of multifunctional flexible and structural supercapacitors based on carbon nanotube fibers.* ETSI Caminos, Canales y Puertos (UPM).

Seshanandan, G., Ravindran, D., & Sornakumar, T. (2016). Mechanical properties of nano titanium oxide particles-hybrid jute-glass FRP composites. *Materials Today: Proceedings*, *3*(6), 1383–1388. doi:10.1016/j.matpr.2016.04.019

Seyhan, A. T., Sun, Z., Deitzel, J., Tanoglu, M., & Heider, D. (2009). Cure kinetics of vapor grown carbon nanofiber (VGCNF) modified epoxy resin suspensions and fracture toughness of their resulting nanocomposites. *Materials Chemistry and Physics*, *118*(1), 234–242. doi:10.1016/j.matchemphys.2009.07.045

Shah, A. U. R., Lee, D. W., Kim, S., Kim, B. S., & Song, J. I. (2013). Tensile and fire retardant properties of nanoclay reinforced Abaca/Polypropylene composite. *Aerospace Science & Engineering (ICASE), 2013 International Conference on IEEE*, 1-5.

Shalin, R. E. (Ed.). (1995). *Polymer Matrix Composites*. Springer Netherlands. doi:10.1007/978-94-011-0515-6

Shanmugharaj, A. M., & Ryu, S. H. (2012). Study on the effect of aminosilane functionalized nanoclay on the curing kinetics of epoxy nanocomposites. *Thermochimica Acta, 546*, 16–23. doi:10.1016/j.tca.2012.07.026

Shanyi, D. (2007). Advanced composite materials and aerospace engineering. *Acta Materiae Compositae Sinica, 1.*

Sharma, V. P., Sharma, U., Chattopadhyay, M., & Shukla, V. N. (2018). Advance Applications of Nanomaterials: A review. *Materials Today: Proceedings*, *5*(2), 6376–6380. doi:10.1016/j.matpr.2017.12.248

Shen, B., Zhai, W., Tao, M., Lu, D., & Zheng, W. (2013a). Enhanced interfacial interaction between polycarbonate and thermally reduced graphene induced by melt blending. *Composites Science and Technology*, *77*, 87–110. doi:10.1016/j.compscitech.2013.01.014

Shen, X. J., Pei, X. Q., Fu, S. Y., & Friedrich, K. (2013b). Significantly modified tribological performance of epoxy nanocomposites at very low graphene oxide content. *Polymer*, *54*(3), 1234–1242. doi:10.1016/j.polymer.2012.12.064

Shim, B. S., Starkovich, J., & Kotov, N. (2006). Multilayer composites from vapor-grown carbon nano-fibers. *Composites Science and Technology*, *66*(9), 1174–1181. doi:10.1016/j.compscitech.2005.11.004

Shinagawa, S., Kumagai, Y., & Urabe, K. (1999). Conductive papers containing metalized polyester fibers for electromagnetic interference shielding. *Journal of Porous Materials*, *6*(3), 185–190. doi:10.1023/A:1009619711017

Shiravand, F., Fraga, I., Cortés, P., Calventus, Y., & Hutchinson, J. M. (2014b). Thermal analysis of polymer layered silicate nanocomposites. Identification of nanostructure development by DSC. *Journal of Thermal Analysis and Calorimetry*, *118*(2), 723–729. doi:10.100710973-014-3709-3

Shiravand, F., Hutchinson, J. M., & Calventus, Y. (2014a). Influence of the isothermal cure temperature on the nanostructure and thermal properties of an epoxy layered silicate nanocomposite. *Polymer Engineering and Science*, *54*(1), 51–58. doi:10.1002/pen.23540

Shiravand, F., Hutchinson, J. M., & Calventus, Y. (2015). Non-isothermal cure and exfoliation of tri-functional epoxy-clay nanocomposites. *Express Polymer Letters*, *9*(8), 695–708. doi:10.3144/expresspolymlett.2015.65

Shokrieh, M. M., Esmkhani, M., Haghighatkhah, A. R., & Zhao, Z. (2014). Flexural fatigue behavior of synthesized graphene/carbon-nanofiber/epoxy hybrid nanocomposites. *Materials & Design (1980-2015), 62*, 401–408. doi:10.1016/j.matdes.2014.05.040

Shokrieh, M., & Rafiee, R. (2010). A review of the mechanical properties of isolated carbon nanotubes and carbon nanotube composites. *Mechanics of Composite Materials*, *46*(2), 155–172. doi:10.100711029-010-9135-0

Singh, M., Lara, S., & Tlali, S. (2017). Effects of Size and Shape on the Specific Heat, Melting Entropy and Enthalpy of Nanomaterials. *Journal of Taibah University for Science Journal*, *11*(6), 922–929. doi:10.1016/j.jtusci.2016.09.011

Singh, P., Babbar, V. K., Razdan, A., Srivastava, S. L., & Puri, R. K. (1999). Complex permeability and permittivity, and microwave absorption studies of Ca(CoTi)Fe12−2O19 hexaferrite composites in X-band microwave frequencies. *Materials Science and Engineering B*, *67*(3), 132–138. doi:10.1016/S0921-5107(99)00328-1

Sinha, R., Liang, V., Paredis, C., & Khosla, P. (2001). A survey of modeling and simulation methods for design of engineering systems. *Journal of Computing and Information Science in Engineering*, *1*(1), 84–91. doi:10.1115/1.1344877

Smalley, R. E. (1992). Self-assembly of the fullerenes. *Accounts of Chemical Research*, *25*(3), 98–105. doi:10.1021/ar00015a001

Smalley, R. E. (1993). From dopyballs to nanowires. *Materials Science and Engineering B*, *19*(1), 1–7. doi:10.1016/0921-5107(93)90156-H

Smith, J. Jr, Connell, J., Delozier, D., Lillehei, P., Watson, K., Lin, Y., ... Sun, Y.-P. (2004). Space durable polymer/carbon nanotube films for electrostatic charge mitigation. *Polymer*, *45*(3), 825–836. doi:10.1016/j.polymer.2003.11.024

Smith, P., Prabhu, S., & Friedman, J. (2007). Best Practices for Establishing a Model-Based Design Culture, 2007-01-0777. *SAE World Congress & Exhibition*, Detroit, MI. 10.4271/2007-01-0777

Sobol, I. M. (1993). Sensitivity Analysis for Nonlinear Mathematical Models. *Mathematical Modeling & Computational Experiment*, *1*, 407–414.

Solov'ev, M. E., Irzhak, T. F., & Irzhak, V. I. Russ. (2015). Kinetics of formation of nanoparticles from first group metal carboxylates. *The Journal of Physical Chemistry A*, *89*(3), 1642–1647.

Song, M., Hourston, D. J., Yao, K. J., Tay, J. K. H., & Ansarifar, M. A. (2003). High performance nanocomposites of polyurethane elastomer and organically modified layered silicate. *Journal of Applied Polymer Science*, *90*(12), 3239–3243. doi:10.1002/app.12979

Song, S. H., Park, K. H., Kim, B. H., Choi, Y. W., Jun, G. H., Lee, D. J., ... Jeo, S. (2013). Enhanced thermal conductivity of epoxy/graphene composites by using non-oxidized graphene flakes with non-covalent functionalization. *Advanced Materials*, *25*(5), 732–737. doi:10.1002/adma.201202736

Song, Y. S., & Youn, J. R. (2005). Influence of dispersion states of carbon nanotubes on physical properties of epoxy nanocomposites. *Carbon*, *43*(7), 1378–1385. doi:10.1016/j.carbon.2005.01.007

Spitalsky, Z., Tasis, D., Papagelis, K., & Galiotis, C. (2010). Carbon nanotube–polymer composites: Chemistry, processing, mechanical and electrical properties. *Progress in Polymer Science*, *35*(3), 357–401. doi:10.1016/j.progpolymsci.2009.09.003

Srinath, G., & Gnanamoorthy, R. (2005). Effect of nanoclay reinforcement on tensile and tribo behaviour of Nylon 6. *Journal of Materials Science*, *40*(11), 2897–2901. doi:10.100710853-005-2439-0

Stan, A., Dinca, I., Ban, C., Ilina, S., Donescu, D., Paven, H., . . . Voicu, I. (2012). Epoxy-layered silicate and epoxy MWCNTs nanocomposites. Applied Mechanics and Materials Trans Tech Publications, 146, 160-169.

Stankovich, S., Dikin, D. A., Dommett, G. H. B., Kohlhaas, K. M., Zimney, E. J., Stach, E. A., ... Ruoff, R. S. (2006). Graphene-based composite materials. *Nature*, *442*(7100), 282–286. doi:10.1038/nature04969 PMID:16855586

Stankovich, S., Dikin, D. A., Piner, R. D., Kohlhaas, K. A., Kleinhammes, A., Jia, Y., ... Ruoff, R. S. (2007). Synthesis of graphene-based nanosheets via chemical reduction of exfoliated graphite oxide. *Carbon*, *45*(7), 1558–1565. doi:10.1016/j.carbon.2007.02.034

Staudenmaier, L. (1898). Verfahren zur Darstellung der Graphitsäure. *Berichte der Deutschen Chemischen Gesellschaft*, *31*(2), 1481–1487. doi:10.1002/cber.18980310237

Stephan, O., Ajayan, P. M., Colliex, C., Redlich, P., Lambert, J. M., Bernier, P., & Lefin, P. (1994). Doping graphitic and carbon nanotube structures with boron and nitrogen. *Science*, *266*(5191), 1683–1685. doi:10.1126cience.266.5191.1683 PMID:17775629

Stevenson, W. J. (2002). Operations Management (7th ed.). Boston: McGraw-Hill/Irwin.

Suenaga, K., Johansson, M. P., Hellgren, N., Broitman, E., Wallenberg, L. R., Colliex, C., ... Hultman, L. (1999). Carbon nitride nanotubulite – densely-packed and well-aligned tubular nanostructures. *Chemical Physics Letters*, *300*(5–6), 695–700. doi:10.1016/S0009-2614(98)01425-0

Su, F., Zhang, Z., & Liu, W. (2006). Mechanical and tribological properties of carbon fabric composites filled with several nano-particulates. *Wear*, *260*(7), 861–868. doi:10.1016/j.wear.2005.04.015

Suresh, S. S., Mohanty, S., & Nayak, S. K. (2017a). Composition analysis and characterization of waste polyvinyl chloride (PVC) recovered from data cables. *Waste Management (New York, N.Y.)*, *60*, 100–111. doi:10.1016/j.wasman.2016.08.033 PMID:27594574

Suresh, S. S., Mohanty, S., & Nayak, S. K. (2017b). Preparation and characterization of recycled blends using poly (vinyl chloride) and poly (methyl methacrylate) recovered from waste electrical and electronic equipments. *Journal of Cleaner Production*, *149*, 863–873. doi:10.1016/j.jclepro.2017.02.057

Susin, S. B., Pistor, V., Amico, S. C., Coelho, L. A. F., Pezzin, S. H., & Zattera, A. J. (2014). Investigation of cure kinetics in epoxy/multi-walled carbon nanotube nanocomposites. *Journal of Applied Polymer Science*, *131*(3), 39857–39862. doi:10.1002/app.39857

Suzdalev, I. P. (2006). *Nanotechnology. Physics and chemistry of nanaocrystals, nanostructures and nanomaterials*. Moscow: Kom Kniga. (in Russian)

Swain, S. S., Unnikrishnan, L., Mohanty, S., & Nayak, S. K. (2017). Carbon nanotubes as potential candidate for separation of H 2 CO 2 gas pairs. *International Journal of Hydrogen Energy*, *42*(49), 29283–29299. doi:10.1016/j.ijhydene.2017.09.152

Szeteiová, K. (2010). *Automotive Materials: Plastics in Automotive Markets Today*. Bratislava: Slovak University of Technology.

Talebi, F., Mahmoudi, A. H., & Shahi, M. (2010). Numerical Study of Mixed Convection Flows in a Square Lid-Driven Cavity Utilizing Nano-Fluid. *International Communications in Heat and Mass Transfer*, *37*(1), 79–90. doi:10.1016/j.icheatmasstransfer.2009.08.013

Tan, H., Huang, Y., Liu, C., & Geubelle, P. H. (2005). The Mori–Tanaka method for composite materials with nonlinear interface debonding. *International Journal of Plasticity*, *21*(10), 1890–1918. doi:10.1016/j.ijplas.2004.10.001

Tan, S. T., Zhang, M. Q., Rong, M. Z., & Zeng, H. (1999). Effect of interfacial modification on metal fiber filled polypropylene composites and property balance. *Polymer Composites*, *20*(3), 406–412. doi:10.1002/pc.10366

Tans, S. J., Verschueren, A. R., & Dekker, C. (1998). Room-temperature transistor based on a single carbon nanotube. *Nature*, *393*(6680), 49–52. doi:10.1038/29954

Tao, K., Yang, S., Grunlan, J. C., Kim, Y.-S., Dang, B., Deng, Y., ... Wei, X. (2006). Effects of carbon nanotube fillers on the curing processes of epoxy resin-based composites. *Journal of Applied Polymer Science*, *102*(6), 5248–5254. doi:10.1002/app.24773

Thawre, M. M., Pandey, K. N., Dubey, A., Verma, K. K., Pehswe, D. R., Paretkar, R. K., ... Manjunatha, C. M. (2015). Fatigue life of a carbon fiber composite T-joint under standard fighter aircraft spectrum load sequence. *Composite Structures*, *127*, 260–266. doi:10.1016/j.compstruct.2015.03.026

Thind, K. S., Singh, J., Saini, J. S., & Bhunia, H. (2015). Mechanical and wear properties of hybrid epoxy nanocomposites. *Indian Journal of Engineering and Materials Sciences*, *22*(4), 421–428.

Thostenson, E. T., Gangloff, J. J. Jr, Li, C., & Byun, J.-H. (2009). Electrical anisotropy in multiscale nanotube/fiber hybrid composites. *Applied Physics Letters*, *95*(7), 073111. doi:10.1063/1.3202788

Thostenson, E. T., Ren, Z., & Chou, T.-W. (2001). Advances in the science and technology of carbon nanotubes and their composites: A review. *Composites Science and Technology*, *61*(13), 1899–1912. doi:10.1016/S0266-3538(01)00094-X

Tibbetts, G. G., Lake, M. L., Strong, K. L., & Rice, B. P. (2007). A review of the fabrication and properties of vapor-grown carbon nanofiber/polymer composites. *Composites Science and Technology*, *67*(7–8), 1709–1718. doi:10.1016/j.compscitech.2006.06.015

Timofeeva, E. T., Routbort, J. L., & Singh, D. (2009). Particle shape effect on thermophysical properties of alumina nanofluids. *Journal of Applied Physics*, *106*(1), 1–10. doi:10.1063/1.3155999

Tiwari, R. K., & Das, M. K. (2007). Heat Transfer Augmentation in a Two-Sided Lid-Driven Differentially Heated Square Cavity Utilizing Nano-Fluids. *International Journal of Heat and Mass Transfer*, *50*(9-10), 2002–2018. doi:10.1016/j.ijheatmasstransfer.2006.09.034

Toldy, A., Szolnoki, B., & Marosi, G. (2011). Flame retardancy of fibre-reinforced epoxy resin composites for aerospace applications. *Polymer Degradation & Stability*, *96*(3), 371–376. doi:10.1016/j.polymdegradstab.2010.03.021

Tong, X. C. (2016). *Advanced materials and design for electromagnetic interference shielding*. CRC Press.

Tripathi, S. N., Saini, P., Gupta, D., & Choudhary, V. (2013). Electrical and mechanical properties of PMMA/reduced graphene oxide nanocomposites prepared via in situ polymerization. *Journal of Materials Science*, *48*(18), 6223–6232. doi:10.100710853-013-7420-8

Turan, O., Poole, R. J., & Chakraborty, N. (2012). Influences of Boundary Conditions on Laminar Natural Convection in Rectangular Enclosures with Differentially Heated Side Walls. *International Journal of Heat and Fluid Flow*, *33*(1), 131–146. doi:10.1016/j.ijheatfluidflow.2011.10.009

Turkevich, J., Stevenson, P. C., & Hillier, J. (1951). A study of the nucleation and growth processes in the synthesis of colloidal gold. *Faraday Discussions of the Chemical Society*, *11*, 55–75. doi:10.1039/df9511100055

Tzeng, S. S., & Chang, F. Y. (2001). EMI shielding effectiveness of metal-coated carbon fiberreinforced ABS composites. *Materials Science and Engineering A*, *302*(2), 258–267. doi:10.1016/S0921-5093(00)01824-4

Ullman, D. (1992). *The Mechanical Design Process*. New York: McGraw-Hill Inc.

Usuki, A., Hasegawa, N., Kato, M., & Kobayashi, S. (2005). Polymer-clay nanocomposites. In *Inorganic polymeric nanocomposites and membranes* (pp. 135–195). Berlin: Springer. doi:10.1007/b104481

Valentini, L., Puglia, D., Carniato, F., Boccaleri, E., Marchese, L., & Kenny, J. M. (2008). Use of plasma fluorinated single-walled carbon nanotubes for the preparation of nanocomposites with epoxy matrix. *Composites Science and Technology*, *68*(3–4), 1008–1014. doi:10.1016/j.compscitech.2007.07.011

Vallittu, P. K. (1999). Flexural properties of acrylic resin polymers reinforced with unidirectional and woven glass fibers. *The Journal of Prosthetic Dentistry*, *81*(3), 318–326. doi:10.1016/S0022-3913(99)70276-3 PMID:10050121

Vao-soongnern, V., Doruker, P., & Mattice, W. L. (2000). Simulation of an amorphous polyethylene nanofiber on a high coordination lattice. *Macromolecular Theory and Simulations*, *9*(1), 1–13. doi:10.1002/(SICI)1521-3919(20000101)9:1<1::AID-MATS1>3.0.CO;2-R

Varrla, E., Venkataraman, S., & Sundara, R. (2011). Functionalized graphene-PVDF foam composites for EMI shielding. *Macromolecular Materials and Engineering*, *296*(10), 894–898. doi:10.1002/mame.201100035

Verdejo, R., Saiz-Arroyo, C., Carretero-Gonzalez, J., Barroso-Bujans, F., Rodriguez-Perez, M. A., & Lopez-Manchado, M. A. (2008). Physical properties of silicone foams filled with carbon nanotubes and functionalized graphene sheets. *European Polymer Journal*, *44*(9), 2790–2797. doi:10.1016/j.eurpolymj.2008.06.033

Vescovo, L., Sangermano, M., Scarazzin, R., & Kortaberria, G. (2010). In-situ-synthetized silver/epoxy nanocomposites: Electrical characterization by means of dielectric spectroscopy. *Macromolecular Chemistry and Physics*, *211*(17), 1933–1939. doi:10.1002/macp.201000138

Vidal, B. C. (2013). Using the FT-IR linear dichroism method for molecular order determination of tendon collagen bundles and nylon 6. *Acta Histochemica*, *115*(7), 686–691. doi:10.1016/j.acthis.2013.02.002 PMID:23506648

Villacorta, B. S., Ogale, A. A., & Hubing, T. H. (2013). Effect of Heat Treatment of Carbon Nanofibers on the Electromagnetic Shielding Effectiveness of Linear Low Density Polyethylene Nanocomposites. *Polymer Engineering and Science*, *53*(2), 417–423. doi:10.1002/pen.23276

Visco, A., Calabrese, L., & Milone, C. (2009). Cure rate and mechanical properties of a DGEBF epoxy resin modified with carbon nanotubes. *Journal of Reinforced Plastics and Composites*, *28*(8), 937–949. doi:10.1177/0731684407087560

Vodicka, R. (1996). *Thermoplastics for airframe applications: a review of the properties and repair methods for thermoplastic composites*. Academic Press.

Völkle, C. M., Gebauer, D., & Cölfen, H. (2015). High-resolution insights into the early stages of silver nucleation and growth. *Faraday Discussions*, *179*, 59–69. doi:10.1039/C4FD00269E

Volmer, M. (1939). *Kinetik der Phasenbildung*. Dresden: Steinkopff.

Vyazovkin, S., & Sbirrzaauoli, N. (1996). Mechanism and kinetics of epoxy-amine cure studied by differential scanning calorimetry. *Macromolecules*, *29*(6), 1867–1873. doi:10.1021/ma951162w

Wagner, Z. (1961). Theorie der Alterung von Niderschlagen durch Umlösen (Ostwald Reifung). *Zeitschrift für Elektrochemie*, *65*, 581–591.

Wang, B., Sun, Y., & Wang, H. (2010). Preparation and properties of electrospun PAN/Fe_3O_4 magnetic nanofibers. *Journal of Applied Polymer Science*, *115*(3), 1781–1786. doi:10.1002/app.31288

Wang, D. H., Sihn, S., Roy, A. K., Baek, J.-B., & Tan, L.-S. (2010). Nanocomposites based on vapor-grown carbon nanofibers and an epoxy: Functionalization, preparation and characterization. *European Polymer Journal*, *46*(7), 1404–1416. doi:10.1016/j.eurpolymj.2010.04.025

Wang, F., Richards, V. N., Shields, S. P., & Buhro, W. E. (2014). Kinetics and mechanisms of aggregative nanocrystal growth. *Chemistry of Materials*, *26*(1), 5–21. doi:10.1021/cm402139r

Wang, G., Chang, Y., Wang, L., Liu, L., & Liu, C. (2013). Facilely preparation and microwave absorption properties of Fe_3O_4 nanoparticles. *Materials Research Bulletin*, *48*(3), 1007–1012. doi:10.1016/j.materresbull.2012.11.089

Wang, K., Kelly, D., & Dutton, S. (2002). Multi-objective optimisation of composite aerospace structures. *Composite Structures*, *57*(1-4), 141–148. doi:10.1016/S0263-8223(02)00078-8

Wang, L., Yu, Y., Chen, P. C., Zhang, D. W., & Chen, C. H. (2008). Electrospinning synthesis of C/Fe_3O_4 composite nanofibers and their application for high performance lithium-ion batteries. *Journal of Power Sources*, *183*(2), 717–723. doi:10.1016/j.jpowsour.2008.05.079

Wang, M., Singh, H., Hatton, T. A., & Rutledge, G. C. (2004). Field-responsive superparamagnetic composite nanofibers by electrospinning. *Polymer*, *45*(16), 5505–5514. doi:10.1016/j.polymer.2004.06.013

Wang, S., Wang, C., Zhang, B., Sun, Z., Li, Z., Jiang, X., & Bai, X. (2010). Preparation of Fe_3O_4/PVA nanofibers via combining in-situ composite with electrospinning. *Materials Letters*, *64*(1), 9–11. doi:10.1016/j.matlet.2009.09.043

Watzky, M. A., & Finke, R. G. (1997). Nanocluster size-control and "magic number" investigations. Experimental tests of the "living-metal polymer" concept and of mechanism-based size-control reductions leading to the syntheses of Ir(0) nanoclusters centering about four sequential magic numbers. *Chemistry of Materials*, *9*(12), 3083–3095. doi:10.1021/cm9704387

Weisman, R. B., & Subramoney, S. (2006). Carbon nanotubes. *The Electrochemical Society Interface*, 43.

Wei, Y., Zhang, X., Song, Y., Han, B., Hu, X., Wang, X., ... Deng, X. (2011). Magnetic biodegradable Fe_3O_4/CS/PVA nanofibrous membranes for bone regeneration. *Biomedical Materials (Bristol, England)*, *6*(5), 1–16. doi:10.1088/1748-6041/6/5/055008 PMID:21893702

Wei, Z., Huagang, X., Shaokai, W., Min, L., & Yizhuo, G. (2015). Electromagnetic characteristics of carbon nanotube film materials. *Chinese Journal of Aeronautics*, *28*(4), 1245–1254. doi:10.1016/j.cja.2015.05.002

Wichmann, M. H., Sumfleth, J., Gojny, F. H., Quaresimin, M., Fiedler, B., & Schulte, K. (2006). Glass-fibre-reinforced composites with enhanced mechanical and electrical properties–benefits and limitations of a nanoparticle modified matrix. *Engineering Fracture Mechanics*, *73*(16), 2346–2359. doi:10.1016/j.engfracmech.2006.05.015

Wicks, S. S., de Villoria, R. G., & Wardle, B. L. (2010). Interlaminar and intralaminar reinforcement of composite laminates with aligned carbon nanotubes. *Composites Science and Technology*, *70*(1), 20–28. doi:10.1016/j.compscitech.2009.09.001

Wilcoxon, J. P., & Abrams, B. L. (2006). Synthesis, structure and properties of metal nanoclusters. *Chemical Society Reviews*, *35*(4), 1162–1194. doi:10.1039/b517312b

Wojciechowski, S. (2000). New trends in the development of mechanical engineering materials. *Journal of Materials Processing Technology, 106*(1-3), 230–235. doi:10.1016/S0924-0136(00)00619-1

Wright, W. W. (1991). Polymers in aerospace applications. *Materials & Design, 12*(4), 222–227. doi:10.1016/0261-3069(91)90169-5

Wu, B., Zhu, D., Zhang, S., Lin, W., Wu, G., & Pan, B. (2015). The photochemistry of carbon nanotubes and its impact on the photo-degradation of dye pollutants in aqueous solutions. *Journal of Colloid and Interface Science, 439*, 98–104. doi:10.1016/j.jcis.2014.10.020 PMID:25463180

Wu, C. L., Zhang, M. Q., Rong, M. Z., & Friedrich, K. (2002). Tensile performance improvement of low nanoparticles filled-polypropylene composites. *Composites Science and Technology, 62*(10-11), 1327–1340. doi:10.1016/S0266-3538(02)00079-9

Wu, G., Zhou, L., Ou, E., Xie, Y., Xiong, Y., & Xu, W. (2010). Preparation and properties of hydroxylated styrene–butadiene–styrene tri-block copolymer/multi-walled carbon nanotubes nanocomposites via covalent bond. *Materials Science and Engineering A, 527*(20), 5280–5286. doi:10.1016/j.msea.2010.04.094

Wuithschick, M., Paul, B., Bienert, R., Sarfraz, A., Vainio, U., Sztucki, M., ... Polte, J. (2013). Size-controlled synthesis of colloidal silver nanoparticles based on mechanistic understanding. *Chemistry of Materials, 25*(24), 4679–4689. doi:10.1021/cm401851g

Wu, J., & Chung, D. D. L. (2002). Increasing the electromagnetic interference shielding effectiveness of carbon fiber polymermatrix composite by using activated carbon fibers. *Carbon, 40*(3), 445–447. doi:10.1016/S0008-6223(01)00133-6

Wu, Y. H., Yu, T., & Shen, Z. X. (2010). Two-dimensional carbon nanostructures: Fundamental properties, synthesis, characterization, and potential applications. *Journal of Applied Physics, 108*(7), 071301. doi:10.1063/1.3460809

Wu, Z. P., Li, M. M., Hu, Y. Y., Li, Y. S., Wang, Z. X., Yin, Y. H., ... Zhou, X. (2011). Electromagnetic interference shielding of carbon nanotube macrofilms. *Scripta Materialia, 64*(9), 809–812. doi:10.1016/j.scriptamat.2011.01.002

Xiao, H., & Liu, S. (2018). 2D Nanomaterials a Lubricant Additive: A Review. *Materials & Design, 135*, 319–332. doi:10.1016/j.matdes.2017.09.029

Xie, H., Liu, B., Yuan, Z., Shen, J., & Cheng, R. (2004). Cure kinetics of carbon nanotube/tetrafunctional epoxy nanocomposites by isothermal differential scanning calorimeter. *Journal of Polymer Science. Part A, Polymer Chemistry, 42*(20), 3701–3712. doi:10.1002/polb.20220

Xie, X.-L., Mai, Y.-W., & Zhou, X.-P. (2005). Dispersion and alignment of carbon nanotubes in polymer matrix: A review. *Materials Science and Engineering R Reports, 49*(4), 89–112. doi:10.1016/j.mser.2005.04.002

Xin, H. L. L., & Zheng, H. M. (2012). In situ observation of oscillatory growth of bismuth nanoparticles. *Nano Letters, 12*(3), 1470–1474. doi:10.1021/nl2041854

Xu, C., Jia, Z., Wu, D., Han, Q., & Meek, T. (2006). Fabrication of nylon-6/carbon nanotube composites. *Journal of Electronic Materials*, *35*(5), 954–957. doi:10.1007/BF02692553

Xue, B., Zhu, J., Liu, N., & Li, Y. (2015). Facile functionalization of graphene oxide with ethylenediamine as a solid base catalyst for Knoevenagel condensation reaction. *Catalysis Communications*, *64*, 105–109. doi:10.1016/j.catcom.2015.02.003

Yadav, M. D., Dasgupta, K., Patwardhan, A. W., & Joshi, J. B. (2017). High Performance Fibers from Carbon Nanotubes: Synthesis, Characterization, and Applications in Composites—A Review. *Industrial & Engineering Chemistry Research*, *56*(44), 12407–12437. doi:10.1021/acs.iecr.7b02269

Yagci, Y., Sahin, O., Ozturk, T., Marchi, S., Grassini, S., & Sangermano, M. (2011). Synthesis of silver/epoxy nanocomposites by visible light sensitization using highly conjugated thiophene derivatives. *Reactive & Functional Polymers*, *71*(7), 857–862. doi:10.1016/j.reactfunctpolym.2011.05.012

Yagci, Y., Sangermano, M., & Rizza, G. (2008a). A visible light photochemical route to silver–epoxy nanocomposites by simultaneous polymerization–reduction approach. *Polymer*, *49*(24), 5195–5198. doi:10.1016/j.polymer.2008.09.068

Yagci, Y., Sangermano, M., & Rizza, G. (2008b). Synthesis and characterization of gold-epoxy nanocomposites by visible light photoinduced electron transfer and cationic polymerization processes. *Macromolecules*, *41*(20), 7268–7270. doi:10.1021/ma801776y

Yamamoto, M., Kashiwagi, Y., & Nakamoto, M. (2006). Size-controlled synthesis of monodispersed silver nanoparticles capped by long-chain alkyl carboxylates from silver carboxylate and tertiary amine. *Langmuir*, *22*(20), 8581–8586. doi:10.1021/la0600245

Yamamoto, M., & Nakamoto, M. (2003). Novel preparation of monodispersed silver nanoparticles via amine adducts derived from insoluble silver myristate in tertiary alkylamine. *Journal of Materials Chemistry*, *13*(9), 2064–2065. doi:10.1039/b307092a

Yang, G., Fu, S.-Y., & Yang, J.-P. (2007). Preparation and mechanical properties of modified epoxy resins with flexible diamines. *Polymer*, *48*(1), 302–310. doi:10.1016/j.polymer.2006.11.031

Yang, K., Gu, M., Jin, Y., Mu, G., & Pan, X. (2008). Influence of surface treated multi-walled carbon nanotubes on cure behavior of epoxy nanocomposites. *Composites. Part A, Applied Science and Manufacturing*, *39*(10), 1670–1678. doi:10.1016/j.compositesa.2008.07.011

Yang, Q., Li, Z., Hong, Y., Zhao, Y., Qiu, S., Wang, C., & Wei, Y. (2004). Influence of solvents on the formation of ultrathin uniform poly (vinyl pyrrolidone) nanofibers with electrospinning. *Journal of Polymer Science. Part B: Polymer Letters*, *42*(20), 3721–3726. doi:10.1002/polb.20222

Yang, S., Lozano, K., Lomeli, A., Foltz, H. D., & Jones, R. (2005). Electromagnetic interference shielding effectiveness of carbon nanofiber/LCP composites. *Composites. Part A, Applied Science and Manufacturing*, *36*(5), 691–697. doi:10.1016/j.compositesa.2004.07.009

Yang, Y., Guo, Z., Zhang, H., Huang, D., Gu, J., Huang, Z., ... Rutledge, G. C. (2013). Electrospun Magnetic Carbon Composite Fibers: Synthesis and Electromagnetic Wave Absorption Characteristics. *Journal of Applied Polymer Science*, *127*(6), 4288–4295. doi:10.1002/app.38027

Yang, Y., Gupta, M. C., Dudley, K. L., & Lawrence, R. W. (2005). A comparative study of EMI shielding properties of carbon nanofiber and multi-walled carbon nanotube filled polymer composites. *Journal of Nanoscience and Nanotechnology*, *5*(6), 927–931. doi:10.1166/jnn.2005.115 PMID:16060155

Yang, Y., Gupta, M. C., Dudley, K. L., & Lawrence, R. W. (2005). Conductive carbon nanofiber-polymer foam structures. *Advanced Materials*, *17*(16), 1999–2003. doi:10.1002/adma.200500615

Yang, Y., Gupta, M. C., Dudley, K. L., & Lawrence, R. W. (2005). Novel carbon nanotube—Polystyrene foam composites for electromagnetic interference shielding. *Nano Letters*, *5*(11), 2131–2134. doi:10.1021/nl051375r PMID:16277439

Yano, K., Usuki, A., & Okada, A. (1997). Synthesis and properties of polyimide-clay hybrid films. *Journal of Polymer Science. Part A, Polymer Chemistry*, *35*(11), 2289–2294. doi:10.1002/(SICI)1099-0518(199708)35:11<2289::AID-POLA20>3.0.CO;2-9

Yano, K., Usuki, A., Okada, A., Kurauchi, T., & Kamigaito, O. (1993). Synthesis and properties of polyimide–clay hybrid. *Journal of Polymer Science. Part A, Polymer Chemistry*, *31*(10), 2493–2498. doi:10.1002/pola.1993.080311009

Yan, S., Wu, Z., Yu, H., Gong, Y., Tan, Y., Du, R., ... Sun, D. (2014). Time-resolved small-angle x-ray scattering study on the growth behavior of silver nanoparticles. *Journal of Physical Chemistry*, *118*(21), 11454–11463.

Yasmin, A., Luo, J. J., Abot, J. L., & Daniel, I. M. (2006). Mechanical and thermal behavior of clay/epoxy nanocomposites. *Composites Science and Technology*, *66*(14), 2415–2422. doi:10.1016/j.compscitech.2006.03.011

Yebassa, D., Balakrishnan, S., & Raghaνan, D. (2003). Next generation aerospace composites through nanotechnology. *Technical Proceedings of the 2003 Nanotechnology Conference*, 246-249.

Ye, L., Lu, Y., Su, Z., & Meng, G. (2005). Functionalized composite structures for new generation airframes: A review. *Composites Science and Technology*, *65*(9), 1436–1446. doi:10.1016/j.compscitech.2004.12.015

Yi, M., & Shen, Z. (2015). A review on mechanical exfoliation for the scalable production of graphene. *Journal of Materials Chemistry. A, Materials for Energy and Sustainability*, *3*(22), 11700–11715. doi:10.1039/C5TA00252D

Young, R. J., Kinloch, I. A., Gong, L., & Novoselov, K. S. (2012). The mechanics of graphene nanocomposites: A review. *Composites Science and Technology*, *72*(12), 1459–1476. doi:10.1016/j.compscitech.2012.05.005

Yu, H., Liu, J., Wen, X., Jiang, Z., Wang, Y., Wang, L., ... Tang, T. (2011). Charing polymer wrapped carbon nanotubes for simultaneously improving the flame retardancy and mechanical properties of epoxy resin. *Polymer*, *52*(21), 4891–4898. doi:10.1016/j.polymer.2011.08.013

Yu, H., Wang, T., Wen, B., Lu, M., Xu, Z., Zhu, C., ... Cao, M. (2012). Graphene/polyaniline nanorod arrays: Synthesis and excellent electromagnetic absorption properties. *Journal of Materials Chemistry*, *22*(40), 21679. doi:10.1039/c2jm34273a

Yumura, M. (1999). Synthesis and Purification of Multi-Walled and Single-Walled Carbon Nanotubes. In K. Tanaka, T. Yamabe, & K. Fukui (Eds.), *The Science and Technology of Carbon Nanotubes* (pp. 2–13). Oxford, UK: Elsevier Science Ltd.; doi:10.1016/B978-008042696-9/50002-1

Yu, W., & Choi, S. U. S. (2003). The Role of Interfacial Layers in the Enhanced Thermal Conductivity of Nanofluids: Arenovated Maxwell Model. *Journal of Nanoparticle Research, 6*(4), 167–171. doi:10.1023/A:1024438603801

Zabihi, O., Hooshafza, A., Moztarzadeh, F., Payravand, H., Afshar, A., & Alizadeh, R. (2012). Isothermal curing behavior and thermo-physical properties of epoxy-based thermoset nanocomposites reinforced with Fe_2O_3 nanoparticles. *Thermochimica Acta, 527*, 190–198. doi:10.1016/j.tca.2011.10.026

Zabihi, O., Mostafavi, S. M., Ravari, F., Khodabandeh, A., Hooshafza, A., Zare, K., & Shahizadeh, M. (2011). The effect of zinc oxide nanoparticles on thermo-physical properties of diglycidyl ether of bisphenol A/2,2-Diamino-1,1-binaphthalene nanocomposites of surface treated multi-walled carbon nanotubes on cure behavior of epoxy nanocomposite. *Thermochimica Acta, 521*(1–2), 49–58. doi:10.1016/j.tca.2011.04.003

Zeng, J., Kumar, S., Iyer, S., Schiraldi, D. A., & Gonzalez, R. I. (2005). Reinforcement of poly (ethylene terephthalate) fibers with polyhedral oligomeric silsesquioxanes (POSS). *High Performance Polymers, 17*(3), 403–424. doi:10.1177/0954008305055562

Zeng, J., Saltysiak, B., Johnson, W. S., Schiraldi, D. A., & Kumar, S. (2004). Processing and properties of poly(methyl methacrylate)/carbon nano fiber composites. *Composites. Part B, Engineering, 35*(2), 173–178. doi:10.1016/S1359-8368(03)00051-9

Zhang, D., Karki, A. B., Rutman, D., Young, D. P., Wang, A., Cocke, D., ... Guo, Z. (2009). Electrospun polyacrylonitrile nanocomposite fibers reinforced with Fe_3O_4 nanoparticles: Fabrication and property analysis. *Polymer, 50*(17), 4189–4198. doi:10.1016/j.polymer.2009.06.062

Zhang, D., Villarreal, M. G., Cabrera, E., Benatar, A., Lee, L. J., & Castro, J. M. (2018). Performance study of ultrasonic assisted processing of CNT nanopaper/solventless epoxy composite. *Composites. Part B, Engineering*.

Zhang, G. D. (2003). Electrical Properties of Nano-ceramics Reinforced with Ropes of Single-Walled Carbon Nanotubes. *Applied Physics Letters, 83,* 1228-1230. doi:10.1063/1.1600511

Zhang, H., Zhang, Z., Friedrich, K., & Eger, C. (2006). Property improvements of in situ epoxy nanocomposites with reduced interparticle distance at high nanosilica content. *Acta Materialia*, *54*(7), 1833–1842. doi:10.1016/j.actamat.2005.12.009

Zhang, L., Li, W., Liu, J., & Ren, B. (1995). Process and performance of electroless Cu/Ni–P double film. Cailiao Gongcheng/ *J. Materials & Design*, *7*, 38–41.

Zhang, R., Zhang, Y., Zhang, Q., Xie, H., Qian, W., & Wei, F. (2013). Growth of half-meter long carbon nanotubes based on Schulz–Flory distribution. *ACS Nano*, *7*(7), 6156–6161. doi:10.1021/nn401995z PMID:23806050

Zhang, W. D., Shen, L., Phang, I. Y., & Liu, T. (2004). Carbon Nanotubes Reinforced Nylon-6 Composite Prepared by Simple Melt-Compounding. *Macromolecules*, *37*(2), 256–259. doi:10.1021/ma035594f

Zhang, Y., Small, J. P., Pontius, W. V., & Kim, P. (2005). Fabrication and electric-field-dependent transport measurements of mesoscopic graphite devices. *Applied Physics Letters*, *86*(7), 073104. doi:10.1063/1.1862334

Zhao, D. L., Qiao, R. H., Wang, C. Z., & Shen, Z. M. (2006). Microstructure and mechanical property of carbon nanotube and continuous carbon fiber reinforced epoxy resin matrix composites. *Advanced Materials Research*, *11*, 517–520. doi:10.4028/www.scientific.net/AMR.11-12.517

Zhi, X., Liu, J., Zhang, H.-B., Hong, S., & Yu, Z.-Z. (2018). Simultaneous enhancements in electrical conductivity and toughness of selectively foamed polycarbonate/polystyrene/carbon nanotube microcellular foams. *Composites. Part B, Engineering*, *143*, 161–167. doi:10.1016/j.compositesb.2018.01.022

Zhou, T., Wang, X., & Wang, T. (2009). Cure reaction of multi-walled carbon nanotubes/diglycidyl ether of bisphenol A/2-ethyl-4-methylimidazole (MWCNTs/DGEBA/EMI-2,4) nanocomposites: Effect of carboxylic functionalization of WCNTs. *Polymer International*, *58*(4), 445–452. doi:10.1002/pi.2558

Zhou, Y., Pervin, F., & Jeelani, S. (2007). Effect vapor grown carbon nanofiber on thermal and mechanical properties of epoxy. *Journal of Materials Science*, *42*(17), 7544–7553. doi:10.100710853-007-1618-6

Zhou, Y., Pervin, F., Jeelani, S., & Mallick, P. K. (2008). Improvement in mechanical properties of carbon fabric–epoxy composite using carbon nanofibers. *Journal of Materials Processing Technology*, *198*(1), 445–453. doi:10.1016/j.jmatprotec.2007.07.028

Zhou, Y., Pervin, F., Rangari, V. K., & Jeelani, S. (2006). Fabrication and evaluation of carbon nano fiber filled carbon/epoxy composite. *Materials Science and Engineering A*, *426*(1), 221–228. doi:10.1016/j.msea.2006.04.031

Zhu, J., Morgan, A. B., Lamelas, F. J., & Wilkie, C. A. (2001). Fire properties of polystyrene–clay nanocomposites. *Chemistry of Materials*, *13*(10), 3774–3780. doi:10.1021/cm000984r

Zhu, J., Wei, S., Ryu, J., Budhathoki, M., Liang, G., & Guo, Z. (2010). In situ stabilized carbon nanofiber (CNF) reinforced epoxy nanocomposites. *Journal of Materials Chemistry*, *20*(23), 4937–4948. doi:10.1039/c0jm00063a

Zilg, C., Mülhaupt, R., & Finter, J. (1999). Morphology and toughness/stiffness balance of nanocomposites based upon anhydride-cured epoxy resins and layered silicates. *Macromolecular Chemistry and Physics*, *200*(3), 661–670. doi:10.1002/(SICI)1521-3935(19990301)200:3<661::AID-MACP661>3.0.CO;2-4

About the Contributors

Noureddine Ramdani received a Bachelor of Engineering at National Preparatory School of Engineers, Ruiba, Algeria in 2007 and joined the Military School Polytechnics, Algeria, where he was an engineer in propellants and energetic materials. He transferred into the College of Chemical Engineering, Harbin Engineering University in 2011, and received a Ph.D. degree in Materials Science at Harbin Engineering University in 2016, under the supervision of Prof. Dr. Wen-bin Liu. He is now Ph.D. supervisor in Materials Science, postgraduate's supervisor in Materials Science and Chemical Engineering at the Harbin Engineering University. In recent years he focused his research on synthesis and relationship between structure and properties of high-performance resins based composites and nano-composites such as polybenzoxazines, phthalonitrile-based composites application, as well as the preparations of novel resins and armor composites.

* * *

Rohit Babhulkar is a 3rd-year B.Tech student at the Department of Metallurgical and Materials Engineering, V.N.I.T. Nagpur.

Swamini Chopra is a Research Scholar at Department of Metallurgical and Materials Engineering, Visvesvaraya National Institute of Technology, Nagpur. She has completed her Graduation (B.E.) in 2013 and M.Tech in Manufacturing Engineering in 2015 from BATU, Lonere. She is Gold medalist throughout her academics. Her areas of research interest are synthesis, post-processing, property testing and material characterization of polymers and polymer composites. She has 17 Journal publications, 2 Books and 1 patent to her credit. She has expertise in handling various sophisticated instruments used for characterization of Materials.

Mohamed Darwish (1979) works at Egyptian Petroleum Research Institute (EPRI). He earned his PhD (Chemical Engineering) from TU Clausthal, Germany, he has a postdoc. at TU Liberec, Czech Republic and at Cambridge University, UK as a visiting scientist. Darwish's specialized in the field of design of functional nanocomposites polymeric materials for advanced applications. Target application areas are biomedical technology, biomaterials, drug delivery, catalysis, magneto-optical materials and in water treatment application.

Kavita Abhay Deshmukh is a Research Associate at Department of Metallurgical and Materials Engineering, Visvesvaraya National Institute of Technology, Nagpur. She has completed her Graduation in 2006 and M.Sc. in Chemistry in 2008 from Nagpur University. She did her Ph.D in the area of Polymers from VNIT, Nagpur in 2013. Besides, she is the recipient of various awards like Late Shri S.L.Shinde memorial award, Late Shri M.J.Pahade memorial award, Late Kashibai Murhar memorial prize. Her areas of interests include Electro and organic Chemistry, Polymer, Composites, Blends, Fatigue, Characterization, Cryogenics, Nanocomposites and Supercapacitors. She has more than 40 Journal publications, 3 Books, 2 Book chapters and 6 Patents to her credit. Beside she has over 9 years experience in handling more than 15 sophisticated instruments. She is associated with many NGO's working for Science.

Kesiya George completed her M.Sc degree in Biopolymer Science from Cochin University of Science and Technology and is currently pursuing her PhD in Polymer Science and Technology at Laboratory for Advanced Research in Polymeric Materials (LARPM), Central Institute of Plastic Engineering and Technology, Bhubaneswar. Her main research interests are the development of elastomeric materials for thermal insulation and polymer nanocomposites etc.

Swaksha Halde is a 3rd-year B.Tech student at the Department of Metallurgical and Materials Engineering, V.N.I.T. Nagpur.

Ezgi Ismar received her MSc in Nanoscience and Nanoengineering from Istanbul Technical University (ITU) in 2013. In 2013 she was hired as a research assistant by the Istanbul Technical University where she is currently working. Also, in the same year, she started her Ph.D. studies at Nanoscience and Nanoengineering Department, ITU. She is studying carbon nanofibers starting from their polymeric precursor, oxidation and end product properties and she has several publications on that topic.

Tamara Irzhak was born in 1947 in Lviv (Ukraine), graduated in 1970 from the Moscow Technological Institute as an economist. Since 1973, she has worked as a programmer at the Institute of Problems of Chemical Physics of the Russian

Academy of Sciences in Chernogolovka. In co-authorship she published more than 60 papers on the kinetics of the formation of polymers and polymer composites. She raised two sons, both physicists, PhD.

Vadim Irzhak is a Doctor of Chemical Sciences, Professor, Chief Researcher of the Institute of Problems of Chemical Physics of the Russian Academy of Sciences, Honored Scientist of the Russian Federation. V. Irzhak was born in Nyandoma, Arkhangelsk region in 1935. In 1958 he graduated from the chemical faculty of the Gorky State University, and in 1962 graduate school of the Institute of Chemical Physics (Moscow). Since then, he works at IPCP. He published about 500 scientific papers including 7 monographs on the kinetics and mechanism of polymerization processes and the topological structure of the polymers formed, as well as on the formation of polymer composites.

Ayesha Kausar is currently employed at National University of Sciences and Technology, Islamabad, Pakistan. She has previously worked for National Centre for Physics and Quaid-i-Azam University, Islamabad, Pakistan. She obtained her PhD from Quaid-i-Azam University, Islamabad/KAIST (Korea Advanced Institute of Science & Technology), Graduate School of EEWS, Daejeon, Republic of Korea. Her research interests include: synthesis, characterization and structure-property relationship of new polymeric materials; synthesis of nanomaterials including organic–inorganic nanocomposites/hybrid materials; polymeric blends via incorporating nanoparticles into thin polymer films; exploration of practical and potential prospects of the novel synthesized materials (new polymers and nanomaterials) counting morphological, mechanical, thermal, electrical, conducting, etc.; flame retardant materials; proton conducting fuel cell membranes; nanocomposites for polymer Li-ion battery electrodes; composites based on electrospun nanofibers; production of various polymeric nanoparticles and their composites for solar cells; polymer/carbon nanotube/nanoparticle composites for water treatment; potential of polymer/graphite nanocomposites; polymer/graphene hybrids; polymer/fullerene nanomaterials; fabrication of epoxy-based nanocomposite for various applications; radar absorbing materials; aerospace relevance, and so on.

Siddhartha Kosti did his engineering in the Mechanical stream from MITS Gwalior, India in 2010. After that he prepared for the GATE exam and secured 667 AIR 2011, then he took admission in IIT Kanpur in Thermal stream and completed his master in 2013. After this, he is in teaching and currently, he is working as an Assistant Professor under TEQIP-III, in the Mechanical Engineering Department of Rajkiya Engineering College, Banda, U.P.

Indradeep Kumar is a Chairman of BIBHUTI EDUCATION AND RESEARCH. He is an Aeronautical Engineer, and till date Published 9 Research paper in International Journal. He also applied for a patent, "Water operated Generator" in 2016. He is also an Entrepreneur and a finalist for Global Student Entrepreneur Award in India and grabbed many other International as well as National prizes. He is a scientific council of the member of IAETSD.

Smita Mohanty is working as a senior scientist at Laboratory for Advanced Research in Polymeric Materials (LARPM), Central Institute of Plastic Engineering and Technology, Bhubaneswar and has more than 12 years of research experience. She has also initiated advanced research at LARPM including E-waste recycling, biopolymers from natural resources, polymer nanocomposites and polymer electrolytes.

Sanjay K. Nayak holds the position of Professor and Chair of R&D wings of Central Institute of Plastic Engineering and Technology with 30 years of experience in teaching and research. He is an active researcher in the field of polymer science and technology, with several publications in peer-reviewed international journals and patented technologies in the area of advanced polymeric materials and characterization.

D. R. Peshwe is Professor at Department of Metallurgical and Materials Engineering Visvesvaraya National Institute of Technology, Nagpur. He did his B.E in Metallurgical Engineering from VRCE, Nagpur University in 1983, Masters in 1985 and achieved his Ph. D from the same institution in 1993. His areas of interests include Physical Metallurgy, Wear of Engineering Materials, Tribology, Polymers & Composites, Cryogenics and Nanomaterials. In his teaching career of 30 years he has guided more than 15 Ph.D scholars and has over 200 publications in various National and International Journals and conferences, 2 Books and 6 Patents. He has coordinated 16 R&D projects of Government of India. He is the recipient of various prestigiosus awards like the Jawaharlal Nehru Memorial Trust Award, Best S&T Innovation award of KVIC and the IIM-SAIL Gold medal of Indian Institute of Metals. In addition he is associated with various technical and social activities.

A. Sezai Sarac received his BSc, and MSc in Chem.Eng.from Istanbul Technical University, PhD. from University of Missouri-Rolla,USA in Chemistry (1976).He held postdoctoral position at University of Leeds –UK (1979-1980).He is a professor at Istanbul Technical University, Department of Chemistry, Polymer Science &Technology and Nanoscience & Nanoengineering graduate programs .He was a visiting professor at Materials and Surface Science Institute University of Limerick in the period of 2002-2003,and was a Head of Physical Chemistry Division, Head of Department of Chemistry between 1991 -1994, and Dean of Faculty of Science and Letters between 1994- 1997,Director of Polymer Science &Technology between 2004-2010. He has authored about 200 scientific publications, and 8 review articles and book chapters, and 2 books, and ~200 conference contributions. His recent book is titled as "Conjugated Polymeric Nanofibers" Pan Stanford Publishing 2016. (http://www.researcherid.com/rid/N-1628-2014) He is awarded a Doctor (Honoris Causa) in Chemical Sciences, Tajikistan Academy of Sciences (2011) and is member of New York Academy of Sciences (1995-2004), Chemical Society of Turkey. He was a visiting Professor at Universitat Regensburg, Germany, Berlin Technical University, Institut für Dünnschichttechnologie und Mikrosensorik e.V.Teltow (IDM-Teltow) and University of Potsdam (between 2006-2010 with different intervals).

S. Sreya is a 3rd-year B.Tech student at the Dept. of Metallurgical and Materials Engineering, V.N.I.T. Nagpur.

Sunil S. Suresh earned M.Phil degree in Marine Chemistry from Cochin University of science and technology (Kerala, India) in 2011, after completing M.Sc. degree in Chemistry from Mahatama Gandhi University (Kerala, India) in 2009. Presently, he is concluding PhD in Chemistry at Laboratory for Advanced Research in Polymeric Materials (LARPM), Central Institute of Plastic Engineering and Technology, Bhubaneswar. His research interests include material recycling, value addition of the recycled plastics, environmental sustainability of the recycled plastics and polymer nanocomposites.

Index

N

O

P

R

S

T

W

www.ingramcontent.com/pod-product-compliance
Lightning Source LLC
LaVergne TN
LVHW081332021026
849161LV00008BA/2091